Journal of Applied Logics - IfCoLog Journal of Logics and their Applications

Volume 5, Number 5

August 2018

Disclaimer
Statements of fact and opinion in the articles in Journal of Applied Logics - IfCoLog Journal of Logics and their Applications (JAL-FLAP) are those of the respective authors and contributors and not of the JAL-FLAP. Neither College Publications nor the JAL-FLAP make any representation, express or implied, in respect of the accuracy of the material in this journal and cannot accept any legal responsibility or liability for any errors or omissions that may be made. The reader should make his/her own evaluation as to the appropriateness or otherwise of any experimental technique described.

© Individual authors and College Publications 2018
All rights reserved.

ISBN 978-1-84890-286-2
ISSN (E) 2055-3714
ISSN (P) 2055-3706

College Publications
Scientific Director: Dov Gabbay
Managing Director: Jane Spurr

http://www.collegepublications.co.uk

All rights reserved. No part of this publication may be reproduced, stored in a retrieval system or transmitted in any form, or by any means, electronic, mechanical, photocopying, recording or otherwise without prior permission, in writing, from the publisher.

Editorial Board

Editors-in-Chief
Dov M. Gabbay and Jörg Siekmann

Marcello D'Agostino	Melvin Fitting	Henri Prade
Natasha Alechina	Michael Gabbay	David Pym
Sandra Alves	Murdoch Gabbay	Ruy de Queiroz
Arnon Avron	Thomas F. Gordon	Ram Ramanujam
Jan Broersen	Wesley H. Holliday	Chrtian Retoré
Martin Caminada	Sara Kalvala	Ulrike Sattler
Balder ten Cate	Shalom Lappin	Jörg Siekmann
Agata Ciabttoni	Beishui Liao	Jane Spurr
Robin Cooper	David Makinson	Kaile Su
Luis Farinas del Cerro	George Metcalfe	Leon van der Torre
Esther David	Claudia Nalon	Yde Venema
Didier Dubois	Valeria de Paiva	Rineke Verbrugge
PM Dung	Jeff Paris	Heinrich Wansing
Amy Felty	David Pearce	Jef Wijsen
David Fernandez Duque	Brigitte Pientka	John Woods
Jan van Eijck	Elaine Pimentel	Michael Wooldridge
		Anna Zamansky

Area Scientific Editors

Philosophical Logic
Johan van Benthem
Lou Goble
Stefano Predelli
Gabriel Sandu

New Applied Logics
Walter Carnielli
David Makinson
Robin Milner
Heinrich Wansing

Logic and category Theory
Samson Abramsky
Joe Goguen
Martin Hyland
Jim Lambek

Proof Theory
Sam Buss
Wolfram Pohlers

Logic and Rewriting
Claude Kirchner
Jose Meseguer

Human Reasoning
Peter Bruza
Niki Pfeifer
John Woods

Modal and Temporal Logic
Carlos Areces
Melvin Fitting
Victor Marek
Mark Reynolds.
Frank Wolter
Michael Zakharyaschev

Automated Inference Systems and Model Checking
Ed Clarke
Ulrich Furbach
Hans Juergen Ohlbach
Volker Sorge
Andrei Voronkov
Toby Walsh

Formal Methods: Specification and Verification
Howard Barringer
David Basin
Dines Bjorner
Kokichi Futatsugi
Yuri Gurevich

Logic and Software Engineering
Manfred Broy
John Fitzgerald
Kung-Kiu Lau
Tom Maibaum
German Puebla

Logic and Constraint Logic Programming
Manuel Hermenegildo
Antonis Kakas
Francesca Rossi
Gert Smolka

Logic and Databases
Jan Chomicki
Enrico Franconi
Georg Gottlob
Leonid Libkin
Franz Wotawa

Logic and Physics (space time. relativity and quantum theory)
Hajnal Andreka
Kurt Engesser
Daniel Lehmann
Istvan Nemeti
Victor Pambuccian

Logic for Knowledge Representation and the Semantic Web
Franz Baader
Anthony Cohn
Pat Hayes
Ian Horrocks
Maurizio Lenzerini
Bernhard Nebel

Tactical Theorem Proving and Proof Planning
Alan Bundy
Amy Felty
Jacques Fleuriot
Dieter Hutter
Manfred Kerber
Christoph Kreitz

Logic and Algebraic Programming
Jan Bergstra
John Tucker

Logic in Mechanical and Electrical Engineering
Rudolf Kruse
Ebrahaim Mamdani

Logic and Law
Jose Carmo
Lars Lindahl
Marek Sergot

Applied Non-classical Logic
Luis Farinas del Cerro
Nicola Olivetti

Mathematical Logic
Wilfrid Hodges
Janos Makowsky

Cognitive Robotics: Actions and Causation
Gerhard Lakemeyer
Michael Thielscher

Type Theory for Theorem Proving Systems
Peter Andrews
Chris Benzmüller
Chad Brown
Dale Miller
Carsten Schlirmann

Logic Applied in Mathematics (including e-Learning Tools for Mathematics and Logic)
Bruno Buchberger
Fairouz Kamareddine
Michael Kohlhase

Logic and Computational Models of Scientific Reasoning
Lorenzo Magnani
Luis Moniz Pereira
Paul Thagard

Logic and Multi-Agent Systems
Michael Fisher
Nick Jennings
Mike Wooldridge

Logic and Neural Networks
Artur d'Avila Garcez
Steffen Holldobler
John G. Taylor

Logic and Planning
Susanne Biundo
Patrick Doherty
Henry Kautz
Paolo Traverso

Algebraic Methods in Logic
Miklos Ferenczi
Rob Goldblatt
Robin Hirsch
Idiko Sain

Non-monotonic Logics and Logics of Change
Jurgen Dix
Vladimir Lifschitz
Donald Nute
David Pearce

Logic and Learning
Luc de Raedt
John Lloyd
Steven Muggleton

Logic and Natural Language Processing
Wojciech Buszkowski
Hans Kamp
Marcus Kracht
Johanna Moore
Michael Moortgat
Manfred Pinkal
Hans Uszkoreit

Fuzzy Logic Uncertainty and Probability
Didier Dubois
Petr Hajek
Jeff Paris
Henri Prade
George Metcalfe
Jon Williamson

Scope and Submissions

This journal considers submission in all areas of pure and applied logic, including:

pure logical systems	dynamic logic
proof theory	quantum logic
constructive logic	algebraic logic
categorical logic	logic and cognition
modal and temporal logic	probabilistic logic
model theory	logic and networks
recursion theory	neuro-logical systems
type theory	complexity
nominal theory	argumentation theory
nonclassical logics	logic and computation
nonmonotonic logic	logic and language
numerical and uncertainty reasoning	logic engineering
logic and AI	knowledge-based systems
foundations of logic programming	automated reasoning
belief revision	knowledge representation
systems of knowledge and belief	logic in hardware and VLSI
logics and semantics of programming	natural language
specification and verification	concurrent computation
agent theory	planning
databases	

This journal will also consider papers on the application of logic in other subject areas: philosophy, cognitive science, physics etc. provided they have some formal content.

Submissions should be sent to Jane Spurr (jane.spurr@kcl.ac.uk) as a pdf file, preferably compiled in LaTeX using the IFCoLog class file.

CONTENTS

ARTICLES

The logical foundations of strategic reasoning: inconsistency-management
    as a test case for logic 945
   John Woods

Reasoning with diagrams and images: observation and imagination as rules
    of inference 987
   John Sowa

Playing with anticipation as abduction: strategic reasoning in an eco-cognitive
    perspective 1061
   Lorenzo Magnani

Abductive cognition: affordance, curation, and chance 1093
   Akinori Abe

Conjectures and abductive reasoning in games 1121
   Ahti-Veikko Pietarinen

Enthymematic interaction in Baduk 1145
   Woosuk Park

When is a strategy in games? 1169
   Woosuk Park

What strategicians might learn from the common law: implicit and tacit understandings of the unwritten **1205**
John Woods

The logical foundations of strategic reasoning: Inconsistency-management as a test case for logic

John Woods
University of British Columbia
john.woods@ubc.ca

Abstract

The notion of foundations has a large presence in theoretical enquiry. The notion of *logical* foundations has an equivocal presence, depending on how logic is to be understood. If taken in its modern mathematical sense, logic's foundational presence can be dated from 1879.[15] [1]

If understood as the systematic analysis of real-life human reasoning in real time, logic has had a foundational significance since Aristotle in the 4th century BC. Whether logical or otherwise (and in whichever sense), enquiry's foundational search for roots is both very old and of enduring importance. In section A on *Foundations* I'll devote some time to tracking the historical lineage of this idea, in the course of which we'll be able to grasp how significant a step it took in 1879 when, in effect, logic decided to detach itself from the business of sorting out the conditions under which real people reason well in real time in the shifting circumstances of human life. Section B on *Cognitive Economics* picks up on the question of what enables beings like us to do as well as we do when we negotiate the flow-through of premissory input to conclusional output. I will suggest that in large measure our successes as reasoners (and our

---

This paper derives from the keynote lecture to the conference on the Logical Foundations of Strategic Reasoning, at the Korean Advanced Institute of Science and Technology, Daejeon, November 3, 2016. Its original title was "Inconsistency-management in big information systems: Tactical and strategic challenges to logic." A companion piece in this same issue of the *IfCoLoG Journal* is entitled "What strategicians might learn from the common law: Implicit and tacit understandings of the unwritten."

[1] The English title in the van Heijenoort volume is *A Formula Language, Modeled upon that of Arithmetic, for Pure Thought.* Notwithstanding Frege's own psychologistic references – thought, judgement, inference, concepts, certitude, and the like – Frege's "formula language" is an artificial one not meant for human speech or as a vehicle of cogitation. In section 6, of *Begriffsschrift,* the "Aristotelian modes of inference" are discussed. However, nothing Aristotle ever wrote about modes of inference are those ones.

failures too) are a function of particular features of the human being's cognitive architecture. This will set the stage for section C on *Strategic Reasoning,* which when practised by beings like us, is practised under the same constraints that our architecture imposes on cognitive processing in general. This will also be a good place to re-pose the logical foundations questions. Is strategic reasoning the kind of enterprise for which the idea of a foundationally realized cognitive security is a plausibly entertainable one? If so, what case, if any, could be made for the doctrine that the foundations of reasoning are in any load-bearing sense *logical*? I will propose that a logic that disregarded the peculiarities and limitations of the human reasoner's cognitive architecture would be ill-suited to the foundational demands its invokers place upon logic.

Section D on *Inconsistency* brings to the fore two concepts of inconsistency, one "robust" and the other "absolute". In section E on *Logic* we will see that when it comes to inconsistency, the absolute appears to dominate over the robust, made so to do by the truth of a long-held logician's thesis about inconsistent systems, known as *ex falso quodlibet*. When we come to section F on *Real World InconsistencyManagement*, it will be time for a suitably naturalized logic of human cognition to start paying its way. In the course of it all, hypotheses will be engendered and passed on to the research community for further study and test. Finally, in section G on *Software Engineers*, inconsistency-robustness makes a brief return, and I'll have some closing words about foundations.

1 Foundations[2]

A safely built house has secure foundations. Anyone who has built one will have a good working knowledge of what its foundations are and how they are assembled. Sometimes people write books on how upstanding houses are built. It would not be startling for such a manual to be entitled *The Foundations of House-Building* or even, with tongue somewhat in cheek, *The Logic of Building Houses*. In the first case, we speak of foundations in a *material* sense. In the second, we speak of them in a *doctrinal* sense. This is the sense in which specified premisses are advanced in support of the claims made by the book. In the preceding sense, foundations are what support the house. They are what keep it from crashing down or tipping over. A third notion has to do with the creation or origins, as when John D. Rockefeller founded Standard Oil or when Aristotle founded systematic logic. Here we have foundations in what we could call the *origination* sense.

When philosophers speak of foundations, they are usually speaking doctrinally. Even so, as the house-example reminds us, it is perfectly possible for entities which

---

[2]I dedicate this essay to the memory of our dearly departed colleague Dale Jacquette (1953-2016).

admit of doctrinally foundational theories to possess foundations in the material sense. Sometimes a good theory contains known falsehoods. Sometimes the falsehoods are derivationally indispensable to the theory is empirical success. Mechanical and electrical engineering are highly successful examples of applied physics. Yet it is known that the physics that underlies these perfectly sound practices is in various respects false.[6] [3] An authoritative theoretical book on engineering might cite some of these falsehoods as doctrinally foundational without thereby making it the case that possessing thereby the factors they misdescribe are somehow materially foundational as well. Of course, having a doctrinal foundationality is no guarantee of a false theory is empirical success. It isn't inconceivable, I suppose, that a theory of house-building could give doctrinal priority to falsehoods about how houses attain and maintain their composure and yet that any house built to the specification of *The Foundations of House-Building* would fall down before the flowers of May come into bloom. Which goes to show that in a competition between the doctrinal and the material, the nod will sometimes go to the material.

It is also possible to have untroubled concurrencies. Aristotle is the originator of logic. It was the foundational core of a wholly general theory of truth-preserving argument. It was foundational in two ways. Its doctrinal foundationality goes without saying. But it – the work itself – also had an originating role. For how could Aristotle have founded a logic for various kinds of truth-preserving argument without having written the *Organon* or at least having had it in his mind for future exposition?

Given the title of the present essay, it behooves us to pay some mind to what it would take for a body of intellectual work to *have* foundations – and, if it did, to what good end – and what it would take to make these foundations *logical*. There are examples of foundational success in mathematics in which, beyond the usual desiderata of rigour, precision and consistency, modern logic plays no role. Since foundation is the older and longer-serving concept, and logic in its mathematical sense a much later entry, I'll begin with foundations *tout court*, in particular with the good that they are supposed to do. Broadly speaking, the foundations of a discipline are comprised of the principles and methods which enable the reduction of the complex to the simple, of the less well-understood to the better-understood, and of the logico-epistemically insecure to the logico-epistemic centre of gravity of the discipline's subject matter. So understood, foundationalizing a discipline is an

---

[3] Also, it is well-known, that in no small measure population genetics owes its substantial empirical success to the embedded falsehood that populations are infinitely large. Its success at the empirical checkout counter confirmed what it said about natural selection on the ground but, needless to say, did nothing to confirm that transfinitely false stipulation about the cardinality of populations.

essentially reductive enterprise in which clarity, understanding and rational security are to be found in the discipline's very roots, under closure conditions that preserve certain of the most valued properties within the roots' secure possession.

The two historically dominant examples of branch-to-root foundationalism are *axiomatization* and *analysis*. Thales (fl. 585 BC) is known to have axiomatized plane geometry. When he wasn't taking care of mathematics, Thales was hypothesizing water as the material foundation of nature. Some two centuries later, Aristotle (384-322 BC) laid the groundwork for the axiomatization of the mature sciences under closure conditions that generated all and only the true propositions of the sciences in question. In Aristotle's hands, axioms reached their apogee as instruments of certainty and generative power. An axiom is a first principle of a discipline or mode of reasoning. It is a proposition that's true, necessary, primary, most intelligible,[4] and neither needful nor susceptible of independent proof or demonstration. As we see, an Aristotelian axiom is an adaptation of the commonplace notion of *premissory support*, or reason for belief or judgement. With axioms, the support is foundational, made so because an axiom is an *unpremissed* premiss, a premiss that neither needs nor allows for premissory support.

Analysis also has deep roots in antiquity. Plato's dialogues, especially the Socratic ones, are replete with attempts at definition. Some fared better than others, and the *Theaetetus*' analysis of knowledge as true belief plus *logos* has flourished from that day to this, where its providence remains surprisingly and I think mistakenly influential.

The antiquity of reductionism is also something to note. There are two (nonaxiomatic) cases of it advanced (but not proved) by Aristotle. In *On Interpretation,* he made the startling claim that anything stateable in Greek is reproducible without relevant loss in the language of categorical propositions. A categorical proposition is a statement in one of four forms: "All A are B", "No A are B", "Some A are B", and "Some A are not B". Let R be any kind of premiss-conclusion reasoning whose closure conditions are both truth-preserving and *discipline-preserving*. A closure is discipline-preserving just in case any proposition in its deductive closure is a proposition with the same disciplinary subject matter. Historically minded readers will know that the conditions that provide the second of these R-outcomes – discipline-preservation – are Aristotle's several conditions on the premisses and/or conclusions of what he would call *syllogisms*. The details need not deter us here. In the interest of time, it will suffice to say that a piece of syllogistic reasoning is R-structured as an ordered triple of categorical propositions, the third of which is the conclusion of the prior two functioning jointly as its premisses, subject to the condition

---

[4]Fully understandable just as it is; intuitively simple and unanalyzable.

that conclusions follow of necessity from their premises, and that further ones are honoured in regulating the entry of propositions to premissory and/or conclusional position. These make for a logically interesting mix of attributes. First there is the requirement that a syllogism be valid, in the sense in which its conclusion follows of necessity from its premises or, in other words, that its premises jointly necessitate it. Aristiotle further stipulates that no premiss be either redundant, or off-topic, or self-inconsistent or inconsistent with the others, or equivalent to the conclusion. Conclusions, in their turn were required to embody only one categorical proposition.[5] Aristotle proffers no definition of the necessitation relation and imposes no further conditions beyond the one that requires it to be truth-preserving. He takes it as given that any neurotypical Greek with an elementary exposure to geometry would know perfectly well how necessitation is to be understood. Accordingly, necessitation was an undefined primitive of his logic.[54][6] In *Prior Analytics* Aristotle proposed that all syllogistic reasoning is reducible without relevant loss to first-figure reasoning or chains thereof. It is a piece of reasoning in first-figure just in case the subject term of the first premiss is the predicate term of the second. If these reduction claims were true, Aristotle would have pulled off a foundationalizing triumph of the highest order. He would have shown that all reasoning whatever of both a truth- and discipline-preserving character is fully reproducible by elementary structures of simplified everyday Greek, as paltry as ordered triples of categorical propositions in first figure in well-structured chains. In a further stroke, Aristotle provided a nearly successful (and repairable) method for discerning with certainty the validity of a piece of reasoning in low finite time and a quasi-mechanical way. Aristotle's logic of syllogisms lacks the technical power and mathematical sophistication of Frege's second-order logic. But Aristotle had the mind of a brilliant logician, well-stocked with very modern-seeming methodological insights. No one has seen better than he the great advantage of handling big problems by framing for the big and solving for the small.

Descartes (1596-1650) reduced plane geometry to algebra, providing a high point of 17th century mathematics. Boole (1815-1864) seriously contemplated the reduction of logic to algebra and the emerging theory of relatives, as did Peirce (1839-1983) later on. Prior to his *Begriffsschrift* of 1879 [15], Frege had satisfied himself that all of mathematics had secure foundations in number theory if number theory were itself foundationally secure. However, by the late 1870s he had decided that number theory was incapable of providing its own foundations. Since number theory was the last foundational stop for mathematics, Frege concluded that if foundations were to

---

[5]It is interesting to note that the founding logic for truth-preserving argument and inference was a relevant, nonmonotonic, paraconsistent, and a fair approximation to an intuitionist one.

[6]A fuller discussion can be found in my *Aristotle's Earlier Logic,*

be found they would have to be found elsewhere. In picking logic as his foundational harbour, Frege was not undoing the effort of his 19th predecessors and contemporaries to "lose" logic in mathematics. He was *re-inventing* logic, in an effort to show us how to "lose" mathematics in logic.

What we have here is the second juncture in the history of Western logic at which a discipline's foundations were expressly proclaimed as logical. In the instance of Frege, the discipline in question was late 19th century mathematics, and the logic in question was the second-order functional calculus of predications. The first time that logic's foundationality was so expressly asserted was by Aristotle. In his case, the discipline was truth-preserving premiss-conclusion human reasoning in real time, and the foundations would be the logic of syllogisms. Aristotle was at least as proud of syllogistic logic as Frege was of second-order logic.

We come now to the nub of this section. The key point about the history of foundationalism is that it has nothing intrinsically to do with what today we recognize as logic. Whereupon we have it that the logical foundations of a discipline are something of an outlier in the foundational universe. Some will fight this, citing the example of Aristotle as a decisive counter. After all, isn't Aristotle logic's founder? Whether the counter holds is of central purchase to the project whose launch I am proposing in this essay. It all comes down to what logic is about. Aristotle would say that logic is a family of theories about various sorts of truth-preserving *reasoning* by on-the-ground human reasoners. A great many modern logicians would say that logic is about the *consequence relation*, never mind people on the ground cite.[7]. Others are of the view that at its heart logic is the study of *logical truth*.[8] Peirce is the independent co-founder of modern quantification theory, and although not working in collaboration with Frege to this same end, assuredly speaks for him in saying: "My proposition is that logic, in this strict [= new] sense of the term, has nothing to do with how you think ...." [39, p. 143] In his Preface to *Methods of Logic,* [45] Quine observes that while logic is an ancient discipline, it is only since 1879 that it's been a great one. Uninformed and rather stupid on their face, these remarks would be neither were the following assumptions to hold true:

- The movement from Aristotle's conception of logic to Frege's was a *paradigm-shift*, in which no prisoners were taken.[9] Quine was an eager next-generation

---

[7] See, for example, Tarski (1901-1983): "the proper concept of consequence must be placed in the foreground" [49, p. 143]; "Logic is about consequence"[49, p. 413] And [1]

[8] See, for example, Quine: "I would say that logic is the systematic study of the logical truths." [43, p. vii]

[9] In marked contrast to the meliorizing efforts of Bacon, Locke and Mill to bring logic to its modern senses, in which the syllogistic was to be given a seat at the table, but wouldn't be the only diner.

participant in this take-over. The mission of the new paradigm would call upon logic to *reconstruct* itself in ways that would enable it to supply the foundations for arithmetic and hence (as Frege believed) for all mathematics. The new paradigm would concretize Peirce's observation that the new logic has nothing directly to do with how in general we think, and would assert that it has everything to do with where arithmetic lies.

Accordingly, from the perspective of the new paradigm, it is entirely correct to say that even in the *Prior Analytics* there is no logic going on. The particular reason why is that beyond making it a defining condition on syllogisms, Aristotle's *Organon* has nothing whatever to say about logical consequence which again, for him, is a simple unanalyzed theoretical primitive.[10]

By the new paradigm's lights, the *Organon* is not logic in any recognizable sense, thus exposing the false continuity of Quine's own slighting remarks in *Methods*. By the *Organon*'s lights, the logic of the new paradigm could be little else than a questionable appendix to its historical predecessor, in which its unanalyzed concept of consequence is explicitly rendered in a made-up language which would have secured no footfall in the *Organon*.[11] I'd venture to say that any strategist who reposed the quality control of his thinking on the theorems of *The Basic Laws of Arithmetic* (1893, 1903) would see his city sacked before nightfall. Strategy is a way of thinking, a way of human thinking in real time, often in conditions of alarm and mortal peril. Knowing the foundations of mathematics avails him nothing when the strategist needs to figure out how to get to Berlin in 1945 before Stalin takes it all. Accordingly, if strategic reasoning has foundations and those foundations are logical, the way to suss them out is by returning to the idea – if not to all its earlier details – that the core business of logic is to regulate the secure flow of human reasoning. In reaching that conclusion, it would take little effort to see how little impact the *Organon* would have on the realities of strategic thinking in real life. The reason is that the *Organon* is focused on truth-preserving reasoning and that, even at its best, strategic reasoning is hardly ever that. Still, the *ways* in which Aristotle catered for the peculiarities of deductive reasoning serve as a model to those of us who seek the ins-and-outs of reasoning at its ampliatively most productive.

Up to now I have been concentrating on deduction and on the logics that have been purpose-built for it. Similarly, the foundations we've so far examined have been foundational for deductively organized systems. Since we all know that a good

---

[10]Frege says from the start that to get to the bottom of numbers it would be necessary to get to the bottom of logical consequence. But it would be wrong to say that the reason Frege produced his logic was his passion to get consequence right. His passion was to get arithmetic right.

[11]Recall that the logics of the *Organon* are centred on premiss-conclusion reasoning *in Greek*.

deal of our most necessary and profitable reasoning is nondeductive — indeed that most human reasoning even at its humanly possible best is nondeductive — why would we pay such mind to deduction? Part of the answer is that, both historically and at present, the contexts in which logic has achieved its greatest successes have been deductively structured. Another is that, in the form in which they have so far evolved, nondeductive logics have been carefully assembled to model some of the most characteristic features of deductive logic, not least its strongly mathematical cast.[12] I will in short order repair my oversight of the nondeductive, but the best place to do it is after reflecting on how the human organism is built for the management of his cognitive agendas in the conditions of real life.

When in 1945 the Western Allies managed to get to Berlin in time to preserve a good chunk of it from Marshall Stalin, we saw the fruits of years of strategic reasoning. The reasoning was done by human beings, reasoning in the ordinary ways of human thinking about matters played out in real-time on the ground. The 1945 matters were complex and dangerous. What was achieved in Berlin was the product of many sectors of the human cognitive economy. It is only natural to ask what human beings would have to be like for that thinking to have occurred and that mainly happy outcome to be achieved. Let's turn to this now.

2 The Cognitive Economy

The money economy is an ecology for the production and circulation of wealth. The cognitive economy is an ecology for the production and circulation of knowledge. The human animal is an information-processor, a knowledge-seeking being and a creature endowed with the wherewithal to achieve it with a steadfast sufficiency for survival, prosperity and the occasional makings of a great civilization. After all, weren't *Prior Analytics* and *Grundgesetze der Arithmetik* achieved at high points on humanity's watch? So was the Standard Model for physics, as were its ancient abumbrations in Thales. Every neurotypical human individual is a belief-having being, responsive to the myriad inducements of belief-change in the face of new information. He is a fallible performer with a good track record in the detection and repair of error after the fact. His belief-revision systems are constantly bombarded with new information, a good deal of which is routinely inconsistent with information already on hand. Whether upon arrival or in anticipation of it, a properly

---

[12]For example, with the exception of autoepistemic logic, defeasibility logics arise from classical logic by attaching to its consequence relation various and shifting families of constraints. As David Makinson observed, nonmonotonic consequence is a relation at two removes from classical consequence, passing from there to superclassical consequence and thence to nonmonotonic consequence. See his [37]

functioning belief-system will try when it can to restore consistency by means of belief-revision. Of course, it can only do this when the inconsistency is somehow spotted. I don't mean consciously spotted, certainly not in the general case. The reason why is that consciousness has a very narrow bandwidth. Taking the sensorium as an example – the juncture at which information arrives from the five sensory modalities – a human being processes $\approx$ 11 million bits of information per second. When it is processed consciously, those 11 million reduce to 40. When processed in linguistically shaped ways, the count falls to 16 bits per second, putting an end to the myth that talk is cheap. Consciousness carries high levels of negative entropy. It is an information-suppressor and a thermodynamically expensive state to be in. [58] It is striking how much the human knower knows on any given occasion, more things certainly than he could even begin to enumerate. For this to be so, cognition must be an *information-thirsty* state. This means that, for wide ranges of cases, knowledge will require more information than the conscious mind can hold at the time the knowledge is acquired and retained. The moral to draw is that most of that information is held unconsciously. Unconscious information-processing has all or most of the following properties, often in varying degrees and harmonies. It is *mechanism-centred, automatic, inattentive, involuntary, semantically and conceptually inert, non-linguistically transacted, deep, parallel* and *computationally luxuriant*.[12, 8] For ease of reference, I'll call this "cognition down-below" – out of sight of the mind's eye, beyond the reach of the heart's command, and nonnegotiable by tongue or pen (or keystroke). This happens to be important for the sciences of cognitive processing. Cognition down-below is not available to introspection. A human being has no direct conscious acquaintance with the workings and conditions of most of what he knows. [13]

This is not to say that science has nothing to say of such matters. A still unsettled case is the attempt by neuroepistemologists (some not all) to model the *brain*'s cognitive productivities on a down-below Bayesian architecture. [11] A long time ago, Gilbert Harman blew the whistle on Bayesian epistemology. He pointed out that the probability conditions on belief-change are too computationally intractable for conscious enactment. For example, if twenty new pieces of information hit a belief-system, a million calculations would be required for the "rational" update of the system. If thirty pieces arrived, a billion computations would be needed. [20] Suppose that the neurotypical human's belief-system harboured a total of 138 logically independent atomic beliefs. A consistency check would require "more time than the twenty billion years from the dawn of the universe to the present."[7] The

---

[13]This means, among other things, that cognitive science by questionnaire has little future but failure and mystification.

reason for this is that all the properties that are distinctive of unconscious activity are oppositely instantiated at the conscious level — *agent-centred, controlled, attentive, voluntary, semantically and conceptually loaded, linguistically expressable, surface, linear* and *computationally feeble*. In particular, not only is consciousness a massive suppressor of information, it is also a massive inhibitor of computational capacity. This prompts the obvious question. Does the human brain have the wherewithal to compute the calculations required by the Bayesian rules? It does not. Consistency checks are computationally intractable.

Cognition down-below is a puzzle. If we accept that knowledge is sometimes *implicit* and *tacit*, there are two assumptions that might strike us as reasonable. One is that when this happens knowledge is stored in nonlinguistic and unconceptualized form. The other is that under the right stimuli, it can surface and, when it does, can *take on* a concentual and linguistic form. A fair hypothesis is that the pure light of consciousness is a bad place for information to dwell.[14] The longer its residency in the cognitive up-above, the less its causal efficacy. This helps us see that most of the information that's most cognitively helpful operates at subconscious levels. Key to this puzzle is the distinction between energy-to-energy transductions, and energy-to-information transitions, and with it the kindred distinction between a blind *reaction* to a causal force and a causal *response* to an information-carrying causal source. The point at which an organism can be causally responsive to information is the point at which nature becomes "readable", without incurring unaffordable thermodynamic costs. What costs? The costs of consciousness, conceptualization and linguistic formulation. "Oh yes?", the reply will come. "How in the world can an information-processor subconsciously *read* experience when it lacks conceptual structure and linguistic form?" How indeed? That is the question. Part of its answer lies in the fact that there is an adaptive advantage in our being able to respond unconsciously to inputs that carry information that enables it to be read.

The idioms of up and down and of surface and depth suggest a vertical and layered organization for information. We might think of the cognitively functioning human individual as an "information-stack" laid, in turn, on layered sheaves of informationless causal nexi. At or near the top would be the information-processing that led to the *Grundgesetze*. Much lower down would be the stimuli that trigger the action of peristalsis. It has been said that the concept of information embodies "a most urgent challenge to philosophical analysis."[23, p. 189]. The reason it does

---

[14]Recent, but far from confirmed, studies speculate on the possibility that electrical neural impulses may embody some of the characteristics of fibre-optics, with photons produced by normal neural metabolism. If this were so, it would give an enormous boost to the speed of information flow. See here Christoff Simon, a University of Calgary quantum physicist in that university's Institute of Science and Technology. Just think of it: Unconscious enlightenment!

is that it lacks an unequivocal meaning. At least four influential notions of information are on offer in the contemporary literature. In what I'll call its "epistemic" sense, information is a propositional representation of what is the case.[15] In its "probabilistic" sense, information is what is channelled from a source to a receiver. The source emits signals with a frequency, and the information picked up by the receiver is seen as the expected reduction of probabilistic uncertainty.[16] In its third or complexity sense, information is a feature of codes. The informational value of a code string is the algorithmic or Kolmogoroff complexity of the string, which is the shortest program that computes it on some fixed universal Turing machine. [43, p. 12] The fourth or "military" sense is typified by the CIA's World Fact Book:

Information is raw data from any source, data that might be fragmentary, contradictory, unreliable, ambiguous, deceptive, or wrong. Intelligence is information that has been collected, integrated, evaluated, analyzed and interpreted.[16, p. 7]

It is difficult to see how there could be a well-unified theory of information in which these four rival and possible incommensurable conceptions come together. But I speculate that there could be an unreductive one in which each conception of it is given load-bearing work to do in the information stacks of the causally responsive human being. It hardly needs saying that these purported distinctions and the ideas that they provoke are anything close to being well-defined or robustly understood. My reaction to this is optimistic. Let's band together and find out. Meanwhile here are some ideas that might repay further consideration.

The fact that most by far of what's processed in the sensorium second by second doesn't make it into consciousness, still less in linguistically expressable ways, suggests that consciousness works as a kind of filtration device screening information for admissibility. The idea here is that much of what isn't admitted wouldn't be helpful to the cognitive ends of the agent at that time. So it would seem that this *irrelevance filter*, if there is one, screens out irrelevant information, unhelpful for the advancement of agendas in the cognitive up-above and in the more implicit and tacit ones a bit lower down. This leaves it open that some of that screened out information is nevertheless causally indispensable for the successes up-above. Furthermore the fact so little is consciously speakable at a given time could also suggest that some of the inadmissible data in the agent's down-below has conceptualized structure, albeit more primitively so, to bear some of the load required for such information-thirsty states as states of knowledge. With it comes some endorsement of the assumption that the information stacks that we humans are admitting of varying degrees of causal epistemic efficacy. Presumably the information that drives peristalsis will

---

[15] See, for example, [13].
[16] See [21, pp. 535-563], [48, pp.623-656]

have no material role in knowing that there is a bluebird in the tree at the back of the garden. But information higher up might well be materially implicated in that knowledge. Whereupon, a interesting possibility comes to mind:

- *Readability*: Subconsciously processed information is readable to the extent of its epistemic materiality.

In *Philosophy of Logic*[43], Quine's conservative strictures about logic led him to draw a red line between the classical logic of truth-preservation and the purported logics of inductive reasoning. Quine's proposal was to send inductive logic to the shakier precincts of epistemology. Writing on page vii of the Preface, Quine avers that "the philosophy of inductive logic ... would be in no way distinguishable from philosophy's mainstream, the theory of knowledge." Quine's 1970 dismissal was not an insulting one, since the year before it had fallen to his own good self to have made epistemology a respectable place in which to limn the lines of solid non-truth-preserving reasoning. In 1969, Quine announced the naturalization of epistemology as a working partnership between philosophy's theories of knowledge and the best-to-date ones of the natural sciences of cognition.[44] As the old saying has it, there are two ways of skinning a cat, or in a less harsh figure, of splitting a difference. Quine's split separated logic from naturalized epistemology. Mine welcomes naturalized epistemology and gives logic a senior partnership within. In this arrangement, logic returns to its historic roots minus the encumbrance of truth-preservation no matter what. In so doing, logic readies itself for the realities of strategic reasoning in human real time. [10, pp. 235-244]

Given a decision to make logic a full partner of epistemology — thereby providing for what we could call "logico-epistemics" — and the further one to make epistemology a working partner of the empirical sciences of cognition, it wouldn't be surprising to see accounts in which the causal character of human knowledge is given due sway. It is a turning which I myself am inclined to take and, in taking it, two working characterizations of matters close to our purpose can be briefly noted.

- *The knowledge as causal hypothesis*: X knows (at some level) that p on information I (at some level) when p is true, X believes p (at some level), and I causally induces X's belief-forming devices to form the belief that p (at that level) X's devices are in good working order and operating here as they should, I is good and well-filtered information, and there is no interference caused by negative externalities.

- *The inference as causal hypothesis*: X infers (at some level) that p from information I (at some level) when in processing p (at some level) X is causally induced to believe (at some level) that p.

Each of these characterizations is only a first pass intended to convey the basic idea. Fuller presentations would require the incorporation of background information, greater specificity about the "at some level" clauses, and a defter handling of good information. (The basic idea is that it not be misinformation or irrelevant.) The reason for mentioning them here is the slack they cut to the idea of knowledge acquired and inferences drawn in the cognitive down-below.

The proposed partnership of logic and epistemology, together with epistemology's partnership with the natural sciences of cognition, has a naturalizing effect upon logic, in which its sometimes rightful leanings toward the mathematical are balanced by the obligations of empirical sensitivity. The corresponding shift of logic's preoccupation with truth-preserving consequence relations back to the founding interest in how human beings manage to think straight in real time helps restore logic to its founding origins as a humanities discipline. All this helps set the stage for a principled discussion of strategic reasoning which, whatever its details, is something that humans do in real time under the press of life's shifting variabilities. Since those involved in it are information-processing beings with cognitive agendas, and the knowledge they achieve is an extraction from information under the right conditions, information is bound to play a foundational role here.[17]

Perhaps the most important lesson to learn from the distinction between energy-to-energy transductions and energy-to-information transitions is that energies that carry information are not thereby *causally* diminished. An information-bearing cause is no less a cause than a blow that blackens an eye. Taken in conjunction with the lately remarked theses of knowledge and inference as causal, supplemented by the thesis that most of knowing and inferring occurs down-below, we can begin to see that knowledge is itself something that has foundations. It is the causal extraction of belief from information under the conditions that qualify belief as knowledge. If the idea has merit, we can see two senses of foundation profitably at work. Knowledge originates in the belief-causing information-processings that produce beliefs that are material elements of it.

Here, too, we have the distinction between knowledge and the theory of knowledge. Often enough, a theory of knowledge postulates foundational principles of knowledge, which are seen as acting in a supporting-premises sort of way. As *The Foundations of House-Building* reminds us, doctrinal foundations aren't always a reliable indicator of material and originating ones, and sometimes can effect a considerable distortion of them, to the point of making them unrecognizable in the doctrinal postulates. This suggests a rule of thumb for theories of subject matters

---

[17] A substantially larger discussion of the place of causal response epistemologies in programmes to naturalize logic can be found in my [53]

that are plausibly taken to have material and/or originating foundations.

- In framing the doctrinal foundations of one's theory of S, it is inadvisable to short-sheet S's material or originating foundations.

Knowledge affords us an attractive example. It can plausibly be said that informationally induced belief is materially foundational for knowledge. A fair candidate for its originating source is the information that triggered the belief in the first place. Yet when we turn to the doctrinal postulations of a good many traditional epistemologists we see scant notice of the fact that more knowledge is produced than the theory's postulates recognize.

It is now time to repair the omissions noted at the end of section one. Most of human reasoning at its best is non-truth preserving. Most of the reasoning that takes on a strategic significance is that way too. If strategic thinking has foundations, and those foundations are in logic, it would repay us to identify the foundations and, if such there be, the logic. There are some obvious places to look, including key works on probabilistic and inductive reasoning, and decision-making. [18] In the selections cited in the footnote below I've restricted myself to the titles in which either "foundations" or "logic" has an occurrence. Of the six, both words occur in the titles of two, and one only of the two in the other four. Five of these works can be considered to be foundational in the sense of being agenda-setters or game-changers, that is, originating. But many more of like influence have titles that flunk the "foundations" and/or "logic" test. I mention this for what it might be worth, beyond noting that a book on the logical foundations of something need not use either word in its title. Readers might wonder why I've decided to pick up the naturalizing option for the locico-epistemics of strategic thinking and practice. I do it for two reasons, one general and the other, as we might now say, strategic. The first is that causally responsive treatments of knowledge and inference in general are superior to rival approaches. The strategic reason is that I want what I say about inference and knowledge to have a decent shot at mattering for some of what I am about to say about strategy.

Let's go back now to Berlin and say something further about strategic reasoning.

---

[18] See, for example, [5, 33, 31, 28, 32, 29]

3 Strategic Reasoning[19]

If logic is an ancient subject, strategics is a century older, arising not in Greece but in China in the 5th century BC. Sun Tzu's *The Art of War* is widely considered its founding document [50]. The name if not its nominatum is Greek, deriving from *stratēgia*, meaning the arts of a troop leader, the office of a general or the exercise of that office. In one of its present-day senses, a strategy is "a comprehensive way to try to pursue political ends, including the threat or actual use of force, in a dialectic of wills" between adversaries. [14] Strategies subsume various more or less distinguishable subsets of skills, including tactics, siegecraft, logistics, operations and the like. In a slightly different usage, strategy is often considered as a plan for problem solving. Its underlying structure is its *kernel*. The kernel subsumes three interacting elements. A *diagnosis* lays out the nature of the problem. A *guiding policy* sets out the ways of meeting the challenges occasioned by it. *Coherent actions* implement them. [46]

Our present question is what a successful theory of strategic reasoning and practice would look like. Would strategics be susceptible to axiomatic organization in the old sense of the word? If strategics *were* foundationally structured, would its foundations be the sorts of thing we could expect to find in logic? If we meant logic in Frege's sense, the answer would be no. Logic in Aristotle's sense would be somewhat closer to the mark, minus the requirement to be deductively truth-preserving. Of all the going nonclassical logics, the ones most sympathetic to Aristotle's in *On Sophistical Refutations* would be logics of games. I admit to some hesitancy about this. Part of the reason why can be found in a deservedly influential book, in which Edward Luttwak writes on the first page of the Preface,

My purpose ... is to uncover *the universal logic* that conditions all forms of war as well as the adversarial dealings of nations even in peace. [36, Emphasis added]

He goes on to say that

... the logic of strategy is *manifest in the outcome* of what is done or not done, and it is by examining those often unintended consequences that the nature and workings of the logic can best be understood. [20]

One infers from this that the logic of strategy is *empirically discernible* in the aftermath of a strategy's actual application. It is an interesting idea: The logic of a strategy has no recognizable presence in the reasoning that grounds it before the

---

[19]Further remarks ab out stragics can be found in my companion piece is this same issue of the *IfCoLoG Journal*.

[20]"Ye shall know them by their fruits. Do men gather grapes of thorns, or figs of thistles? Even so, every good tree bringeth forth good fruit; but a corrupt tree bringeth forth evil fruit .... By their fruit shall ye know them."[19, pp. 16-20](Emphasis added)

fact. It becomes recognizable only after the fact. On the face of it, this is a puzzling and implausible thing to say. Does Luttwak really mean that strategists don't know what they are presently doing but only what they *did* do then? I'm inclined towards a less dramatic reading. What Luttwak might have had in mind is that the logic of a strategy is something of which its planners have only *implicit* awareness before the fact, and *empirical* awareness afterwards. If before-the-fact awareness is implicit in the causal response terms of the preceding section, then it is not then then and there amenable to express formulation. It is not articulable awareness, made so by the fact that it takes place in the cognitive down-below, subconsciously, non-linguistically and so on. There is no room for such awareness in the thermodynamically suppressive conscious mind. It is not, by the way, foreclosed that *ex post facto* awareness of the logic is subject to express and formidable awareness, but it should not be forgotten, that not everything empirically discernible to us can be expressly formulated. Nothing close.

It would be quite wrong to think of the idea of implicit and tacit knowledge as a quirk of Luttwakian strategics. Knowledge that resists explicitization and articulated formulation is a founding epistemological principle of the English common law, evolving from policy chances imposed by the Normans after the Battle of Hastings in 1066. There are several examples of the idea, none more important than in the understanding of unwritten judge-made laws of juridical decisions, in which in the reasons for judgement produce rules of law called *precedents*. When an appeal court's decision on the particular facts before it is rendered, the finding is binding on all courts of the same level and all courts below in all subsequent cases whose facts bear a sufficiency of relevant similarity to the facts of prior one. Although the reasons for decision from which the precedent originates is usually set out in great detail in texts running a great many pages long, the rule of law that it establishes is not written down, not explicitly laid out. There is in this reluctance the stirrings of two ancient principles of common law epistemology. One is its distrust of universalization, the other is distrust of codification. The reason, each time, is similar. If we universalize, we over-generalize a *bona fide* generalization. If we put it into words, we'll misstate it. Some things are best left unsaid. Accordingly, the judge-made rules of law that drive the engines of common law jurisprudence are known and intelligently applied implicitly and tacitly.[21]

In *Strategy*'s Part I, "The Logic of Discovery" [36], Luttwak reflects on what he takes to be the "paradoxical" character of strategy, indeed on "the blatant contradiction" that lies within:

---

[21] I say more of this in my companion piece, "What strategicians might learn from the common law: Inplicit and tacit understandings of the unwritten", in this issue of the journal.

Consider the absurdity of equivalent advice in any sphere of life but the strategic: if you want *A*, strive for *B*, its opposite, as in "if you want to lose weight, eat more" or "if you want to become rich, earn less" – surely we would reject all such." (pp. 1-2)

If Luttwak's attributions of paradox and blatant contradictions are meant literally – I mean literally in a logician's sense – they are attributions of a kind that will send a chill through the body logical. In sections D and E I'll consider both the literalness and the chill.

For the present, I want to take note of a further, and I think essential, feature of Luttwak's approach. It is interesting to note that, while he seeks for the logic of strategy, there is no discussion of the place of game theoretic logic in Luttwak's own. I take this to be telling. Luttwak believes Carl von Clausewitz to have been "the greatest student of strategy who ever lived." (p. 267)[24] Yet he also notes that Clausewitz "was simply uninterested in defining things in generic, abstract terms; he regarded all such attempts as futile and pedantic." What this suggests is a Clausewitzian resistance to formalized approaches to strategic practice, certainly to the idea that the best theoretical language for strategies is a formal one whose formulae carry no propositional content. Frege's way of foundationalizing arithmetic would be the wrong kind of way to approach the theory of strategic reasoning. This matters for game theoretic logics. Although not tethered to Frege's mission, they are tethered to the mathematical treatment of entities constructed from uninterpreted formal languages. For Clausewitz, this would be a step too far. It would harness us to the idea that the concreta of Eisenhower's just-in-time arrival in 1945 Berlin are best understood in the abstracta of generic mathematical models.[22] If this matters for Clausewitz, it also matters for Luttwak. It matters for Luttwak because he models himself on Clausewitz. It also matters for anyone who follows Mintzberg that he identifies a strategy as "a pattern in a stream of decisions", rather than as a kind of overt planning.[38][23]

Of course, it would be wrong to overlook the importance and routineness of a strategist's conscious contemplation of options in his planning space. It would be equal folly to downplay the pressure that falls on strategic planners to get it down on paper, as completely and coherently as possible. There is nothing in Mintzberg to gainsay these observations or diminish their importance. Still, Mintzberg is onto

---

[22] Even so, big-information systems are perpetually and pervasively inconsistent. Perhaps they could be thought to model well, or at least in a helpful way, the inconsistencies that inhere in the information systems embodied in Sun Tzu's *The Art of War* and von Clausewitz's *On War*, Michael Howard and Peter Paret, editors, Princeton: Princeton University Press, 1984; originally published in German as *Vom Krieg* in 1832. I discuss bit-information systems later in this section.

[23] Henry Mintzberg, "Patterns in strategy formation", *Management Science,* 24 (1978), 934-948.

something important, and is so in a way that helps explain Luttwak's respect for Clausewitz. When he says that strategy is discernible in a pattern of decisions, what Mintzberg is suggesting is that strategy is implicit and unvoiced in the interactive dynamics of decision-sequences, some of which might be accessible to expression and consideration in the *historian*'s and *strategician*'s cognitively conscious up-above.

The question that now presses is whether the strategy that inheres in decision-chains can be causally efficacious to good cognitive end without being consciously discernible to its beneficiaries. I am bound to say that the answer is yes, but I'll settle for saying that if the answer were no I'd have a hard time seeing why or how. It now becomes explicable why the greatest strategic thinker ever to draw breath would resist the idea that the canons of good strategic practice would be found in a well-articulated abstract theory of decision-making, in which the strategist's behaviour on the ground is seen as an implementation of its laws. Why would he resist this? Couldn't the real-world decision maker be implementing those laws unconsciously? I suspect, without knowing it, that Clausewitz's answer might resemble my own. Mine is that in the absence of independently available evidence to the contrary, the very idea that the everyday flesh and blood decision makers implement a complex piece of highly idealized mathematics doesn't bear thinking about. There are three reasons why. One has to do with computational intractability. Another pertains to normativity. The third concerns approximation. Beginning with the first, no human being is remotely capable of making the computations laid down by establishment theorists in his conscious up-above, or in his unconscious down-below either.

In response to this, it is frequently assumed that, while these mathematical models fail empirically, they succeed normatively. If this were so, most human reasoning would lack rationality to the degree that it failed the purported ideals. But we see no reflection of this scepticism in the everyday behaviour of the idealizers themselves after a hard day's work at the think tank. That's the trouble with scepticism. Not even its boosters believe it. Then, too, there is the question of how the empirically unperformable idealizations become normatively authoritative for *us*? It is often said that empirical discomportment with the model's idealizations isn't outright irrationality, and that an on-the-ground human is rational to the degree that his reasoning approximates to the reasoning mandated by the model. Not only is the notion of approximation not defined for these contexts, many of the standard idealizations can't be approximated to at all.[24]

I suppose that there might be a concept of foundations for a discipline or practice

---

[24]Classical decision theory requires that belief be closed under consequence, hence that the rational believer would have infinitely many beliefs for any given belief he had. Some earthlings believe more things than some others. But none of them in any finite degree approximates to the required number more closely than any of the others no matter how brilliant.

which is delivered by any book on the discipline's practice's subject matter by an expert. In this sense, *Strategy* would deliver the goods, or would do so as Luttwak sees them. This makes what he doesn't say there as important as what he does say. What he says suggests that the logic of strategy is implicit in its aftermath and empirically discernible there. He does not say that, even in its empirically recognizables state, the logic of strategy is expressly formulable as a body of consequences of some well-articulated principles. It bears notice, however, that there is no indexical reference in this book to logic, foundations, axioms, rules of inference, implications, or anything we could call logic in Frege or Peirce's sense.[25] What this suggests is that if there were a comfortable home in logic for Luttwakian insights, it would likely be found in a naturalized logic for real-time human inference, within which the implicity and tacity of successful strategic reasoning is given a reasonable chance of receiving their due.

Forty-one years ago Edward Luttwak published his ground-breaking work [51], *The Grand Strategy of the Roman Empire: From The First Century CE to the Third*,[26] which "was instantly recognized as the most coherent and compelling account of Roman frontier policy ever produced."[?][27] As Peter Thonemann observes, "*The Grand Strategy of the Roman Empire* [51] has a good claim to being the single most influential book ever written on Roman military history." (25) He continues:

"The book's influence was, if anything, even greater outside the Roman world. The 1970s and 80s saw a dramatic proliferation of Institutes of Defence and Strategic Studies, particularly (but not only) in the United States."

He adds:

"The new discipline of Strategic Studies was, for obvious reasons, narrowly programmatic and empirical, short on "isms" on specific contemporary problems (whether to use force to remove a particular country's nuclear programme, and so forth)." (25-26)

These four decades later, it is startling to see how careworn this classic book has become, and how difficult it has been for Luttwak to restitch it for another forty-year run. Here again is Thonemann:

"A more fundamental problem with Luttwak's approach is what we might call the watchmaker fallacy. Eyeballing maps, he believed that he could make out gen-

---

[25] Neither of whom, by the way, would have had the slightest inclination to say that Luttwak's book would have secured a safer anchorage had it been possible to find it a home in second-order predicate logic or the quantified logic of relatives.

[26] Johns Hopkins University Press, 1976.

[27] Peter Thonemann, "A man, a plan: a canard. Why the Romans were never quite as strategic as was previously thought", *Times Literary Supplement,* October 14, 2016, pages 25-26, a review of the revised and updated edition, also from Hopkins, 2010. Quotations here are from Thonemann.

eral strategic patterns in Roman frontier policy, which could only be the result of intelligent design by generations of successive emperors. The trouble is is that we have not a scrap of *evidence* for this kind of conscious and deliberate long-term planning, and it goes against everything we know about the ways in which policy was actually formulated by individual emperors (ad hoc, reactive, and driven by short-term considerations of prestige and profit.)" (26).

If my causal-response remarks of two or more pages ago held water, we'd have reason to say that this objection largely misses the mark. For on the causal response model, the question of conscious awareness of the strategy before the fact draws a negative answer. Even so, Thonemann, too, is onto something important. Up to now, our logico-epistemic reflections have concentrated on the solo cognitive agent, whether Ike or George or Caesar or anyone else. But, as Thonemann reminds us, much of human knowledge is achieved not by individual agents operating in the advancement of their own particular cognitive agendas, but rather by *multi-agents*, not just the ones that generations of emperors gave constitutive rise to, but the ones that established SHAEF in 1943. The Supreme Headquarters Allied Expeditionary Force commanded the largest number of formations ever assigned to a given operation on the Western Front: including First Airborne Army, British 21st Army Group (First Canadian Army and Second British Army), American 12th Army Group, and American 6th Army Group (French First and American Seventh). It was purposed to discharge Operation Overlord against occupied France. SHAEF was a large and complex multi-agent: A multi-agent is a composite of sub-agents, often themselves multi-agents in their own right, working interactively according to some operational agenda or in fulfilment of some conventional arrangement. In some cases, multiagency is an additive combination of its separate parts. In others, it is an emergent fusion of subsets of its parts, a cohesion of "the mangle of practice."[40][28] From a Mintzbergian perspective, Operation Overlord's strategy is the patterning of nodes in complex and interacting decision-chains ensuing from General Sir Frederick E. Morgan's original plan moulded into its final version in mid-March 1943 and executed on June 6 in the following year.

SHAEF was disbanded in July 1945. The documentation produced in these scant years is immense, much of it produced by individuals and more of it by multiagents operating within the superagency. The immensity of the documentation bespeaks the immensity of the information which drove Operation Overlord. SHAEF operated what computer scientists call "Big information systems" (an understatement). The story of the closure of Overlord's agenda is told in the aftermath of what happened on

---

[28] Andy Perkins, *The Mangle of Practice: Time, Agency and Science,* Chicago: University of Chicago Press, 1985.

June 6th until the just-in-time arrival in Berlin several months later. The European Allies – the U.S., Britain, Canada, Free France, Poland, Norway, Australia, New Zealand, Netherlands, Belgium and Czechoslovakia – fought and won the war in Europe, in complex causal chains extending from Camp Griffiss to the last man's fall on May 7th, 1945. The Alliance was a mammoth multi-agent. The multi-agent won the battle of Europe because it found a way to do it. In that sense, the Allies knew what they were doing. If they knew it explicitly, it might have done it sooner. Had they known it explicitly they could have written it down. When it came right down to it, they knew but didn't know how to say what they knew. From which we may conclude that the Allies knew it unconsciously.

In a startling and now classic paper, John Hardwig discusses a multiauthored contribution to *Physical Review Letters,* entitled "Charm photoproduction cross section at 20 GeV".[29] The letter reports experimental results in particle physics by a widely scattered team of co-authors of varying backgrounds and expertise. The experiment was successful and important enough to appear in this top journal. The diversity of the this expertise was such that

"...no one person would have done this experiment – in fact ... no one university or natural laboratory could have done it – and many of the authors of an article like this will not even know a given number in the article was arrived at." (p. 345)

In a footnote, William Bugg is reported as saying that although a few persons –

"the persons most actively working on the data and who therefore understood most about it – wrote up the experiment ..., they really only prepared a draft for revisions and corrections by the other authors."

All this raises the question of what the charm photoproduction multi-agent knew if several of its expert numbers didn't know what other experts knew and couldn't understand it if they were told it. The group, was a multi-agent of ninety-nine agents, whose own knowledge is largely derived from other multi-agents, by way of information they're told by "sayso" manifolds of knowledge distributions in the home disciplines. By numerical comparison, SHAEF is a multi-agent that dwarfs the charmers. Even so, the same questions arise. Did SHAEF know anything? Assuming that some of its sub-agents did, does their knowledge compose to the superagency? If not, what then? If so, in virtue of what? Did SHAEF know what it knows implicitly and tacitly. We are now in a position to advance a characterization of *strategics.*

- Strategics is the logico-epistemics of multi-agent causal response to information geared to advance large-scale agendas for quelling "dialectics of the will" by

---

[29] John Hardwig, "Epistemic dependence", *Journal of Philosophy*, 82 (1985), 335-349. [18]

force of arms or the threat of them (or other conditions of unwelcome force and threat).

Most scholarly work on the cognitive workings of multi-agency is abstract in ways that would give Clausewitzian offence.[30] There is much to admire in the formal semantics of interactive intelligent agency of all stripes. But I would revert to a lately noted admonition. In foundationalizing the workings of multi-agent reasoning, it would be ill-advised to overlook the material and originating foundations of multi-agency itself, and the role they play in accommodating them in the broader embrace of a causal response logico-epistemics.

Where Thonemann is almost certainly right is in the observation that there is no empirical evidence that field commanders saw themselves as comporting with the patternings of multi-agent decision-sequences both past, present and yet to come. Multi-agency is part of the problem, and incomplete decision runs are another. The problem posed by multi-agency is intra-agency blindness and other forms of cognitive alienation. Individual agents need not know who the others are and, if they did, would frequently be unable to understand them. The problem posed by decision-chains that haven't reached their end is a problem for pattern-recognition. It is a projection problem, a central challenge for trend analysis. Even if decisions taken up to now show a certain patterning, projecting the pattern could meet with recalcitrant future decisions. It is a nasty version of Hume's problem, thanks to the fact that the next war, like all of them, is *sui generis*.

Where Thonemann may be wrong is in thinking (or suggesting) that patterning couldn't be discernible to anyone looking for it after the fact. Suppose that Luttwak's reading of the Roman patterns is at least approximately accurate. That would help *us* in seeing the structure of strategic decision-sequences in a way that might not have done the war-planners on the ground a lick of good. The war-fighting pattern-makers are in the field *in medias res*, and Luttwak the pattern reader peacefully reposes in a condition of after-the-fact scholarly reflection. There is little that's new in this suggestion. In what is sometimes called "the historian's conceit", we find it said that history is not understood by the people who make it, but only after the fact by people who had no hand in it, namely, the people who are smart enough not to be warriors and to be historians instead.[31]

---

[30] Distributive problem-solving, non-cooperative game theory, multiagent communication and learning, social choice theory, mechanism design, mechanism design theory, conditional game theory, mathematical programming and Markov decision processes, among others.

[31] Of course, there are notable exceptions, Caesar being one of them. Rommel wrote up his own victories and George Patton was one of its readers. Then Patton used what Rommel told him in crushing his boastful adversary in North Africa.

In these days of AI, it is only natural to wonder whether the SHAEF information as set out in the massive documentation of its operations could be *computerized*. Could the software engineers computationally model it? Could computers know what we earthlings can't? Efforts of similar scope have been attempted with various levels of success. But big information-systems are problematic. The problem is their internal inconsistency. Big systems are "inconsistency-robust".

4 Inconsistency

The name "inconsistency-robustness" was coined by the MIT computer scientist Carl Hewitt, presently affiliated with Stanford University.[32] Five Eyes, a joint product of the governments of the United States, Canada, the United Kingdom, Australia and New Zealand, is the world's largest information-gathering system in matters of national and international security; the biggest and many would say the most invasive.[33] Five Eyes employs inconsistency-robust reasoning protocols of limited effectiveness, made so by deficiencies in the system's automation procedures. Inconsistency-robustness is strictly speaking a property of Five Eyes' *practices*. But there would be no harm in attributing the property to the system itself, as a kind of expository short-cut.[34] Needless to say, Overlord's information-system was also robustly inconsistent.

An information-system is inconsistency-robust when it is big in ways that require multiple millions of lines of code to computerize, as with climate modelling and modelling of the human neural system.[35] Its inconsistencies are perpetual, pervasive, expungeable in localized contexts but irremovable over-all. Although IR systems do indispensable practical work, they are imperfect and costly. Over-zealous efforts to spare them inconsistency's ignominy seriously damage their practical utility.

Five Eyes plays an indispensable role in supporting the strategic thinking of the

---

[32] For want of alternatives to date, his and my edited book is the publication of record for inconsistency-robustness. See Carl Hewitt and John Woods, editors, *Inconsistency Robustness*, volume 52 of Studies in Logic, London: College Publications, 2015. Second revised edition to appear in 2018. [22]

[33] The participating entities are the U.S.'s National Security Agency (NSA), the Communication Security Establishment of Canada (CSEC), the British Government Communications Headquarters (GCHQ), the Australian Signals Directorate (ASD), and New Zealand's Government Communications Security Bureau (GCSB).

[34] Hewitt to Woods, personal communication September, 2016. The emphasis on practice is easily explained. Even if the system itself *were* free of inconsistency, it couldn't, as far as is now known, be put to productive use in an inconsistency-free manner.

[35] Even a top-end sedan of the Mercedes E series has 30 million lines of code in its computer systems.

Defence Departments of all five participating governments. If the Hewitt assumption holds true, national and multi-national defence establishments are riddled with inconsistencies without anything approximating to destructive consequence. No one seriously supposes that the Pentagon has been incapacitated for successful work by the inconsistencies within, many of them supplied by the inconsistencies that lurk in information in the military sense typified by the CIA World Book. No one thinks either that cleansing Five Eyes of its inconsistencies comes close to making practical sense. This is a point to hang on to for discussion in the section to follow.

The theoretical core of the Hewitt-Woods book is Hewitt's Inconsistency Robust Direct Logic,[36] or IRDL for short. IRDL embodies some formidable heavy-equipment mathematical machinery, and is still very much a work in progress. There is no need here to absorb its many technicalities. It is perfectly possible to reflect on its importance for logic without going into the engineering nuts and bolts. As we have it now, inconsistency-robustness has a large but still quite selective providence. In recent writings, I have suggested that it is a property that travels well, rather in the way that Swiss wine is believed not to.[37] What I mean by this is that it is a property that appears to be fruitfully applicable to inconsistent systems that might not be as big as Five Eyes is, or as national systems of banking or health-care either. Most information-systems that aren't at all small aren't big in the Five Eyes sense. All the same, they can be a lot bigger than we might think.

The IR project is founded on assumptions which many logicians and epistemologists would take to be unpersuasive. How, they will ask, is it known that big systems inherently harbour widespread inconsistency or at least do so with a very high probability? How is it known that these inconsistencies aren't expungeable without serious damage to their practical utility? Even granting that these assumptions are common knowledge in various precincts of informatica, couldn't we have some supporting evidence? These are fair and necessary questions, for which we'll have no time here. Here is how I propose to proceed meantime. I shall begin by accepting the IR assumptions as working assumptions, and in due course attempt to show that the massive inconsistency hypothesis has legs regardless of whether IRDL also does.[38]

It is widely believed that a human being's deep memory is inconsistent.[39] It

---

[36] Hewitt has a fondness for caps but tends to slight hyphens. I admit to reverse preferences. But nothing of substance depends on that.

[37] "How robust can inconsistency get?" *IfCoLoG Journal of Logics and Their Applications*, 1 (2014), 177-216 [55]. See also "Inconsistency: Its present impacts and future prospects", in Hewitt and Woods, 158-194.

[38] My thanks to Harvey Siegel for constructive correspondence on these points.

[39] A. Collins and M. Quillam, "Retrieval time from semantic memory", *Journal of Verbal Learn-*

too is very big, although not perhaps Five Eyeswise so. As already mentioned, it has also been known for some time that a human being's belief system is not even subject to truth-functional consistency checks. This means that inconsistency is not a systematically recognizable property of belief-systems, leading many researchers to accept inconsistent belief-sets as a matter of course. These same attributions are plausibly applied to the background information of scientific theories and to systems of common knowledge. When I said just now that IR is a concept that travels, what I meant is that, big or not that big, all these other systems are also inconsistent. They are systems of indispensable value to the cognitive economy. Their inconsistencies are pervasive and, while not themselves algorithmically recognizable, the plain fact of their inconsistency is an empirically discernible one. We know that systems of this sort are open to, and bettered by, the repair of localized inconsistency. We might not know that any wholesale ethnic cleansing would actually be a system-wrecker. Even so, as of now, we know of no methods that pull the heavy load of system-wide purification for memory and belief. Consider, too the pervasiveness with which newly arrived information contradicts informations currently resident in the processor's belief-box. At each point of such contact, inconsistency strikes. There is little doubt that to a certain extent these are dealt with at source, what with the rejecting of the newly arrived information on the erasure of something already *in situ*. Even so, new information frequently arises without notice, engendering inconsistencies that are likewise overlooked and undetected by consistency checks we're too computationally feeble to run. Given causal efficacies of unconsciously operational information, the impact of flows of unannounced inconsistency upon belief-formation would be a matter of course. From which we would have it that along with the inconsistencies of deep memory come the pervasive inconsistencies of its close kin, deep belief.

Here briefly are two further and different kinds of case to consider. The Newton-Leibniz calculus is inconsistent, and yet played an indispensable role in the theory that revolutionized physics. Old Quantum theory, also thought to be inconsistent, made a major contribution to quantum physics. In each case, the whole theory was inconsistent and, in informational terms, big. But in neither case, did physics go into bankruptcy and announce the close of business. Newton thought that the inconsistency of his treatment of infinitesmals could safely be expunged (and he was right.) Bohr never thought that the inconsistency of his model would disable quantum physics (and he was right). What is different about these cases is that the

---

ing and Verbal Behavior, 8 (1969), 240-249 [9]; M. Howe, *Introduction to Human Memory*, New York: Harper & Row, 1970 [25]; A. Klatzley, *Human Memory: Structures and Processes*, San Francisco, W. H. Freeman, 1975 [30]; and P. Lindsay and D. Norman, *Human Information Processing*, New York: Academic Press, 1977 [35].

inconsistencies struck their abettors as neither pervasive nor ineliminable.

5 Logic

In this section widely pervasive inconsistency robustness will temporarily cede centre-stage to its seemingly slighter sibling, *single-sited* inconsistency. The reason why is that, all on its own, one-shot inconsistency is a very large problem. Unless we know how to handle it, we'll have no chance of keeping robust inconsistency from going off its tracks. Or so it would appear. The strategic challenge that inconsistency poses for logic is to figure out how best to deal with it. The tactical challenge is to meet the strategic target without making matters worse. The best place to start is with a theorem called *ex falso quodlibet* which, loosely translated, says that from a logical falsehood every statement whatever follows of necessity. If we accept that B follows of necessity from A just in case it is not logically possible for A to be true and B not, then *ex falso* easily drops out. If A is logically false, it can't possibly be true. So it is impossible for both A to be true and any B conjointly false. This is a widely accepted definition. If it's the correct definition then it lies in the very nature of logical implication that inconsistent premisses logically imply every statement whatever of the language in which the premisses themselves are expressible.

An information-system in which the negation of a derivable sentence is also derivable is said to be "negation-inconsistent". In the 1920s, Emil Post showed that any negation-inconsistent system that conforms to *ex falso* is also "absolutely inconsistent", in the sense that each and every sentence of this system also follows, as does its own negation. The reverse implication is plain to see. From this it follows that any system with even a smidgeon of inconsistency "detonates" into inconsistency everywhere.[40][3] Inconsistency goes viral.

There are a hardly any logics of inconsistency-management. That is to say, there are scant few of them describing how the real-life neurotypical human agent copes with his inconsistencies without falling into cognitive impotence. What these logics

---

[40]"Detonation" is a charming play on words. The words are the title of Russell's famous 1905 paper, "On denoting", in *Mind*. See Peter Schotch and Ray Jennings, "On detonating", in Graham Priest, Richard Routley and Jean Norman, editors, *Paraconsistent Logic*, pages 306-327 [47], Munich: Philosophia Verlag, 1989. *Paraconsistent Logic* was an authoritative source-book in 1989, and still is. Also important, and more recent, is Jean-Yves Béziau, Walter Carnielli and Dov Gabbay, editors, *Handbook of Paraconsistency*, volume 9 of Studies in Logic, London: College Publications, 2007 [2]. An excellent collection on preservationist paraconsistency is Peter Schotch, Bryson Brown and Raymond Jennings, editors, *On Preserving: Essays on Preservationism and Paraconsistent Logic*, Toronto: University of Toronto Press, 2009.

try to do is describe how inconsistent formal logistic systems, not flesh-and-blood people, manage to spare themselves the perceived chaos of absolute inconsistency. In so doing, in some quarters it is thought that "paraconsistent" models formally represent how you and I should do the same. Most, by far, evade the question of whether its measures are consciously implementable in neurotypical human practice, as opposed to enactable in the down-below, and whether they are implementable even there.[41][41] Many carry the unearned assumption that any real-life agent not implementing the system's routines would be less than rational. This, for me, is a question of a more general methodological significance. It is the question of how mathematically contrived models can formally represent properties of real-world interest without making them unrecognizable in the models. Because time is limited, I'll not take up that question here. I must say, however, that, in more instances than not, the distortion-beyond-recognition problem remains largely unsolved.[42]

Virtually without exception, a paraconsistent logic is one in which *ex falso* fails.[43] The name "paraconsistent" was coined by Miró Quesada and Newton da Costa in 1976, but important paraconsistent logics preceded their baptism as such. Jaśkowski's contradictory deductive systems in 1948 and Ackermann's system of *Strenge Implikation* in 1958 are influential examples, as are the relevant logics of Pittsburgh and Bloomington, and also Canberra and Melbourne.[44] In all these logics, the disposal of *ex falso* is a primary objective, but an even more

pressing one was the destruction of the 1932 Lewis-Langford proof of the dread theorem.[45] Here it is schematically rendered:

---

[41] A possible exception is Graham Priest in his 1987 book *In Contradiction: A Study of the Transconsistent,* Dordrecht: Kluwer, second expanded edition, Oxford: Clarendon Press, 2006; pages references to the second edition. At page 19 Priest writes (correctly, in my opinion): "There is, as far as I am aware, no linguistic or grammatical evidence at all that the English predicate 'is true' does typically ambiguous duty for an infinite hierarchy of predicates at the deep level." Even so, he thinks that logic *should* capture the "deep structure of natural language, or at least part of it." (p. 74). When we make this investigation, presumably we'll learn something about the deep structure of natural language truth and, whatever we find, we won't find it to be transfinitely ambiguous. My chief departure from Priest, and by no means a slighting one, lies among other things in a preference for logic to investigate the deep structure of real-life human *cognition*, all the while not ignoring how conscious knowing also goes. See the section to follow. My thanks to Serban Dragulin for correspondence on this point.

[42] For a larger discussion, readers could consult my "Does changing the subject from A to B really provide an enlarged understanding of A?", *Logic Journal of IGPL,* (2016), 24, 456-480 [56].

[43] There are some paraconsistent logics in which *ex falso* holds. Apparently the intention is for it to hold in a limited way. It holds for entailment but fails for inference. I thank Gillman Payette for helpful conversations about this.

[44] And indeed, in the fourth century B.C. Aristotle's logic of the syllogism was paraconsistent at its very founding.

[45] C. I. Lewis and C. H. Langford, *Symbolic Logic,* New York: Dover, 1959, pp. 288-289 [34].

1. $A \wedge \sim A$ by assumption
2. A 1, by $\wedge$-elimination
3. $A \vee B$ 2, by $\vee$-introduction for arbitrary B
4. $\sim A$ 1, by $\wedge$-elimination
5. B 3, 4, by disjunctive syllogism (DS)

By far the most prominent point of attack by paraconsistent critics has been the validity of the disjunctive syllogism rule DS. Others have questioned the joint validity of $\wedge$-intro and DS. A third complaint is the failure of the truth functional connectives to capture the real meanings of their natural language counterparts. A fourth is the refusal of the conjunction rule.

There is no need to go into the details of how well or badly the Lewis-Langford proof has weathered the paraconsistent pressures of rival logistic systems. Suffice it to say that, as presented in *Symbolic Logic,* the proof was a construction within the truth-functional propositional calculus in relation to the uninterpreted language of that system. Even so, its authors went on to claim that their proof conformed to the ordinary meanings of "proof" and "inference", that is, to their meanings in English. My reaction to this assertion is that if it were true, it would only stand to reason that there'd be an *informal* proof of *ex falso* that is valid for English, and in which no truth-functional formal connective need appear. Let S schematize an English declarative statement S, and let the "not" of "not-S" be taken as sentence-negation. Then

1. S and not-S.(by assumption)

2. If S and not-S, then it is true that S and not-S. (Condition T)[46]

3. If it is true that S and not-S then, then on the principle that if both of two things hold true so does each, S is true.

4. If S is true then, on the principle that for any pair of sentences containing S at least one of them is true, at least one of S, X is true for arbitrary X.

5. If not-S is true then, on the negation principle, S is false and therefore by bivalence is not true.[47]

First issued in New York by Appleton Century-Croft, 1932.

[46] Advanced by Tarski in "The concept of truth in formalized languages" as a condition of "material adequacy" for any theory of natural language truth. The full condition asserts that "S" is true if and only if S ("Snow is white" is true if and only if snow is white). The condition's biconditional structure provides that if S then it is true that S, which is the form in which we have line (2).

[47] Why, then, retain bivalence. If we lost it, we'd still have excluded middle, and negation would flip the truth value of a true statement to one that's not true. If it doesn't flip to falsity, non-truth functions as a third truth value, and negation would retain its negational force.

6. If at least one of S, X is true and S is not, then, on the principle that if at least one of two particular sentences is true and this one is not, it's the other one that's true, X is true.

7. Since each of these steps save the first arises in a truth-preserving way from prior such lines, we have it that ours is a valid conditional proof of the statement that contradictions logically imply the negations of anything they imply.[48]

It is interesting to note that lines (1) to (5) give a proof by contradiction that no contradiction is true.

I needn't remind logician readers that the proof of *ex falso* is a hotly contested one. I won't take the time to litigate the issue here, beyond noticing what bothers these critics and trying to mitigate their concerns. The key question is whether at (6) "not-S" can exclude S from the choice-space between S and X in light of the fact that we already have it that S itself verifies the assumption that at least one of S, X is true. The nub of this question – the deep centre of it – is this. At what point of the proof does "not-" lose its *negational* potency? If it loses its power at line (1), we'll be landed in the "cancellationist" camp, in which a contradictory pair of propositions say nothing at all, and will thereby have dealt a nasty blow to mathematics (which would lose proof by contradiction). This alone is a good reason for thinking that "not-"'s negational authority is untrifled with at line (1).

Very well, then, suppose that the "not-" of "not-S" has full negational potency with regard to S. If it *lacked* this feature at line (1), we'd lose all interest in it. From which I conclude that (1)'s interest is wholly centred on "not-"'s negational powers. The question that now presses in why would "not-" lose its negational potency lower down the proof's chain of truth-preserving reasoning? The fact that at line (6) it verifies *ex falso* strikes me as no reason at all to think that the S of line (1) doesn't negate the "not-S" of the same line, or that lower down the proof goes off the negational track.

Why would I think so? In its present form, the proof centres on the powers of the negation-operator, whose role in life is to flip truth values. If S is true then not-S is false. If S is false, then not-S is true. Giving the proof this focus helps us see that what's really on the line here is whether "not-" retains its truth-value flipping powers under the assumption of a contradictory conjunction.[49]

---

[48]Let's also note that the proof contains no occurrence of the contested word "or", and makes no use of the transformation rules of the propositional calculus.

[49]In earlier versions of the proof, for example, in my paper for the Schotch volume, there is no mention of truth values, hence no occasion to consider whether "not-" always flips them. Regrettably, this omission helps disguise the fact that flipping is the principal issue of the proof.

(i) In approaching the S, X pair at line (4), it is necessary to bear in mind that we already have it independently that if "S and not-S" is true, so is S. In approaching the S, X pair at line (6), it is necessary to bear in mind that we already have it independently on this same assumption that S is not true.

(ii) In the general case in which we have it by assumption that at least one of two statements S*, X* is true and that S* is not true, we default to the conclusion that X* is the true member of the pair.

(iii) However, ours is not the general case. It is the quite particular case in which on the assumption of the proof we have it independently that if at least one of S, X is true, one of them is S, without the necessity of the other one also being true. On the other hand, this is a case in which on that same assumption we also have it independently that if at least one of S, X is true and yet S isn't true, then X is the one that is.

What we have here is the appearance of a standoff. At different validly derived stages of the proof S's truth makes it the case that at least one of the pair S, X is true, and also that S's non-truth makes it the case that X is. The question is whether under this assumption we can have it both ways. My answer is the safe one. Either we can have it both ways or we can't. If we can, *ex falso* is secured by a wholly safe conditional proof. If we cannot, we have made negation unrecognizable. We have the word "not" but there is no truth-value flipping operator it signals. And if that were so line (1), like death, would have no sting. Thinking otherwise, I take it that (1) possesses the sting of negation, that its sting is not erased down the proof's truth-preserving line, and that therefore any inconsistent system, big or small, theoretical or everyday, as a validly derivable negation for each of its derivable sentences, indeed for each of the sentences of the system's language.[50] If so, every proposition in SHAEF's information-system has a validly implied negation. In one of Leonard Cohen's songs, it is proposed that "first we take Manhattan and then we take Berlin." How in the world would an absolutely inconsistent SHAEF take

---

[50] For the languages of formal systems – e.g. languages of the first-order functional calculus – it will depend on whether the system's formal representations of the sentences of natural language and of its properties of interest are sufficiently tight to reflect properties of the formal system's linguistic items back onto their natural language counterparts. If so, absolute inconsistency passes to the formally represented natural language system. If not, the cognitive formal representations can't have been of much utility for natural language in the first place. In the case of information-flows in the down-below, it matters whether unconceptualized and nonlinguistic items of information can stand to one another in any relation of incompatibility sufficient to trigger of *ex falso*'s proof. Final answers aren't yet in, but for now I'll give the nod to a qualified "Yes". The qualification is this. That subconscious and unconceptualized information which is materially efficacious for conscious awareness can bear other such information in like analogues of incompatibility to one another.

Berlin?

6 Real World Inconsistency Management

In this section I'll try to marshal two facts into an instructive coalition. The one fact is a logical fact. The other is an empirical one. To help box our compass, consider again all the knowledge of astrophysics that was gleaned from *Philosophiae Naturalis Principia Mathematica* of 1687. Consider all the knowledge of the quantum world conveyed by Old Quantum Theory associated with Bohr's work in the 'teens of the last century.[51] The logical fact is that all negation-inconsistent systems are absolutely inconsistent, including Newton's and Bohr's. The empirical fact is that people who know this aren't much troubled by it and don't see it as an impediment to knowledge acquisition. The two facts lie in fateful conjunction. Given that we know these facts to be true, there is a pressing need to explain how this came to be so. With that comes the necessity of determining how knowledge is extracted from information-systems within which the negations of all its sentences are validly derivable. We know that these systems detonate *derivationally*. But it is also evident that they do *not* detonate *epistemically*. This sets up a key question. By what mechanism do we determine that certain of these detonationally derived sentences are *truth-tracking*, whereas others are not? Similarly, how does it come to pass that there is much to learn, much knowledge to be gleaned, from Newton's mechanics and Bohr's quantum theory? How is it possible for all of what we know of such things to have come from true beliefs, each of which has a validly derived negation within?

Every system, consistent or not, has a countable infinity of deductive consequences. Even a simple system with a mere scatter of sentences and the sentential connective for or-introduction has an infinite closure. Any theory which advances our knowledge of its subject matter is like this too. Peano arithmetic is thought to be consistent. It has an axiom which says that zero is a number, from which it follows that (i) zero is a number or it is nice in Nice in November, and (ii) that either water is the very same thing as H_2O or not. Both these propositions are true and follow from the Peano axiom in a truth-preserving manner, but neither tells the truth in a way that advances our knowledge of the natural numbers. Neither of them is a truth of Peano arithmetic. The moral is that truth-preservation is no guarantor

---

[51]There is some uncertainty about whether and in what sense Bohr's theory is actually inconsistent, but none at all about the adiabatic principle. See here Bryson Brown, "Old Quantum Theory: A paraconsistent approach," *PSA 1992*, 2, 397-411 [4], and Peter Vickers, *Understanding Inconsistent Science*, Oxford: Oxford University Press, 2013, chapter 3.

of *subject-matter* preservation. This is a point worthy of some official notice:

(1) limitation theorem for PA truth: There are infinitely many true consequences of the Peano axioms that aren't truths of Peano arithmetic.

When Peano wrote down his axioms – actually an updated version of Dedekind's – his object was to capture all the true propositions of natural number theory. For that objective to be fulfilled, the theory would have to be governed by something like a built-in irrelevance filter, whereby the demonstrative output of the axioms would link to true sentences of number theory.[52] To the best of my knowledge, there has been little recognition of this device and virtually no theoretical working-up of how it functions or how it made footfall in the economics of human cognition.[53]

Let's turn now to the best known inconsistent theory in the history of modern logic. Frege's axiomatization of sets was making the rounds of working mathematics in the closing two decades of the nineteenth century. It turned out that the axioms harboured a contradiction, sending shockwaves throughout some sectors of the set-theoretic community. People abandoned "intuitive" set theory in droves.[54] The neurotypical person at large might not be ruffled by an occasional inconsistency – especially if he didn't realize its provisions for omniderivability – but among mathematicians both then and now it is a house-rule that a system blighted by inconsistent axioms cannot be allowed to stand. We might think that set theorists gave up on Frege's set theory for good. The truth is that some did and others didn't. People who still do set theory Frege's way are fewer than those who don't. But the

---

[52] Along the lines of the irrelevance-filter hypothesized in section B to screen out information from conscious awareness because of its unhelpfulness in effecting cognitive awareness in the conscious up-above.

[53] See, however, Dov M. Gabbay and John Woods, *Agenda Relevance: A Study in Formal Pragmatics*, volume 1 of their *A Practical Logic of Cognitive Systems*, Amsterdam: North-Holland, 2003 [17], and my *Errors of Reasoning: Naturalizing the Logic of Inference* [53].In both those works, the hypothesis of irrelevance filters was arrived at abductively, using methods most recently reviewed in my "Reorienting the logic of abduction" [57], in Lorenzo Magnani, and Tommaso Bertolotti , editors *Handbook of Model-Based Science* to appear with Springer in 2017. It updates the position I took in "Cognitive economics and the logic of abduction", the *Review of Symbolic Logic*, 5 (2012),148-161 [52]. But in none of those places is there any systematic exposure of how the filter actually works. This remains an open problem for the research programmes of human cognition.

[54] Save for dialethic logicians, who think that some paradoxical sentences are actually true without violating the Law of Non-Contradiction, hence not triggering *ex falso's* proof. A good state of the art exposition of dialethism is Graham Priest, "Paraconsistency and dialetheism" [42], in Dov M. Gabbay and John Woods, editors, *The Many Valued and Nonmonotonic Turn in Logic*, volume 8 of Gabbay and Woods, editors, *Handbook of the History of Logic*, pages 129-204, Amsterdam: North-Holland, 2007

fact remains that those who do it Frege's way manage to convey a lot of perfectly solid knowledge of sets.

Roughly speaking, the majority is made up by set theorists, whereas the minority is made up of people who either teach introductory set theory or use it as a tool for advancing non-set-theoretic agendas.[55] A recent teaching example, is Guam Bezhanishvili and Eachan Landreth, *An Introduction to Elementary Set Theory* available on the MAA100 website of the Mathematical Association of America at `http://www.maa.org/press/periodicals/convergece/an-introduction-to-elementary-set-theory`. In this text, the Russell paradox is clearly identified and briefly discussed. It is then set aside with the explanation that its further treatment is unnecessary for the purposes of the course. Almost certainly these authors are classically minded about logic, and must know about *ex falso*. Every true sentence of their book has a validly derived negation, and yet there is a lot of solid knowledge about sets which the book imparts to students. Closer to home is the set theory I taught my children when they were in elementary school. I used the old axioms knowing *ex falso* to be true. I told them about it, and invited them not to worry about it. They learned a good deal about sets on those Saturday mornings, followed at lunch time by hamburgers at the best place for them on the Pacific coast.

In the matter of unwanted consequences, the Frege-situation steals a step on the Peano. Frege's logic has an infinite closure of the sort that Peano's axioms has, but it also included in that closure the negation of each of the sentences in it. An irrelevance-filter might keep the consistent parts of that closure from conveying non-set-theoretic truths – "∅ is a set or Nice is nice in November" is not a true proposition of set theory. But can it also deal with the falsehood "∅ is not a set" on grounds of irrelevance? The answer, I think, is that the Gabbay-Woods notion of *agenda-relevance* might have a shot at providing this service. The set theorist's agenda is to pick out the true propositions of set theory from the transfinite closure of Frege's axioms. Should his axioms be inconsistent, his agenda needn't change. Even if every truth of set theory has its negation in the axioms' closure, the fact remains that none of them will be true. On the G-W model, agenda relevance is a *causal* relation defined over ordered quadruples $\langle X, I, K, A \rangle$, where X is a cognitive agent, I is information, K is X's current and/or background information together with his present cognitive capacities, and A his cognitive agenda. Then, as a first rough pass,

G-W agenda-relevance: I is relevant for X with respect to A iff in processing I, X is put in a state of mind contextualized by K which advances or closes A.[56]

---

[55] For example, to understand model theory, it is necessary to have some grasp of sets. It is not necessary that it be a post-paradox understanding of them.

[56] Further details can be found in *Errors of Reasoning* at pages 290-291. For an indication of how

In the situation presently in view, X is set theorist or teacher, Information I is made up of the intuitive axioms, and K is a working knowledge of mathematical practice. A is a desire to get at what these axioms tell us about sets. If K also includes an agenda-irrelevance filter, his agenda would progressively advance towards agenda-closure. That alone may be reason enough to propose the agenda-irrelevance filter thesis as an abduced working hypothesis. If such a filter exists, it lends some operational significance to the metaphor of an investigator's "nose for the truth", which Peirce recognized as a cognitive instinct. Accordingly,

(2) *Inconsistency is no bar to knowledge*: An absolutely inconsistent theory can be a true theory and a knowledge-acquiring and knowledge-advancing one, if it is equipped with an agenda-irrelevance filter that enables subsets of its deductive closure to be truth-tracking. These would be subsets sufficient for agenda-closure

Call these theories "detonated but epistemically productive" ones – DEP theories, for short.

Let me now close this section with a further abduction.

(3) *The down-below abduction*: The hypothesis that best fits the available empirical evidence is that the cognitive system of the individual human being embodies a DEP architecture, operating for the most part in the cognitive down below.[57]

Of course, critics will cavil, and up to a point rightly. I've speculated on the operational roles of two directly unevidenced cognitive filters, which screen out the true irrelevancies of consistent systems and the falsities of inconsistent ones. I have offered no head-on experimental evidence for its existence. So why isn't this just hopeful smoke-blowing? I respond as follows: (i) It is no intrinsic condition on the soundness of an abductive inference that it provide any evidential support for its successfully abduced hypotheses. (ii) Since most of the conjectured workings of DEP systems happens in the unintrospectable down-below – DEP performance is depth performance – would my critics do me the kindness of explaining how the absence of directly confirming evidence weighs so heavily against my hypothesis? For ease of reference, let's agree to call the present account the causal response theory of

---

the agenda-relevance approach helps with *inference*, see also pages 292-293, where consideration is given to how agendas operate in the cognitive down-below.

[57] How we can abductively advance to *knowledge* is explained in greater detail in my "Reorienting the logic of abduction"[57], which prior to editorial excision had been preceded by its main title, "Knowledge without evidence".

inconsistency-management. There is a twofold reason to accept the baptism. One is that the data-responsiveness of CR epistemology. The other is the CR approach adopts the filtration devices on offer here.

SHAEF's information system was robustly inconsistent and, absolutely so. It detonated derivationally, but not epistemically. SHAEF's system retained its truth-tracking capacities, in the absence of which it could not imaginably taken Berlin in 1945. The Alliance that got to Berlin in time before losing it all to Stalin embodied a DEP architecture operating for the most part in the cognitive down-below, thanks in no small part to the causal efficacy of devices that separate derivationally secure falsehoods from derivationally secure true ones. Whether there are logics that describe (or model) this architecture remains to be seen.

7 Software Engineers

Suppose that the causal-response account we've been developing for inconsistency-management in the cognitive ecologies of beings like us were interesting enough for logicians to pause over and turn their minds to. One of the questions that might arise is whether it is feasible or even possible to make a computer model of how things actually go were the causal response theory epistemologically true. Given the similarities between my detonated information systems and Hewitt's robustly inconsistent ones, especially regarding the extent of their respective inconsistencies, it seems natural to ask whether the logic on offer for Hewitt's perpetually, pervasively and ineradicably inconsistencies could be adapted to provide for my absolutely inconsistent systems which happen to be the information systems of neurtypical humanity at large. Hewitt's logic for his inconsistent systems is IRDL, and in IRDL *ex falso* certainly fails. But since *ex falso* is certainly true for English, why not ask Hewitt whether IRDL might be expanded to encompass the transfinitely more capacious infinities under CR management to produce IRDL+? How would Hewitt respond to this? He would likely talk to some software engineers. He probably would talk to the ones he talked to in the process of working up IRDL. Suppose in those earlier conversation, that he encountered considerable pushback concerning the varying throng of (negation) inconsistency-tolerant alternatives. For ease of exposition, let's label these logics as NIT_1, NIT_2, ..., NIT_n. A common complaint might have been that such systems aren't amenable to computer modelling, owing among others to problems of computational complexity.[58]

---

[58]For example, "Deep Learning can require learning exponentially many parameters in the size of an input feature set, which can require exponential time if many outputs depend on numerous overlapping sets of input features" quoted from a draft of Carl Hewitt's contribution to the 2018

- *Computability as an adequacy condition*: Is computer-modellability a necessary condition on the accuracy of a logic's provisions for inconsistency-management?

The second is:

- *The computability of IRDL*: Is the IRDL management system for *robust* inconsistency amenable to computer modelling?

The third is:

- *The computability of IRDL+*: Is the IRDL+ management system for *absolute* inconsistency amenable to computer modelling?

The fourth is:

- *The computability of CR manageability*: Is the CR management system for absolute inconsistency amenable to computer modelling?

At this juncture of our enquiry, I would say that if the IRDL+ system were computable and the CR system were not, that would provide some grounds for saying:

- *Logical foundationality*: If computability were an adequacy condition on logics of inconsistency management, then IRDL would supply the logical foundations of at least a large part of strategic reasoning.

Needless to say, the question posed by the logical foundationality's subjunctive conditional is whether there is weighty reason to think that its antecedent is true. Let's call this the Open Question.

When I myself have had occasion to talk to software engineers about material of current interest (to me), often the response, although always friendly, is not always encouraging. The data are imprecisely specified[59], working assumptions are unclear,

---

edition of *Inconsistency Robustness.[22]*. It is also widely known that defeasibility logics are combinatorial nightmares. Other problems have to do with large numbers of parameters and the high levels of parametric interaction. If true, it would raise a question of fundamental importance for logic, actually four of them.

[59] Any first-year economics student soon learns that data don't speak for themselves. They have to be interpreted and modelled. This is too fast. It is true that data are of no use if they can't be understood. Sometimes data that are difficult to understand have to be interpreted or conceptualized in a certain way. But it is a big stretch to say that conceptualization is modelling. Sometimes adequately understood data simply can't be modelled. The computer science friend who told me that there was too much muck in what I had shown him also told me that he couldn't "see" my data.

comparative considerations aren't quantifiable, many of the assumptions are hostile to commonly used procedures (e.g. probabilities don't carry real-numbered values), the account at hand is too undeveloped (e.g. postulated filters are much too slightly described), and so on. The trouble, I'm told, is there isn't enough in what I've shown them to go on. In the words of one of my software friends, "too much muck".[60] In its current state of development that is the kind of response I'd expect the CR inconsistency-management account to draw. Our open question is whether computability is an adequacy condition on theories of inconsistency-management. If so, the bad news from software engineers would put the CR-account out of business. This inclines me to close the Open Question question with a negative answer.

A few closing words about foundations. The most influential foundational accomplishments by logicians in the history of logic are Aristotle's logic of syllogisms and Frege's second-order functional calculus. In each case, a master reductionist was driving the action. Each was in the service of framing for the big and solving for the small. To achieve the goal of a comprehensive theory of truth-preserving argument, Aristotle relied on key reductions. Let's briefly return to them now. One purported that anything stateable in Greek could be reproduced without relevant loss is a language of four types of categorical propositions. A second one provided that every truth-preserving argument in Greek can be reproduced without relevant loss in the language of categorical propositions. A third proposed that every syllogism can be infallibly be reduced in finite time to the scant few of first figure syllogisms. A fourth asserted that any non-obvious syllogism can be made obvious in low finite time, completely reliably and in a quasi-mechanical way. The first two of these reductionist claims are untrue. The third is much closer to the mark. The fourth was almost proved by Aristotle, and is easily proved now. Whereupon we have it that the logic of syllogisms does not serve the needs of a wholly general theory of truth-preserving argument. We also have it that Aristotle had the soul of a great logician.

The second-order functional calculus was in large measure of Frege's own making. It was a logic purpose-built to provide foundational security for the arithmetic of the natural numbers. In his *habilitation* dissertation had done the same for imaginary numbers, reducing all that's true of them to what's true of the naturals. Frege's reductions to pure logic would be truth-preserving, moving a theory of foundationally insecure truths to the foundational security of the new logic for arithmetic. Of course, the trouble was that the new logic itself was foundationless owing to the

---

[60]Consider also the view that most published statistical research findings are false. See, for example, John Ioannidis, "Why most published research findings are false", *PLOS Medicine*, 2 (2005), e124 [26], and "Why most published research findings are false: Author's reply to Goodman and Greenland", *PLOS Medicine*, 4 (2007), e215 [27].

Russell paradox.[61] Whether Frege's logic is a reasonable one is, to some extent, still a debatable issue. But the fact remains that, like Aristotle, Frege had the soul of a great logician.

The ties that bind these great logicians is that foundations cannot be secured until problems framed for the large are solved for the small and the way to bring this about is by reducing the large to the small and the insecure to the secure, and then take it from there. If we took this to be itself a foundational principle for foundationalism, it would lend at least present and foreseeable discouragement of the idea that strategic reasoning *en large* even has foundations, logical, game-theoretic, or whatever you like.

Acknowledgement

For valuable pre-conference suggestions, I warmly thank Carl Hewitt, Jean-Yves Beziau, Gillman Payette, Woosuk Park, Lorenzo Magnani, Dale Jacquette, Harvey Siegel and Serban Dragulin, and for fruitful post-conference and conference resistance I am grateful to Ahti-Veikko Pietarinen and Paul Bartha. Less resistant, but equally fruitful, were further suggestions from Park, Magnani and Hewitt, and more recently from Dom Lopes and Alirio Rosales.

References

[1] Jeffrey C Beall and Greg Restall. *Logical pluralism*. Oxford University Press on Demand, 2006.

[2] Jean-Yves Béziau, Walter Alexandre Carnielli, and Dov M Gabbay. *Handbook of paraconsistency*. College publications, 2007.

[3] Manuel Bremer. Peter schotch, bryson brown, and raymond jennings, eds., on preserving: Essays on preservationism and paraconsistent logic. reviewed by. *Philosophy in Review*, 30(6):430–431, 2011.

[4] Bryson Brown. Old quantum theory: A paraconsistent approach. In *PSA: Proceedings of the Biennial Meeting of the Philosophy of Science Association*, volume 1992, pages 397–411. Philosophy of Science Association, 1992.

[5] Rudolph Carnap. Logical foundation of probability. *Chicago, Illinois*, 1950.

[6] Nancy Cartwright. How the laws of physics lie, 1983.

[7] Christopher Cherniak. Computational complexity and the universal acceptance of logic. *The Journal of Philosophy*, 81(12):739–758, 1984.

[8] Jonathan D Cohen and Jonathan W Schooler. Scientific approaches to consciousness. 1997.

---

[61] And, as some would later say, also on account of the Gödel incompleteness result.

[9] Allan M Collins and M Ross Quillian. Retrieval time from semantic memory. *Journal of verbal learning and verbal behavior*, 8(2):240–247, 1969.

[10] Michael DePaul. Foundationalism. *Bernecker and Pritchard*, pages 235–44, 2011.

[11] Kenji Doya, Shin Ishii, Alexandre Pouget, and Rajesh P.N. Rao. *Bayesian brain: Probabilistic approaches to neural coding*. MIT press, 2011.

[12] Jonathan St BT Evans et al. *Hypothetical thinking: Dual processes in reasoning and judgement*, volume 3. Psychology Press, 2007.

[13] Luciano Floridi. Blackwell guide to the philosophy of computing and information. 2003.

[14] Lawrence. Freedman and EBSCOhost. *Strategy : a history / Lawrence Freedman*. Oxford University Press Oxford ; New York, 2013.

[15] Gottlob Frege. Begriffsschrift, a formula language, modeled upon that of arithmetic, for pure thought. *From Frege to Gödel: A source book in mathematical logic*, 1931:1–82, 1879.

[16] Dov M Gabbay, Paul Thagard, John Woods, Pieter Adriaans, and Johan FAK van Benthem. *Philosophy of information*. Elsevier, 2008.

[17] Dov M Gabbay and John Woods. *Agenda relevance: A study in formal pragmatics*, volume 1. Elsevier, 2003.

[18] John Hardwig. Epistemic dependence. *The Journal of philosophy*, 82(7):335–349, 1985.

[19] Douglas Hare. *Matthew*. Westminster John Knox Press, 2009.

[20] Gilbert H. Harman. *Induction*, pages 83–99. Springer Netherlands, Dordrecht, 1970.

[21] Ralph VL Hartley. Transmission of information. *Bell Labs Technical Journal*, 7(3):535–563, 1928.

[22] Carl Hewitt. John woods assisted by jane spurr, editors. *Inconsistency Robustness*.

[23] Jaakko Hintikka. *Socratic epistemology: Explorations of knowledge-seeking by questioning*. Cambridge University Press, 2007.

[24] Michael Howard, Peter Paret, and Rosalie West. *Carl Von Clausewitz: On War*. Princeton University Press, 1984.

[25] Michael JA Howe. *Introduction to human memory: A psychological approach*. New York: Harper & Row, 1970.

[26] John PA Ioannidis. Why most published research findings are false. *PLoS medicine*, 2(8):e124, 2005.

[27] John PA Ioannidis. Why most published research findings are false: author's reply to goodman and greenland. *PLoS medicine*, 4(6):e215, 2007.

[28] Richard C Jeffrey. *The logic of decision*. University of Chicago Press, 1990.

[29] James M Joyce. *The foundations of causal decision theory*. Cambridge University Press, 1999.

[30] Roberta L Klatzky. *Human memory: Structures and processes*. WH Freeman, 1975.

[31] AN Kolmogorov. Foundations of the theory of probability. chelsea, new york, 1933. *Trans. N. Morrison*, 1956.

[32] David Krantz, Duncan Luce, Patrick Suppes, and Amos Tversky. Foundations of mea-

surement theory. 1971.

[33] J Leonard. Savage: The foundations of statistics. *New York*, 1954.

[34] Clarence Irving Lewis, Cooper Harold Langford, and P Lamprecht. *Symbolic logic*. Dover Publications New York, 1959.

[35] PH Lindsay and DA Norman. Human information processing. 1977.

[36] Edward Luttwak. *Strategy: the logic of war and peace*. Harvard University Press, 2001.

[37] David Makinson. *Bridges from classical to nonmonotonic logic*. King's College, 2005.

[38] Henry Mintzberg. Patterns in strategy formation. *Management science*, 24(9):934–948, 1978.

[39] Charles Sanders Peirce. *Reasoning and the logic of things: The Cambridge conferences lectures of 1898*. Harvard University Press, 1992.

[40] Andrew Pickering. *The mangle of practice: Time, agency, and science*. University of Chicago Press, 1995.

[41] Graham Priest. In contradiction: A study of the transconsistent, 2nd expanded ed, 2006.

[42] Graham Priest. Paraconsistency and dialetheism. *The Many Valued and Nonmonotonic Turn in Logic*, 8:129–204, 2007.

[43] Willard V Quine. *Philosophy of logic*. Harvard University Press, 1986.

[44] Willard Van Orman Quine. *Ontological relativity and other essays*. Number 1. Columbia University Press, 1969.

[45] Willard Van Orman Quine. *Methods of logic*. Harvard University Press, 1982.

[46] Richard P Rumelt. Good strategy/bad strategy. crown business. Technical report, ISBN 978-0-307-88623-1, 2011.

[47] Peter K Schotch, Raymond E Jennings, et al. On detonating. *G. Priest, R. Routley, and J. Norman, editors, Paraconsistent Logic: Essays On The Inconsistent*, pages 306–327, 1989.

[48] CE Shannon. A mathematical theory of communication. *Bell System Technical Journal*, 27:379–423 & 623–656, July & October, 1948.

[49] Alfred Tarski. *Logic, semantics, metamathematics: papers from 1923 to 1938*. Hackett Publishing, 1983.

[50] Sun Tzu. *The art of war*. e-artnow, 2012.

[51] Edith Mary Wightman. The grand strategy of the roman empire from the first century ad to the third, 1978.

[52] John Woods. Cognitive economics and the logic of abduction. *The Review of Symbolic Logic*, 5(1):148–161, 2012.

[53] John Woods. *Errors of reasoning: Naturalizing the logic of inference*. College Publications, 2013.

[54] John Woods. Aristotle's earlier logic. 2014.

[55] John Woods. How robust can inconsistency get? *IfCoLog Journal of Logics and their Applications*, 1(1):177–216, 2014.

[56] John Woods. Does changing the subject from a to b really provide an enlarged understanding of a? *Logic Journal of the IGPL*, 24(4):456–480, 2016.

[57] John Woods. Reorienting the logic of abduction. In *Springer Handbook of Model-Based Science*, pages 137–150. Springer, 2017.

[58] Manfred Zimmermann. The nervous system in the context of information theory. In *Human physiology*, pages 166–173. Springer, 1989.

Reasoning with Diagrams and Images

John F. Sowa

sowa@bestweb.net

Abstract

Visualization and analogy are the heart and soul of mathematics. Long before they write a formal proof, mathematicians train their intuition on diagrams, visualize novel patterns, and discover creative analogies. For over two millennia, Euclid's diagrammatic methods set the standard for rigorous proof. But the abstract algebra of the 19th century led many mathematicians to claim that all formal reasoning must be algebraic. Yet Peirce and Polya recognized that Euclid's diagrammatic reasoning is a better match to human thought patterns than the algebraic rules. A combination of Peirce's graph logic, Polya's heuristics, and Euclid's diagrams is a better candidate for a natural logic than any algebraic formalization. This combination justifies Peirce's claim that his graph logic generates "a moving picture of the action of the mind in thought." Formally, it can be specified as rigorously as any algebraic proof in mathematics or computer science.

With his existential graphs (EGs), Peirce developed a graphic version of logic that was as precise and expressive as his linear notation for predicate calculus. He also hinted at the possibility of generalizing EGs to accommodate complex diagrams in more than two dimensions. This article develops Peirce's hints by adding two rules of inference: observation and imagination. By observation, any fact that may be observed in a diagram may be asserted by a formula in any version of logic, linear or graphic. By imagination, a statement of the form "If diagram, then formula" may be asserted if the formula is true of the diagram. For diagrams as simple and precise as Euclid's, these rules are sound: they preserve truth. For more complex diagrams, the rules are as dependable or fallible as the methods of observation and testing.

1 Languages, Logics, and Imagery

Is language based on logic? Is logic based on language? Or are they both based on imagery? These questions have been debated since antiquity, and modern cognitive science is beginning to provide some answers. To illustrate the issues, consider sentences spoken by a child named Laura shortly before her third birthday [32]:

> Here's a seat. It must be mine if it's a little one.
> I want this doll because she's big.
> When I was a little girl I could go "geek-geek" like that. But now I can go "this is a chair."

Laura's content words express concrete images, directly related to her actions. But her syntax and function words express a surprising amount of complex logic: possibility, necessity, tenses, indexicals, conditionals, causality, quotations, and metalanguage about her own language. As another example, a mother was talking with her son, who was about the same age as Laura [27]:

> Mother: Which of your animal friends will come to school today?
> Son: Big Bunny, because Bear and Platypus are eating.

The mother looked in his room, where the stuffed bear and the platypus were sitting in a chair and "eating". The boy had imagined a situation, built a model of it, and based his reasoning on it: The bear and the platypus are eating. They can't eat and go to school at the same time. Big Bunny isn't doing anything. Therefore, Big Bunny is available.

The reasoning abilities of children have challenged the syntax-based theories of logic and linguistics. The semantic aspects of imagery, action, and feelings appear to be more important than syntax. This point is supported by studies of infants with one deaf parent and one speaking parent. At every stage of development, they have equal ability to express themselves in one-dimensional speech or in moving, three-dimensional gestures [49]. In fact, infants with two deaf parents babble with their hands, not with vocal sounds. The neuroscientist Antonio Damasio [2]) summarized the issues:

> The distinctive feature of brains such as the one we own is their uncanny ability to create maps... But when brains make maps, they are also creating images, the main currency of our minds. Ultimately consciousness allows us to experience maps as images, to manipulate those images, and to apply reasoning to them.

The maps and images form mental models of the real world or of the imaginary worlds in our hopes, fears, plans, and desires. They provide a "model theoretic" semantics for language that uses perception and action for testing models against reality. Like Tarski's models, they define the criteria for truth, but they are flexible, dynamic, and situated in the daily drama of life.

The logician, mathematician, scientist, and philosopher Charles Sanders Peirce would agree. Although he invented the algebraic notation for predicate calculus [43],

Peirce claimed that all reasoning is based on a "a concrete, but possibly changing, mental image" that may be aided by "a drawing or a model":

> All necessary reasoning without exception is diagrammatic. That is, we construct an icon of our hypothetical state of things and proceed to observe it. This observation leads us to suspect that something is true, which we may or may not be able to formulate with precision, and we proceed to inquire whether it is true or not. For this purpose it is necessary to form a plan of investigation, and this is the most difficult part of the whole operation. We not only have to select the features of the diagram which it will be pertinent to pay attention to, but it is also of great importance to return again and again to certain features. [47, Vol 2, p.21]

> The word *diagram* is here used in the peculiar sense of a concrete, but possibly changing, mental image of such a thing as it represents. A drawing or model may be employed to aid the imagination; but the essential thing to be performed is the act of imagining. Mathematical diagrams are of two kinds; 1st, the geometrical, which are composed of lines (for even the image of a body having a curved surface without edges, what is mainly seen by the mind's eye as it is turned about, is its generating lines, such as its varying outline); and 2nd, the algebraical, which are arrays of letters and other characters whose interrelations are represented partly by their arrangement and partly by repetitions. If these change, it is by instantaneous metamorphosis. [48, Vol 4, p. 219]

For informal reasoning, stuffed animals or objects of any kind can aid the imagination in building models and reasoning about them. For mathematical reasoning, Euclid's diagrams are the classical examples. But moving or changing images can be even more suggestive. Figure 1 shows a diagram that inspired Archimedes to imagine polygons converging to a circle. As the number of sides increases, the perimeter of the inner polygon expands, and the perimeter of the outer polygon shrinks. The limit for both is the circumference of the circle. If the diameter of the circle is 1.0, the limit is π. By using 96-agons, Archimedes approximated π as $22/7 = 3.142857...$, an error of 0.004%.

Figure 1 embodies two creative insights. First, Archimedes recognized that the perimeters of the outer polygon and the inner polygon are upper and lower bounds on the circumference of the circle. Second, the two bounds come closer together as the number of sides increases. With those two insights, a good mathematician with enough time and enough paper could compute π to any desired approximation.

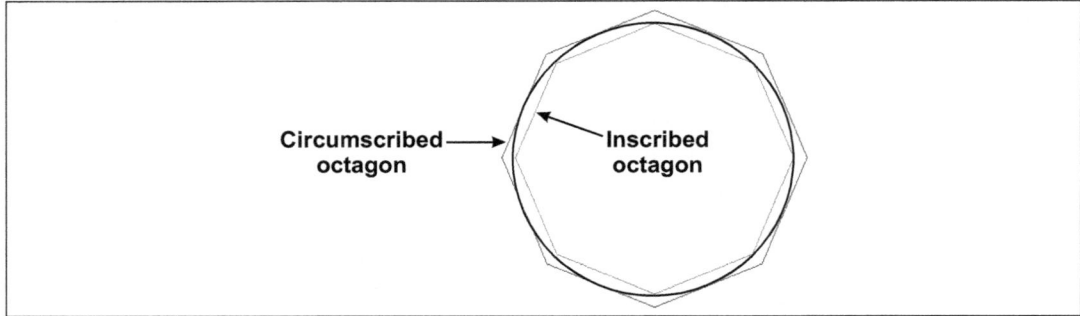

Figure 1: Approximating π by polygons converging to a circle

There were many good mathematicians before Archimedes, but none of them had those two insights. The mathematician Paul Halmos [13] said that the essence of creativity consists of visualizing novel analogies and guessing which ones are the most promising to pursue:

> Mathematics — this may surprise or shock some — is never deductive in its creation. The mathematician at work makes vague guesses, visualizes broad generalizations, and jumps to unwarranted conclusions. He arranges and rearranges his ideas, and becomes convinced of their truth long before he can write down a logical proof... the deductive stage, writing the results down, and writing its rigorous proof are relatively trivial once the real insight arrives; it is more the draftsman's work not the architect's.

To study and compare the thought processes of mathematicians, Jacques Hadamard [12] asked some of the most creative to answer a few questions. Their responses were similar to the comments by Peirce and Halmos. Einstein emphasized visual and muscular images and their combinations:

> The words or the language, as they are written or spoken, do not seem to play any role in my mechanism of thought. The psychical entities which seem to serve as elements in thought are certain signs and more or less clear images which can be voluntarily reproduced and combined... The above-mentioned elements are, in my case, of visual and some of muscular type. Conventional words or other signs have to be sought for laboriously only in a secondary stage, when the mentioned associative play is sufficiently established and can be reproduced at will.

For teaching students how to think, visualize, and guess, George Polya [51] presented *heuristics* or methods of plausible reasoning that promote the insight:

> You have to guess a mathematical theorem before you prove it; you have to guess the idea of a proof before you carry through the details. You have to combine observations and follow analogies; you have to try and try again. The results of the mathematician's creative work is demonstrative reasoning, a proof; but the proof is discovered by plausible reasoning, by guessing.

Since Peirce understood the importance of visualization for discovering a proof, he experimented with graph notations that used visual methods in the proof itself. His first version, the *relational graphs* of 1882, could express a subset of first-order logic. But like the early semantic networks in artificial intelligence, Peirce's relational graphs couldn't delimit the scope of quantifiers and negations. A dozen years later, he extended his relational graphs to *existential graphs* (EGs) by using nested ovals to represent scope. Following is an EG for the sentence "A cat is on a mat" and its translation to predicate calculus (Peirce's algebra with Peano's choice of symbols):

```
Cat—On—Mat
```
$\exists x \exists y (\mathrm{Cat}(x) \land \mathrm{Mat}(y) \land \mathrm{On}(x,y))$

In existential graphs, a line by itself represents something that exists. The labels 'Cat', 'On', and 'Mat' represent relations. In combination, the EG states "Some cat is on some mat." The predicate calculus states "There is an x, there is a y, x is a cat and y is a mat and x is on y." For the image of a cat on a mat, English, EG, and predicate calculus could all be called diagrammatic because their mapping is simple and direct: the labels map to the two objects and the relation between them. But the predicate calculus is less digrammatic than the others because its syntactic details are more complex.

Before Peirce published his algebraic notation, Frege [9] had published a tree notation for FOL, which he called *Begriffsschrift* (concept writing). His goal was to show the reasoning steps explicitly, not to represent imagery or natural language. Figure 2 shows the Begriffsschrift for the cat sentence.

The symbols in Figure 2, reading from left to right, consist of a vertical bar for assertion, a short vertical bar for negation, two cup symbols for universal quantifiers with the variables x and y, two hooks for implication, another negation, and the same three relation labels as the EG or the predicate calculus. In English, the Begriffsschrift may be read "Assert: it is false that for every x, for every y, if x is a cat then if y is a mat, then it is false that x is on y." It is equivalent to the following predicate calculus: $\sim \forall x \forall y (\mathrm{Cat}(x) \supset (\mathrm{Mat}(y) \supset\, \sim \mathrm{On}(x,y)))$

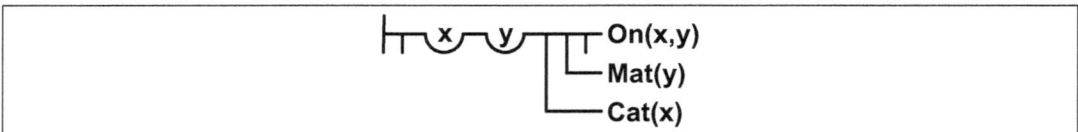

Figure 2: Frege's Begriffsschrift for "A cat is on a mat."

Although many logicians considered Begriffsschrift unreadable, it could be considered diagrammatic by Peirce's criteria. Frege designed the notation to highlight the relationship between his logical operators and his rules of inference. Occurrences of the three operators ∀, ⊃, and ∼ trigger the three rules of inference: universal instantiation, modus ponens, and double negation. But the operators of existence and conjunction have a more direct mapping to imagery and ordinary language. If Frege had added a square cup for existence and the symbol & for conjunction, an extended Begriffsschrift would resemble predicate calculus (Figure 3).

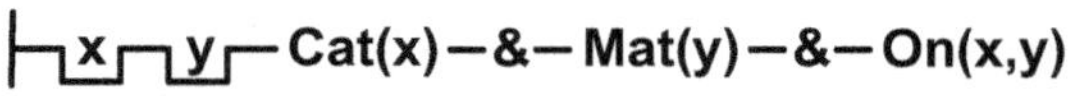

Figure 3: Extended Begriffsschrift with operators for existence and conjunction

But Frege refused to add more operators to Begriffsschrift, which he regarded as "a device invented for certain scientific purposes, and one must not condemn it because it is not suited to others." He had no desire to simplify the mapping to the language of everyday life (*Sprache des Lebens*). Instead, he hoped "to break the domination of the word over the human spirit by laying bare the misconceptions that through the use of language often almost unavoidably arise concerning the relations between concepts."

Peirce, however, considered language and logic as aspects of the broader field of semiotic. Instead of a sharp dichotomy, he emphasized the continuum of formal and informal methods. The logic-based methods of induction, abduction, and deduction are based on the same kinds of analogies used in observation and imagination. That view led Peirce to a better balance of readability and proof theory. His existential graphs not only have a simpler mapping to and from language and imagery, they also have a direct mapping to simpler rules of inference with an elegant version of model theory. For these reasons, Peirce called EGs his "chef d'oeuvre" and claimed that their rules of inference generate "a moving picture of the mind in thought." After a detailed comparison of Peirce's EGs to current theories about mental models, the psychologist Johnson-Laird [22] agreed:

> Peirce's existential graphs are remarkable. They establish the feasibility of a diagrammatic system of reasoning equivalent to the first-order

predicate calculus. They anticipate the theory of mental models in many respects, including their iconic and symbolic components, their eschewal of variables, and their fundamental operations of insertion and deletion. Much is known about the psychology of reasoning. But we still lack a comprehensive account of how individuals represent multiply-quantified assertions, and so the graphs may provide a guide to the future development of psychological theory.

The remainder of this article develops these ideas. Section 2 summarizes the syntax of existential graphs and their mapping to the linear Existential Graph Interchange Format (EGIF). Section 3 presents the EG rules of inference, and Section 4 uses Peirce's *endoporeutic* to show that the rules are sound. Section 5 shows how arbitrary images can be inserted in any EG and used in a proof. As an example, Euclid's proofs can be translated line-by-line to EGs that use Peirce's rules of inference to construct Euclid's diagrams. Section 6 presents research issues that are raised and sometimes solved by these methods. Finally, Section 7 presents evidence for the claim that EGs supplemented with images are a promising candidate for a natural logic.

2 Existential Graph Syntax

Charles Sanders Peirce was a pioneer in logic. Although Frege published the first complete system of first-order logic in 1879, no one else adopted his notation. Peirce published the algebraic version of FOL and HOL in 1885. With a change of symbols by Peano and some extensions by Whitehead and Russell, Peirce-Peano algebra is still the most widely used logic today [52]. But as early as 1882, Peirce experimented with graph notations to express "the atoms and molecules of logic." In 1897, he developed existential graphs (EGs) as a notation that expressed the semantics of first-order predicate calculus with equality. For the first-order subset of EGs, Peirce used the same structure and rules of inference in every manuscript from 1897 to 1911. But he experimented with variations, extensions, and semantic foundations [54]. For the graphics, this article uses the notation he preferred after 1906. Unless otherwise cited, all quotations by Peirce are from the *New Elements of Mathematics*, pages 3:162 to 3:169.

As an example, the EG on the left of Figure 4 asserts that there is a phoenix. The line, which he called a *line of identity*, represents existence. By itself, the line asserts "There is something." The word phoenix is the *name* of a relation type. The line attached to the name asserts "There exists a phoenix." As Peirce explained,

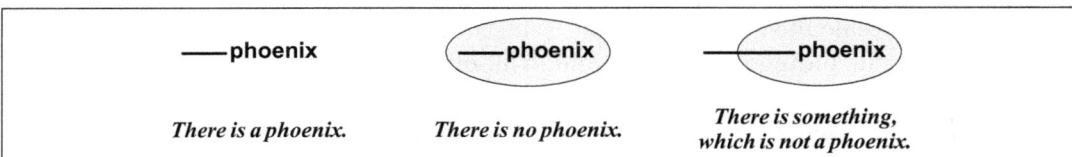

Figure 4: Three existential graphs

	EGIF	Predicate Calculus
Left	(phoenix *x)	$\exists x\ \text{phoenix}(x)$
Middle	~[(phoenix *x)]	$\sim \exists\ \text{phoenix}(x)$
Right	[*x] ~ [(phoenix x)]	$\exists x \sim \text{phoenix}(x)$

Table 1: EGIF and predicate calculus for Figure 4

To deny that there is any phoenix, we shade that assertion which we deny as a whole [EG on the left of Figure 4]. Thus what I have just scribed [EG in the middle] means "It is false that there is a phoenix." But the [EG on the right] only means "There is something that is not identical with any phoenix."

To indicate negation in his early EGs, Peirce used an unshaded oval enclosure, which he called a *cut* or a *sep* because it cut or separated the *sheet of assertion* into a positive (outer) area and a negative (inner) area. In the later versions, he added shading to highlight the distinction between positive and negative areas. Table 1 shows the Existential Graph Interchange Format (EGIF) for each of the graphs in Figure 4 and the corresponding formula in predicate calculus (Peirce-Peano algebra).

In the EGIF for the graph on the left of Figure 4, the parentheses enclose the name phoenix of a *monad* (monadic relation) and a *defining label* *x. The defining label represents the beginning of a line of identity in the graphic EG. It is mapped to an existentially quantified variable $\exists x$ in predicate calculus. For the EG in the middle, the shaded oval is represented in EGIF by a tilde ~ for negation and a pair of brackets ~[ ]. For the EG on the right, the beginning of the line of identity is outside the shaded oval. Since there is no relation outside the negation, the defining label is contained in a *coreference node* [*x] in front of the negation. Inside the parentheses of the relation, the label x without an asterisk is a *bound label* that is within the *scope* of the defining label *x on the outside.

When an oval is drawn inside another oval, the doubly nested area is positive (unshaded), as in Figure 5. Any area nested inside an odd number of ovals is shaded, and any area inside an even number of ovals (possibly zero) is unshaded. As Peirce

Reasoning with Diagrams and Images

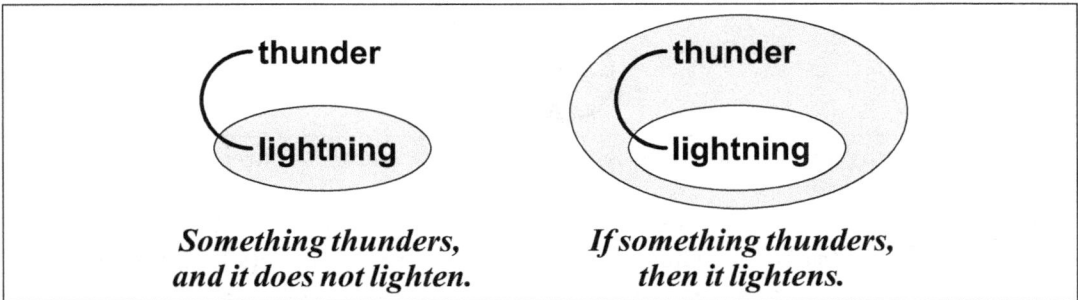

Figure 5: An EG with one negation and an EG with two negations

	EGIF	Predicate Calculus
Left	(thunder *x) ~[(lightning x)]	$\exists x(\text{thunder}(x) \land$ $\sim \text{lightning}(x))$
Right	~[(thunder *x) ~[(lightning x)]]	$\sim \exists x(\text{thunder}(x) \land$ $\sim \text{lightning}(x))$
Optional	[If (thunder *x) [Then (lightning x)]]	$\forall x(\text{thunder}(x) \supset \text{lightning}(x))$

Table 2: EGIF and Predicate Calculus for Figure 5

said, "The graph [on the left] asserts that it thunders without lightning... a denial shades the unshaded and unshades the shaded. Consequently [the graph on the right] means 'If it thunders, it lightens'."

In Table 2 the conjunction (AND operator) is implicit in EGIF; no symbol is required. But the formula in predicate calculus requires the symbol $\land$. In the last line of the table, the EGIF uses the option of replacing the two negations with the keywords **If** and **Then**; the predicate calculus uses the universal quantifier $\forall$ and the implication $\supset$.

In EG and EGIF, a nest of two negations represents an if-then statement in English. Peirce called that combination a *scroll*. To improve readability, the optional line in the table shows how the tildes for the two negations may be replaced by the keywords **If** and **Then**. In predicate calculus, $\sim \exists$ is equivalent to $\forall \sim$. With that conversion, the formula for Figure 5 becomes $\forall x \sim (\text{thunder}(x) \land \sim \text{lightning}(x))$. By the definition of the implication operator $\supset$, the formula $\sim (p \land \sim q)$ is equivalent to $p \supset q$. Therefore, the formula for the EG on the right of Figure 5 may be rewritten as $\forall x(\text{thunder}(x) \supset \text{lightning}(x))$. In English, this formula may be read "For every x, if x thunders, then x lightens." This example shows that the EG and EGIF for if-then, has a more direct mapping to English than the operator $\supset$ in predicate

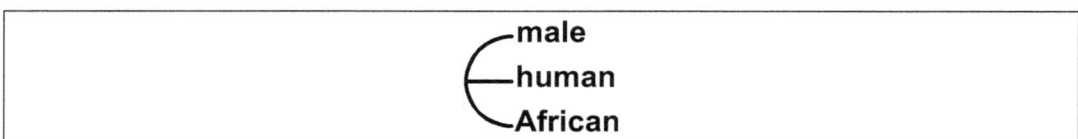

Figure 6: A teridentity is a ligature that connects three lines

	EGIF	Predicate Calculus
Teridentity	(male *x)(human *y) (African *z) [x y z]	$\exists x \exists y \exists z (\text{male}(x) \land \text{human}(y) \land \text{african}(z) \land x = y \land y = z)$
One label	(male *x) (human x) (African x)	$\exists x (\text{male}(x) \land \text{human}(x) \land \text{african}(x))$

Table 3: EGIF and Predicate Calculus for Figure 6

calculus. In fact, EGs are isomorphic to the *discourse representation structures*, which Kamp and Reyle [24] developed for representing natural language semantics.

> A graph may be complex or indivisible. Thus [Figure 6 shows] a graph instance composed of instances of three indivisible graphs which assert 'there is a male', 'there is something human', and 'there is something African'. The syntactic junction or point of *teridentity* asserts the identity of something denoted by all three.

In EGIF, Figure 6 may be represented by three indivisible nodes followed by a coreference node [x y z], which shows that that three lines of identity refer to the same individual. Peirce called such a junction a ligature. Since all three lines refer to the same thing, the defining labels *y and *z may be replaced by bound labels for *x. Then the coreference node is no longer needed, and it may be deleted, as shown in the following table:

In modern terminology, Peirce's indivisible graphs are called *atoms*. In predicate calculus, each atom consists of a single relation followed by its list of *arguments*, which Peirce called logical subjects. In EGs, an N-adic relation has N pegs, each of which is attached to a line of identity for its logical subject. In EGIF, each atom is represented by a pair of parentheses that enclose a relation name followed by a list of defining labels or bound labels for its logical subjects..

Peirce represented a *proposition* as a relation with zero pegs. He called it a *medad*; the prefix me- comes from the Greek μη for not. In EGIF, a proposition p is represented as a relation with no logical subjects: (p). In early versions of EGs, Peirce distinguished two subsets: Alpha for propositional logic and Beta for first-order logic. By treating medads as relations, he avoided the need to distinguish Alpha from Beta. The same rules of inference apply to both.

Peirce continued, "Every indivisible graph instance must be wholly contained in a single area. The line of identity may be regarded as a graph composed of any number of dyads '—is—' or as a single dyad." To illustrate that option, consider the graph man — African, which may be read "There is an African man." Replacing the line with two copies of —is— would break the line into three segments: man—is—is—African. This EG may be read, "There is a man that is something that is something African." Following is the EGIF and predicate calculus:

(man *x) (is x*y) (is y*z) (African z)
$\exists x \exists y \exists z (\mathrm{man}(x) \wedge x = y \wedge y = z \wedge \mathrm{African}(z))$

The two copies of the dyad is are equivalent to two coreference nodes [x *y] [y *z], which may be replaced by a single coreference node [x *y *z]. Since all three labels are coreferent, the bound label x may replace all occurrences of the other defining labels and bound labels. The result is (man *x) (African x).

With these examples, Peirce presented EG syntax in less than three printed pages. In another four pages, he presented the rules of inference, a brief summary of endoporeutic, and an example that shows how an Aristotelian syllogism can be translated to an EG and proved by the EG rules of inference. As he showed, EGs have only two explicit operators: a line to represent the existential quantifier and an oval to represent negation. Conjunction is an implicit operator, expressed by drawing any number of graphs in the same area. Equality is expressed by joining lines.

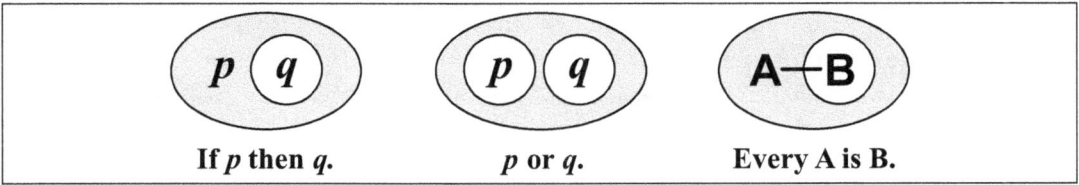

If p then q. p or q.   Every A is B.

Figure 7: Three logical operators expressed as EGs

All other operators of first-order logic are represented by combining these primitives. Figure 7 shows three composite operators defined in terms of nested negations. Although these three operators are composite, their graphic patterns are just as readable as the algebraic formulas with the special symbols $\supset$, $\vee$, and $\forall$. In fact, the explicit nesting of EG ovals directly shows the scope of quantifiers. But the usual Boolean operators $\wedge$, $\vee$, and $\supset$ look so similar that students find it hard to

	EGIF	Predicate Calculus
If p, Then q.	[If (p) [Then (q)]]	$p \supset q$
p or q.	~[~[(p)]~[(q)]]	$p \vee q$
Every A is B	[If (A *x) [Then (B x)]]	$\forall x(A(x) \supset B(x))$

Table 4: EGIF and Predicate Calculus for Figure 7

	EGIF	Predicate Calculus
Left	[*x][*y] ~[[x,y]]	$\exists x \exists y \sim (x = y)$
Middle	[*x][*y]~[(is x y)]	$\exists x \exists y \sim \text{is}(x, y)$
Right	(P *x) (P*y)~[(is x y)]	$\exists x \exists y (P(x) \wedge P(y) \wedge \sim \text{is}(x, y))$

Table 5: EGIF and Predicate Calculus for Figure 8

remember that each one has a different effect on the scope of quantifiers. In EGIF, the propositions or medads (p) and (q) are enclosed in parentheses.

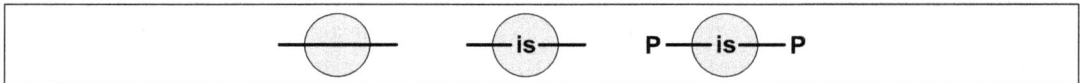

Figure 8: Stating that two things are not identical

Sometimes, the dyad —is— can clarify the translation of an EG to a sentence or formula. For example, Figure 8 shows three ways of saying that there exist two things. In the EG on the left, the shaded area negates the junction of the lines of identity on either side. To emphasize what is being negated, the EG in the middle replaces part of the line with the dyad —is—. Therefore, that EG may be read "There is something x that is not something y." The graph on the right says that there exist two different things with the property P or simply "There are two Ps."

As these examples show, EGs express identity by joining two lines and deny identity with an oval on a line. The linear notations require named labels and symbols such as x=y or x ≠ y. Without labels on lines, graphic EGs do not need special rules for replacing one label with another. But Peirce said that a complex graph with criss-crossing lines may be clearer if some connections are shown by labels, which he called *selectives* [45, 4.460]. When all the lines are replaced by labels, the EG becomes a nest of nodes, isomorphic to EGIF.

In summary, EGIF has the same primitives and conventions as EGs, but with the adaptations necessary to linearize the graphs. An oval for negation is represented by a tilde ~ followed by a pair of square brackets to enclose the EGIF for the subgraph inside the oval. A line of identity is represented by a defining label and zero or one bound label. (See Figure 4 for examples of defining labels with no bound labels.) The graphic options for connecting and extending lines are shown by coreference nodes. A ligature is represented by a coreference node with two or more labels (defining or bound). Many coreference nodes may be eliminated by replacing several defining and bound labels with a single defining label and multiple bound labels. (See Figure 6.) The only coreference nodes that cannot be eliminated are those that are enclosed in a negation and show a junction of two or more lines whose defining labels are outside the negation. (See Figure 8.)

3 EG Rules of Inference

All proofs in Peirce's system are based on "permissions" or "formal rules... by which one graph may be transformed into another without danger of passing from truth to falsity and without referring to any interpretation of the graphs" (CP 4.423). Peirce presented the permissions as three pairs of rules, one of which states conditions for inserting a graph, and the other states the inverse conditions for erasing a graph. Unlike Frege's rules, Peirce's rules are symmetric: any change by one rule can be undone by its inverse. This property enables some remarkable metalevel proofs (shown in Section 6) that are not possible with the common proof procedures. As before, all quotations by Peirce, unless otherwise cited, are from [48, 3:162-169]:

> There are three simple rules for modifying premises when they have once been scribed in order to get any sound necessary conclusion from them.... I will now state what modifications are permissible in any graph we may have scribed.

Peirce's rules are a generalization and simplification of the rules for *natural deduction*, which Gentzen [11] independently discovered many years later. For both Peirce and Gentzen, the rules are grouped in pairs, one of which inserts an operator, which the other erases. For both of them, the only axiom is a blank sheet of paper: anything that can be proved without any prior assumptions is a theorem. Section 6 presents a more detailed comparison with Gentzen's method.

1st Permission. Any graph-instance on an unshaded area may be erased; and on a shaded area that already exists, any graph-instance may be inserted. This includes the right to cut any line of identity on an unshaded area, and to prolong one or join two on a shaded area. (The shading itself must not be erased of course, because it is not a graph-instance.)

The proof of soundness depends on the fact that erasing a graph reduces the number of options that might be false, and inserting a graph increases the number of options that might be false. Rule 1e, which permits erasures in an unshaded (positive) area, cannot make a true statement false; therefore, that area must be at least as true as it was before. Conversely, Rule 1i, which permits insertions in a shaded (negative), area cannot make a false statement true; therefore, the negation of that false area must be at least as true as it was before. A formal proof of soundness requires a version of model theory. Section 4 uses Peirce's model-theoretic semantics, which he called *endoporeutic* for "outside-in evaluation."

These rules apply equally well to propositional logic and predicate logic. Since EGs have no named variables, the algebraic rules for dealing with variables are replaced by rules for cutting or joining lines of identity (which correspond to erasing or inserting an equality or the graph —**is**—). Cutting a line in a positive area has the effect of *existential generalization*, because the newly separated ends of the line may represent different existentially quantified variables. Joining two lines in a negative area has the effect of *universal instantiation*, because it replaces a universally quantified variable with an arbitrary term.

> **2nd Permission**. Any graph-instance may be **iterated** (i.e. duplicated) in the same area or in any area enclosed within that, provided the new lines of identity so introduced have identically the same connexions they had before the iteration. And if any graph-instance is already duplicated in the same area or in two areas one of which is included (whether immediately or not) within the other, their connexions being identical, then the inner of the instances (or either of them if they are in the same area) may be erased. This is called the Rule of Iteration and Deiteration.

In other writings, Peirce stated more detail about the effect of these these rules on lines of identity. Iteration (2i) prolongs a line from the outside inward: any line of identity may be prolonged in the same area or into any enclosed area. Deiteration (2e) retracts a line from the inside outward: any line of identity that is not attached to anything may be erased, starting from the innermost area in which it occurs. Peirce also said that no graph may be copied into any area within itself; it is permissible, however, to copy a graph and then make a copy of the new graph in some area of the original graph.

The proof of soundness of iteration (2i) and deiteration (2e) shows that they are equivalence relations: they can never change the truth value of a graph. First, note that a copy of a graph p in the same area is equivalent to the conjunction $p \wedge p$; inserting a copy of p by Rule 2i or erasing it by 2e cannot change the truth value. For a copy of a subgraph into a nested area, the method of endoporeutic shows that the subgraph makes its contribution to the truth value at its first (outermost) occurrence. The presence or absence of a more deeply nested copy is irrelevant.

> **3rd Permission.** Any ring-shaped area which is entirely vacant may be suppressed by extending the areas within and without it so that they form one. And a vacant ring-shaped area may be created in any area by shading or by obliterating shading so as to separate two parts of any area by the new ring shaped area.

A vacant ring-shaped area corresponds to a *double negation*: two negations with nothing between them. The third permission says that a double negation may be erased (3e) or inserted (3i) around any graph on any area, shaded or unshaded. Note that Peirce considered the *empty graph*, represented by a blank area on a sheet of paper, as a valid existential graph; therefore, a double negation may be drawn or erased around a blank. An important qualification, which Peirce discussed elsewhere, is that such a ring is considered vacant even if it contains lines of identity, provided that the lines begin outside the ring and continue to the area enclosed by the ring without having any connections to one another in the area of the ring. Both rules 3e and 3i are equivalence rules, as endoporeutic shows.

In discussing these rules, Peirce said that their purpose is "to dissect the reasoning into the greatest possible number of distinct steps and so to force attention to every requisite of the reasoning." In that comment, Peirce was not comparing the EG rules to the rules for predicate calculus, because EG proofs are often shorter than other proof procedures for FOL. Instead, he was comparing the EG rules to Aristotle's syllogisms. To illustrate the issues, he proved the syllogism named Barbara:

> I will now, by way of an example of the way of working with this syntax, show how by successive steps of inference to pass from the premises of a simple syllogism to its conclusion. [Figure 9] shows the two premises "Any M is P" and "Any S is M." The first step consists in passing to [Figure 10] by the 2nd Permission.

The rule 2i allows the graph on the left of Figure 4 to be iterated (copied) into the unshaded area of the graph on the right. In translating EGs to EGIF, any labels

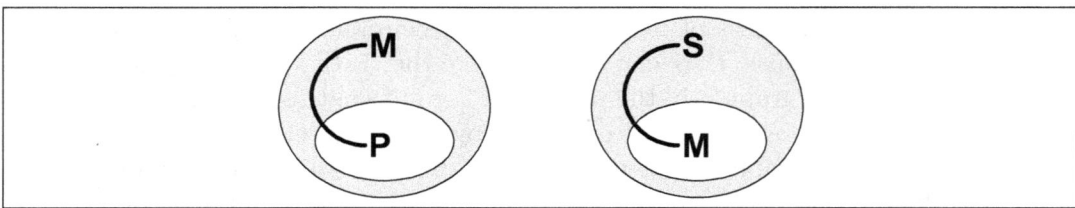

Figure 9: Two premises of the syllogism named Barbara

may be chosen for the lines of identity. To minimize the name clashes in EGIF proofs, each line should be assigned a unique label. For example, *x may be used as the defining label for the graph on the left of Figure 9, and *y for the graph on the right. Following are the EGIF statements for Figure 9 and for Figure 10. Since these graphs have the form of implications, the keywords If and Then are used to represent the negations:

```
[If (M *x) [Then (P x)]]  [If (S *y) [Then (M y)]]
[If (M *x) [Then (P x)]]  [If (S *y) [Then [If (M *x)
   [Then (P x)]] (M y)]]
```

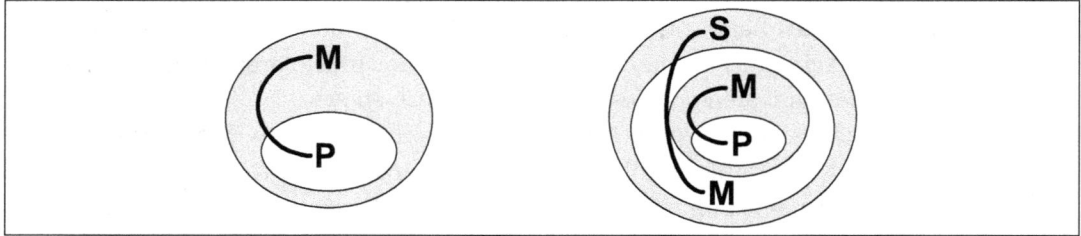

Figure 10: After iterating the left EG into the right EG

The second step is simply to erase "Any M is P" by the 1st Permission. The third step is to join the two ligatures by the 1st Permission as shown in [Figure 11]. It will be observed that in iterating the major premise, I had a right to put the new graph instance at any part of the area into which I put it; and I took care to have the ligature of the minor premise **touch** the shaded area of the iterated graph instance. Now by the 1st Permission I have a right to insert what I please into a shaded area, and without making the new line of junction leave the shaded area, I make it touch the unshaded line of identity of the minor premise.

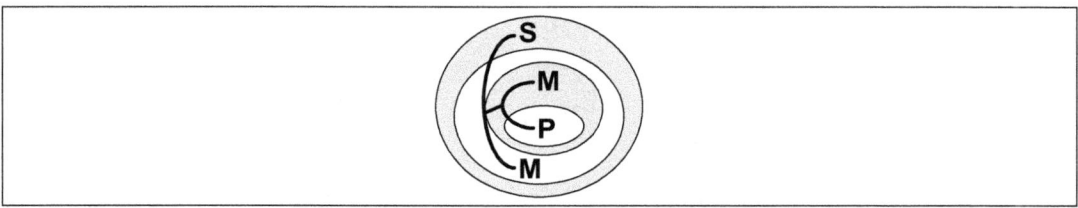

Figure 11: After joining the two lines

Peirce's explanations of lines that touch a boundary depend on two-dimensional features of the drawing. If he had not made the boundary of the iterated graph instance touch the line of identity, the join would take two steps: prolong the outer line into the shaded area by Rule 2i, and join the two lines by Rule 1i. Since EGIF cannot represent lines touching an oval, both steps are required. Therefore, the EGIF proof from Figure 10 to Figure 11 takes three steps: by 1e, erase the copy of the outer EG that had been iterated; by 2i, prolong the line that connects S to M into the shaded area by inserting the coreference node [y]; and by 1i, show the join of that line to the one that connects M to P by replacing the labels *x and x with y:

```
[If (S *y) [Then [If (M *x) [Then (P x)]] (M y)]]
[If (S *y) [Then [If [y] (M *x) [Then (P x)]] (M y)]]
[If (S *y) [Then [If (M y) [Then (P y)]] (M y)]]
```

Before the two lines were joined, the inner copy of M could not be erased by deiteration because the two copies of M were attached to different ligatures. But after the join, both copies of M are attached to the same ligature, and the inner copy of M can be erased by Rule 2e.

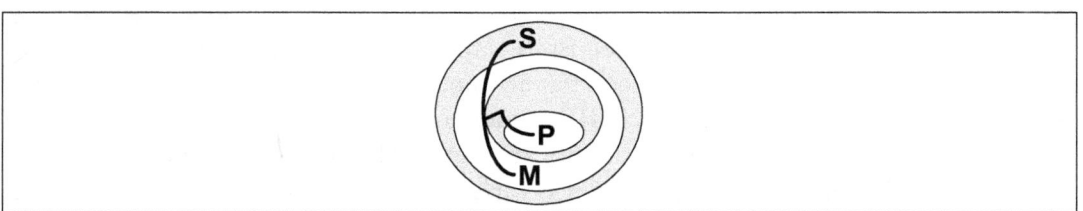

Figure 12: After deiterating (erasing) the inner copy of M

This gives me a right in the fourth step to deiterate M so as to give [Figure 12] by the second permission. The fifth step is to delete the M on an unshaded field giving [Figure 13].

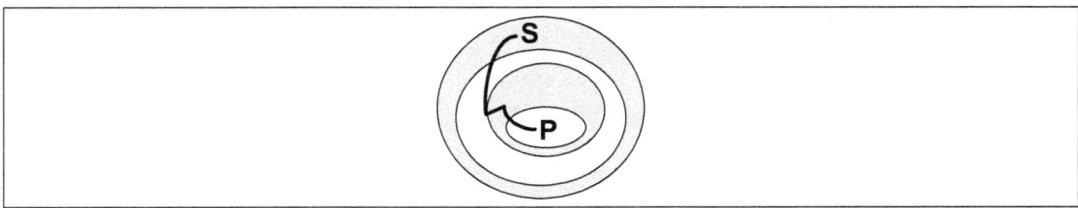

Figure 13: After erasing M in a shaded area

The Sixth step authorized by permission the third consists in getting rid of the empty ring shaped shaded area round the P, giving [Figure 14].

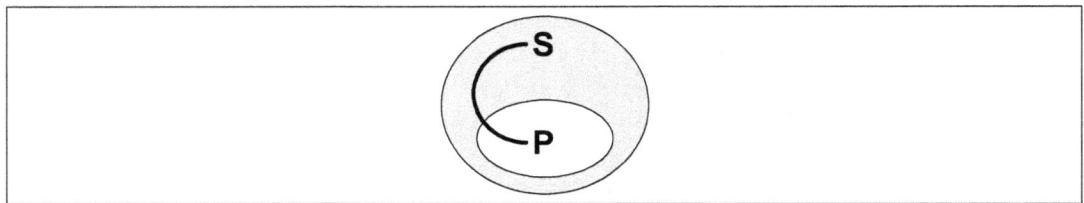

Figure 14: Conclusion of the syllogism

Peirce's rules are fundamentally semantic: each one inserts or erases one or more meaningful units. The differences between an EG proof and an EGIF proof result from syntactic details. In the proof of Barbara, the major differences result from using labels instead of lines of identity. Labels simplify the issues about lines crossing or touching borders, but they may require additional steps when lines of identity are joined.

For propositional logic, which Peirce called Alpha graphs, the meaningful units are relations and negations. For first-order logic with equality, Beta graphs add lines of identity and ligatures of lines. For EGIF, all meaningful units are expressed by *nodes*: relations, negations, and coreference nodes. Peirce treated functions as a special case of relations, but the EGIF grammar presented in the appendix adds nodes called *functions* to support the full semantics of Common Logic.

Since EGIF does not use shading, a positive area is defined by an even number of negations, and a negative area by an odd number of negations. The conditions for first-order logic with equality include those conditions with further constraints and operations on the labels for lines of identity. Although Peirce did not suggest a linear notation for EGs, he recognized that labels on the lines would be useful for complex graphs crossing lines of identity. He called those labels *selectives* and used capital letters as identifiers:

in any case in which the lines of identity become too intricate to be perspicuous, it is advantageous to replace some of them by signs of a sort that in this system are called *selectives*. [45, 4.460]

Any ligature may be replaced by replicas of one selective placed at every hook and also in the outermost area that it enters. In the interpretation, it is necessary to refer to the outermost replica of each selective first, and generally to proceed in the interpretation from the outside to the inside of all [ovals]. [45, 4.408]

Although Peirce did not suggest a linear notation for EGs, these two quotations summarize the crucial first step. But they omit one important step: If two or more ligatures are joined in an area that is nested inside the areas of their outermost points (defining nodes), their juncture must be represented by a coreference node. After that step, the lines are gone, and the EG becomes a nest of nodes. Ovals become nodes that contain other nodes. The only arbitrary choices are the symbols. Inside each oval, the order of listing the nodes is logically irrelevant. But the convention of putting defining labels before their bound labels is a convenience for anyone (human or computer) searching the lists.

The method for translating a well-formed EG to EGIF guarantees that the EGIF is also well formed: all syntactic and semantic constraints on the EGIF are satisfied. If all rules of inference are applied to the graphic EGs before the translation to EGIF, no other constraints need to be stated. But the EG rules combined with the method of translating EG to EGIF imply additional constraints on proofs that use EGIF directly.

These constraints show why human insight is important: the EG rules are easier to see than to describe. But ordinary computer programs cannot "see" the drawings. Therefore, some human who understands EGs and EGIF must design software to enforce the constraints. Before stating them, some definitions are necessary: the *scope* of a defining label; the EGIF equivalents for Peirce's verbs *prolong*, *join*, and *cut*; and another verb *simplify* for reducing the number of labels. Note to the reader: If you always derive EGIF from a well-formed EG, you can skip the next page and a half. Go to Figure 15 and its explanation.

- If a defining label d occurs in an area a, the *scope* of d is the area a and any area directly or indirectly enclosed by any negation in a. Any bound label in the scope of d that has the same identifier as d is said to be *bound* to d. The node that contains d must precede (occur to the left of) all nodes that directly or indirectly contain labels bound to d.

- No area may contain two or more defining labels with the same identifier. If the result of an insertion (by rule 1i or 2i) would violate that constraint, the identifier of the newly inserted defining label and all its bound labels shall be replaced by some identifier that does not violate the constraint.

- To *prolong* a line of identity with a defining label d into any area a in the scope of d is to insert a coreference node in a that has a single bound label that is bound to d. If the area a contains a defining label e with the same identifier as d, the identifier of e and all its bound labels shall be replaced by an identifier that is distinct from all other labels in the same EG.

- To *join* two lines of identity with defining labels d and e in any area a that is in the scope of both d and e is to insert a coreference node in a that contains bound labels for d and e and no others.

- To *cut* a line of identity with defining node d in an area a in the scope of d is to insert a defining node e whose defining label is distinct from all other labels whose scope includes a and to replace some or all of the labels bound to d that are in the scope of e with labels bound to e.

- To *simplify* the coreference nodes in an area a is to perform the following operations as often as they are applicable:

  1. If a coreference node in a contains two or more bound labels with the same identifier, erase all but one of them.
  2. If two coreference nodes in a each contain a bound label with the same identifier, they are *merged*: erase one of the coreference nodes, move all its bound labels to the other, and remove any duplicates according to operation 1.
  3. If a defining node d in a has a bound label in a coreference node c in a and c also contains a bound label b that is not bound to d, then the defining node d is erased, the label in c that was bound to d is erased, and every other label that was bound to d is replaced with a copy of b.
  4. If a coreference node c in area a contains exactly one bound label b, and area a also contains another node with a defining label or a bound label with the same identifier, then node c may be erased.

The basic constraint on erasing or inserting nodes is that no operation may leave or insert a bound label outside the scope of its defining label. Following are further conditions for Peirce's first and second permissions; no more conditions are needed for the third permission:

1. (a) In a negative area, no node that contains a bound label may be inserted in an area that is not in the scope of its defining label. No defining node may be inserted in an area that is in the scope of another defining label that has the same identifier.

    (e) In a positive area, no defining node that has one or more bound labels may be erased, unless all the nodes that contain those bound labels are erased in the same operation.

2. (a) If a defining node is iterated, the copy must be converted to a coreference node that contains a single bound label with the same identifier. A defining node that is enclosed in a negation, however, may remain unchanged when the negation is copied; but to avoid possible conflicts with future operations, the identifier of that defining label and all its bound labels should be replaced with an identifier that is not otherwise used.

    (e) Any nodes that could have been derived by rule 2i may be erased. (Whether or not a node had previously been derived by 2i is irrelevant.)

Peirce's meticulous attention to the smallest steps of reasoning enabled him to prove the soundness of every rule. Frege [9] assumed nine unprovable and non-obvious axioms. Whitehead and Russell [61] assumed five axiom schemata, of which one was redundant, but nobody discovered the redundancy for another 16 years. For EGs, only one axiom is necessary: a blank sheet of assertion, from which all the axioms and rules of inference by Frege, Whitehead, and Russell can be proved by Peirce's rules.

As an example, Figure 15 shows a proof of Frege's first axiom $a \supset (b \supset a)$. With the key words If and Then, that axiom in EGIF would be [If (a) [Then [If (b) [Then (a)]]]], but the proof is easier to see when the negations are shown explicitly.

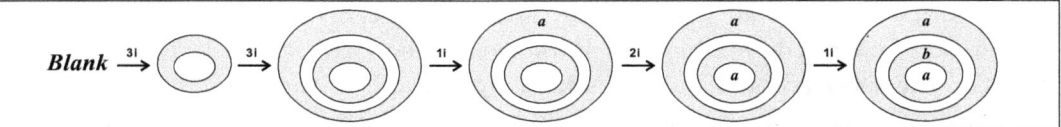

Figure 15: Proof of Frege's first axiom

The axiom in predicate calculus has five symbols, and each step of the EG proof inserts one symbol in its proper place. In EGIF, propositions are represented as relations with zero arguments: (a) and (b). Since there are no lines of identity, the EGIF proof takes the same five steps:

1. By rule 3i, Insert a double negation around a blank: ~[~[ ]]

2. By 3i, insert a double negation around the previous one:
   ~ [~ [~ [~[ ]]]]

3. By 1i, insert (a): ~[(a) ~[~[~[ ]]]

4. By 2i, iterate (a): ~[(a)~[~[~[(a)]]]]

5. By 1i, insert (b): ~[(a)~[~[(b)~[(a)]]]]

Frege's axiom contains five symbols, and each step of the proof inserts one symbol into its proper place in the final result. Frege's two rules of inference were *modus ponens* and *universal instantiation*. Figure 16 shows a proof of *modus ponens*, which derives q from a statement p and an implication $p \supset q$:

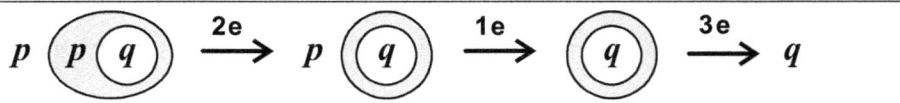

Figure 16: Proof of modus ponens

Proof in EGIF:

0. Starting graphs: (p) ~[(p)~ [(q)]]

1. By 2e, erase the nested copy of (p): (p) ~[~[(q)]]

2. By 1e, erase (p): ~[~[(q)]]

3. By 3e, erase the double negation: (q)

The rule of *universal instantiation* allows any term t to be substituted for a universally quantified variable in a statement of the form $(\forall x)P(x)$ to derive $P(t)$. In EGs, the term t would be represented by a graph of the form —t, in which t represents some monadic relation defined by a graph attached to the line of identity. The formal definition of EGIF in the appendix also supports functions; in EGs, a function could be represented with an arrowhead at the end of the line: ←t.

The universal quantifier $\forall$ corresponds to $\sim\exists\sim$, which is represented by a line whose outermost part occurs in a negative area. Since a pure graph has no named variables, there is no notion of substitution. Instead, the proof in Figure 17 performs the equivalent operation by joining two lines.

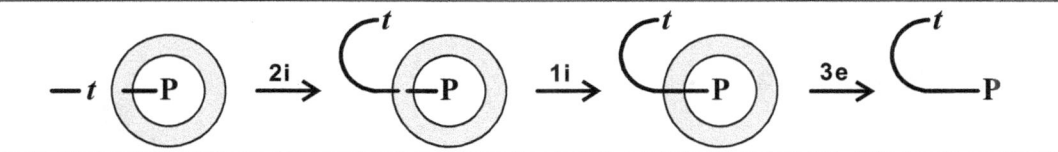

Figure 17: Proof of universal instantiation

In predicate calculus, the proof from (∃x)t(x) and (∀y)P(y) to (∃x)(t(x) ∧ P(x)) would take one step. But the three EG steps in Figure 17 are easy to see and explain. In EGIF, the step of joining two lies by 2i requires two steps to relabel identifiers and simplify the result: In predicate calculus, the process of relabeling is just as complex, but it's not called a separate step.

0. Starting graphs: (t *x) ~[[*y] ~[(P y)]]

1. By 2i, iterate *x (prolong the line) by inserting [x]:
   (t *x) ~[[x] [*y] ~[(P y)]]

2. By 1i, join x and *y by replacing *y and every occurrence of y with x:
   (t *x) ~[[x] [x] ~[(P x)]]

3. By 2e, delete the two unnecessary copies of [x]:
   (t *x) ~[~[(P x)]]

4. By 3e, erase the double negation: (t *x) (P x)

In the *Principia Mathematica*, Whitehead and Russell proved a theorem that Leibniz called the *Praeclarum Theorema* (Splendid Theorem): $((p \supset x) \land (q \supset s)) \supset ((p \land q) \supset (r \land s))$. Starting with five axioms, their proof took a total of 43 steps. With Peirce's rules, this theorem can be proved in just seven steps, starting with a blank sheet of paper (Figure 18).

The first three steps of Figure 18 illustrate a proof procedure that can be used to prove every theorem of mathematics: start with a blank sheet of paper, draw a double negation around a blank area, insert the hypothesis in the shaded area, and copy the hypothesis, or subgraphs of it, into the unshaded area. Section 6 shows that these steps are a simplification and generalization of Gentzen's system of natural deduction.

All five proofs in this section suggest a useful heuristic for guiding the proof: Every proof proceeds in a straight line from premises to conclusion, and most steps tend to make the current EG look more like the desired conclusion. Therefore, begin a proof by drawing an EG for the conclusion and select options that can transform

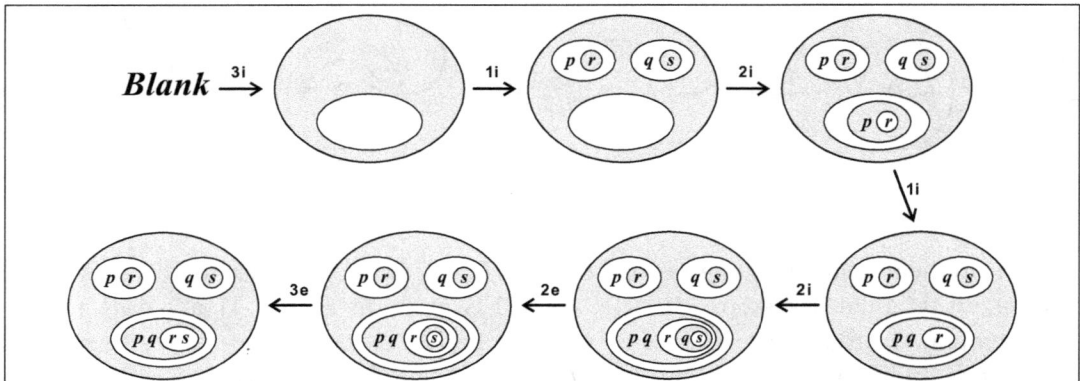

Figure 18: Proof of Leibniz's Praeclarum Theorema

the current EG to one that looks more like the conclusion. After only four steps, the graph in Figure 18 looks almost like the desired conclusion, except for a missing copy of s in the innermost area. Since that area is unshaded, the only way to get s in there is to iterate some graph or subgraph that contains s and erase the parts that are not needed. Step five iterates (by rule 2i) the only subgraph that contains s, and the last two steps erase parts that are not needed. The EGIF proof takes the same seven steps, but they're harder to visualize:

1. By 3i, draw a double negation around the blank: ~[~[ ]]

2. By 1i, insert the hypothesis in the negative area:
   ~[~[(p) ~[(r)]] ~[(q) ~[(s)]] ~[ ]]

3. By 2i, iterate the left part of the hypothesis into the positive area:
   ~[~[(p) ~[(r)]] ~[(q) ~[(s)]] ~[~[(p) ~[(r)]]]]

4. By 1i, insert (q):
   ~[~[(p) ~[(r)]] ~[(q) ~[(s)]] ~[~[(p) (q) ~[(r)]]]]

5. By 2i, iterate the right part of the hypothesis into the innermost area:
   ~[~[(p) ~[(r)]] ~[(q) ~[(s)]]
       ~[~[(p) (q) ~[(r)] ~[(q) ~[(s)]]]]]

6. By 2e, deiterate (q): ~[~[(p) ~[(r)]] ~[(q) ~[(s)]] ~[~[(p) (q) ~[(r)] ~[~[(s)]]]]]

7. By 3e, erase the double negation to generate the conclusion:
   ~[~[(p) ~[(r)]] ~[(q) ~[(s)]] ~[~[(p) (q) ~[(r)] (s)]]]

The visual heuristics that help teachers explain an EG proof can also help students discover proofs by themselves. These heuristics are especially evident in Section 5, which shows how Euclid's diagrams can be inserted or erased in the areas of an existential graph.

4 EG Semantics: Endoporeutic

In his first publication on model theory, Tarski (1933) quoted Aristotle as a precedent. The medieval Scholastics developed Aristotle's insights further, and Ockham (1323) presented a model-theoretic analysis of Latin semantics. Although Ockham wasn't as formal as Tarski, he covered the Latin equivalents of the Boolean connectives, the existential quantifier (*aliquis*), the universal quantifier (omnis), and even modal, temporal, and causal terms. Peirce lectured on Ockham's logic at Harvard, and he defined logic as "the formal science of the conditions of the truth of representations" [45, 2.220]. To see the similarity, compare quotations by Peirce and Tarski:

- Peirce [42]: "All that the formal logician has to say is, that if facts capable of expression in such and such forms of words are true, another fact whose expression is related in a certain way to the expression of these others is also true.... The proposition 'If A, then B' may conveniently be regarded as equivalent to 'Every case of the truth of A is a case of the truth of B'."

- Tarski [59]: "In terms of these concepts [of model], we can define the concept of logical consequence as follows: *The sentence X follows logically from the sentences of the class K if and only if every model of the class K is also a model of the class X.*"

Peirce's most important contribution to model theory was the method of *endoporeutic*, which he defined as an "outside-in" method for evaluating the truth of an existential graph. Before presenting his rules of inference, Peirce briefly summarized endoporeutic [48, 3:165], which he used to justify each rule:

> The rule of interpretation which necessarily follows from the diagrammatization is that the interpretation is "endoporeutic" (or proceeds inwardly) that is to say a ligature denotes "something" or "anything not" according as its **outermost part** lies on an unshaded or a shaded area respectively.

For every model M and proposition p, endoporeutic determines the same truth value as Tarski's definition. But Peirce's method is based on a two-person game between

Graphist, who proposes an EG, and Grapheus, a skeptic who demands a proof. In fact, endoporeutic is a version of *game theoretical* semantics (GTS), which was developed by Hintikka [19]. The similarity between endoporeutic and Hintikka's GTS was first discovered by Hilpinen [16]. For their introduction to model theory, Barwise and Etchemendy [1] chose the title *Tarski's World*, but what they presented was GTS.

In modern terminology, endoporeutic can be defined as a *two-person zero-sum perfect-information game*, of the same genre as board games like chess, checkers, and tic-tac-toe. Unlike those games, which frequently end in a draw, every finite EG determines a game that must end in a win for one of the players. But Henkin [15], the first modern logician to rediscover the GTS methods, showed that it could even evaluate the denotation of some infinitely long formulas in a finite number of steps. Peirce also considered the possibility of infinite EGs: "A graph with an endless nest of [ovals] is essentially of doubtful meaning, except in special cases" [45, 4.494]. As an example, he showed an infinite graph (one in which a certain pattern is repeated endlessly) for which endoporeutic would stop in just 2 or 3 steps. The version of endoporeutic presented here is based on Peirce's writings supplemented with ideas adapted from Hintikka, Hilpinen, Pietarinen [50], and the game-playing methods of artificial intelligence.

In the game of endoporeutic, Graphist proposes an existential graph g and claims that g is true. But Grapheus is a skeptic or devil's advocate who tries to show that g is false. The game begins with some state of affairs M, which corresponds to a Tarski-style model $M = (D, R)$: the domain D is a set of individuals, and R is a set of relations defined over D. The model M, like any Tarski-style model, can be represented as an EG with with no negations: each individual in D is represented by a line of identity; each n-tuple of each relation r in R is represented by a copy of the string that names r with n pegs attached to n lines of identity. The game proceeds according to a recursive procedure with Graphist as the initial proposer and Grapheus as the initial skeptic. During the game, they switch sides as they peel off each negation:

1. If g contains no negations, then the game is over, and the winner is determined by one of three possible cases:

    - If g is an empty graph, it is true because it says nothing false. The current proposer wins.
    - Else if there is a mapping (graph homomorphism) of g to M, the current proposer wins. The mapping need not be an isomorphism, since multiple lines of identity in g may map to the same line in M. Each relation node in g must map to a relation node in M with the same name.

- Else the current skeptic wins.

2. Else if g consists of just a single negation, the graph inside the oval becomes the new g, and the shading of each area in g is reversed. Then the two players switch sides: the proposer becomes the new skeptic; the skeptic becomes the new proposer; and the game continues with the new g and the new roles for both players.

3. Else if g consists of two or more negations, the skeptic chooses any one of the negations as the new g, and the game continues.

4. Else g consists of two or more parts: a subgraph g_0, which has no negations, and one or more negations: $g_1, ..., g_n$. The graph g_0 may contain some lines of identity that are joined to lines inside one or more of the negations. There are now two possibilities:

    - If there is no mapping of g_0 to M, the game is over, and the skeptic wins.
    - Else one or more mappings are possible, and the proposer may choose any one. That choice maps every line of identity x in g_0 to some line y in M. If x is joined to any line z in any negation g_i, then z must remain mapped to y for the duration of the game. Then the subgraph g_0 is erased, leaving g as a conjunction of one or more negations, and the game continues.

Since each pass through this procedure reduces the size of g, any game that starts with a finite graph must terminate in a finite number of moves. Since none of the ending conditions results in a draw, either Graphist, the original proposer, or Grapheus, the original skeptic, must have a winning strategy for any g and model M. If Graphist has a winning strategy, g is true of M. If Grapheus has a winning strategy, g is false of M.

These rules can evaluate Alpha graphs (propositional logic) in exactly the same way as Beta graphs. A model M for propositional logic would be a set of medads, such as {(p), (q), (t)}. A graph g with no negations is true of M if the set of medads in g is a subset of the medads in M. As an exercise, evaluate the EG for the Praeclarum Theorema in terms of some set of medads. Since it is a theorem, it should be true of every model, including the empty set. To illustrate these rules for Beta graphs (FOL), let the EG in Figure 19 be the model M. The three subgraphs state "There is lightning at 1 pm, there is thunder and lightning at 2 pm, and there is thunder at 3pm."

As the first example, let the graph g be `lightning—at—2pm`. Since g has no negations, Graphist, the original proposer, wins because there is a mapping from g

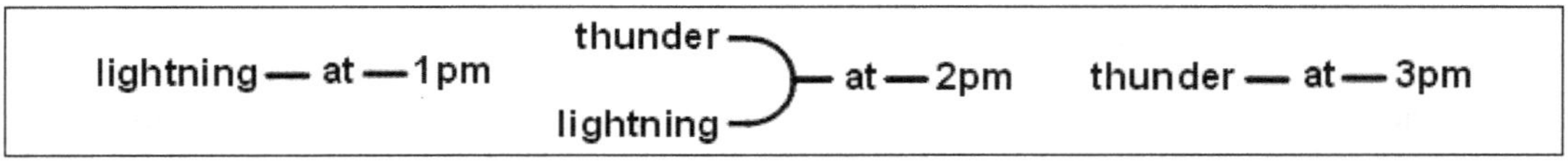

![Figure 19](lightning—at—1pm, thunder/lightning—at—2pm, thunder—at—3pm)

Figure 19: An EG used as the model M

to the middle part of M. Therefore, that graph is true. But if g happened to be `lightning`—at—3pm, then Grapheus, the original skeptic, would win because this g cannot be mapped to M. Therefore, it is false.

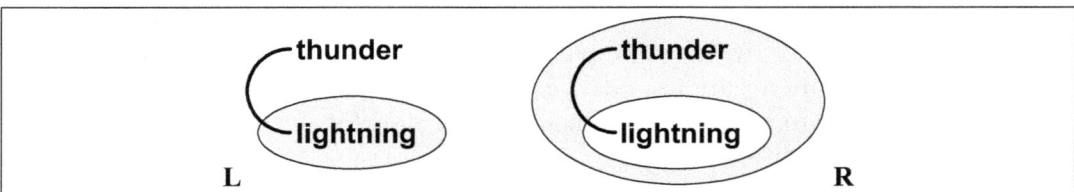

Figure 20: Two EGs, L and R, to be evaluated in terms of M

As an example that requires more than one pass through the procedure, consider the graph L on the left of Figure 20. L represents the sentence "There is thunder, and not lightning," and it meets the conditions for Step 4 of endoporeutic: the subgraph g_0 for the part outside the negation is `-thunder`. Graphist, the proposer, can choose any of two possible mappings of g_0 to M: one for thunder at 2 pm and the other for thunder at 3 pm; this choice also determines the mapping of the line inside the negation. If Graphist chooses 3 pm, the subgraph g_0 is erased, and the game continues. At Step 2, the negation is erased, causing the new g to become `-lightning`, but with the line forced to be mapped to the subgraph of M at 3 pm. Then the two players change sides, making Graphist the new skeptic. The game continues at Step 1, where g cannot be mapped to M because lightning occurred at 2 pm, not 3 pm. Therefore, the skeptic wins because this subgraph with this mapping is false. But the current skeptic is Graphist, who had been the original proposer for the graph L, which is therefore true. Note that if Graphist had made a mistake in the original choice of mapping, then Graphist would have lost. The truth of a graph, however, is not determined by mistakes; it depends only on the existence of a winning strategy if both players make the best choice at each option. As a final example, the graph R on the right adds one more negation. Therefore, R meets the condition for Step 2 of the game, which removes the negation and causes the two players to change sides. That makes Grapheus the proposer for a graph that is now identical to L, which has a winning strategy. Therefore, Grapheus wins the game that started with R. Since a win by Grapheus implies a loss by Graphist, the graph

R is false.

To prove that Peirce's rules of inference are sound, it is necessary to show that each rule preserves truth, i.e., if Graphist has a winning strategy with a given EG, no rule of inference can transform it to a graph that allows Grapheus to win. Therefore, the rules must monotonically increase the winning options for Graphist and monotonically decrease the winning options for Grapheus. Since the two players switch sides when a negation is removed (Step 2), Graphist is the proposer in every positive area and the skeptic in every negative area. The two steps of endoporeutic that depend on the number of options are Step 3, where the skeptic chooses one of the negations, and Step 4, where the proposer chooses one of the possible mappings of g_0 to M.

1. Rule 1e erases an arbitrary subgraph in a positive area. Erasing a negation decreases the number of options for Grapheus, who is the skeptic in a positive area. Erasing all or part of a relational graph reduces the constraints on the mapping to the model M and thereby increases the options for Graphist. Rule 1i, which inserts an arbitrary subgraph in a negative area, has the opposite effects: If negations are added, Graphist, who is the skeptic, has more options to choose. Adding a subgraph to a relational graph adds more constraints on the mapping to M and thereby reduces the options for Grapheus.

2. Rules 2e (deiteration) and 2i (iteration) have no effect on the truth value of any graph because any subgraph g that could be iterated or deiterated has its effect on the evaluation at its first occurrence. There are two possibilities to consider: g is true, or g is false. If g is true, a copy of g inserted or erased from any area has no effect on the truth value of that area. If g is false, the original area in which g occurs is false; the truth value of any nested area is irrelevant.

3. The only effect of evaluating a double negation is to cause the proposer and skeptic to reverse their roles twice. Therefore, Rules 3e (erasing a double negation) and 3i (inserting a double negation) can have no effect on the truth value of any graph.

In summary, rules 1e and 1i monotonically increase the options for Graphist and monotonically decrease the options for Grapheus. Rules 2e, 2i, 3e, and 3i have no effect on the winning strategies for either side. Therefore, if an EG is true, the EG that results from applying these rules will also be true. Therefore, all the rules are *sound*, because they preserve truth. Peirce's rules are also complete for FOL because the rules for other complete systems can be proved in terms of them See Figures 15, 16 and 17.

What distinguishes the game-theoretical method from Tarski's approach is its procedural nature. Unlike Tarski's definition, which maps all variables in a formula to individuals in a model M, endoporeutic is a "lazy" method that avoids mapping a line to M until the proposer chooses it in Step 4. Any subgraph that is not chosen is never evaluated; some very large or even infinite subgraphs can often be ignored.

One reason why Peirce had such difficulty in explaining it is that he and his readers lacked the terminology of the game-playing algorithms of artificial intelligence. No one clearly deciphered Peirce's writings until Hilpinen noticed the similarity to GTS. The discussion in this section is a reconstruction and clarification in modern terminology.

5 Formalizing Euclid

To formalize Euclid's geometry in existential graphs, map the diagrams to EGs, represent the EGs in EGIF, and justify every step of Euclid's proofs by EG rules of inference supplemented with two special rules: observation and imagination. In his writings on diagrammatic reasoning, Peirce described diagrams as "mainly" icons, but he noted that they may also contain symbols (symbolide features) and "features approaching the nature of indices" such as names or labels:

> A Diagram is mainly an Icon, and an Icon of intelligible relations... Now since a diagram, though it will ordinarily have Symbolide Features, as well as features approaching the nature of Indices, is nevertheless in the main an Icon of the forms of relations in the constitution of its Object, the appropriateness of it for the representation of necessary inference is easily seen. [45, 4.531]

For mathematics, Peirce distinguished two kinds of diagrams: geometrical and algebraic. He also discussed physical models, stereoscopic images, and motion pictures as aids to reasoning. Modern discussions of theorem proving ignore diagrams and never talk about observation or imagination. But every computer proof by any algorithm makes observations (searches for patterns) and performs experiments in the imagination (tentative constructions that may succeed or fail in later steps of the proof). In fact, observation and imagination are fundamental for any method of reasoning, formal or informal: (1) observe the current diagram (physical or mental); (2) imagine or construct the next one. As Peirce said,

> all deductive reasoning, even simple syllogism, involves an element of observation; namely deduction consists in constructing an icon or diagram the relation of whose parts shall present a complete analogy with

those of the parts of the object of reasoning, of experimenting upon this image in the imagination, and of observing the result so as to discover unnoticed and hidden relations among the parts. [45, 3.363]

While rereading Euclid's *Elements*, Peirce discovered a distinction between a theorem and a corollary. For every major theorem, Euclid drew a new diagram. But he used the same diagram to observe the patterns stated as corollaries. Mathematicians have used those terms for centuries, but Peirce was the first to state a clear distinction between them: the proof of a theorem requires "an experiment in the imagination"; the proof of a corollary just makes observations about an already given or imagined diagram. Yet logicians today are still searching for syntactic criteria to distinguish theorems and corollaries. But Peirce's distinction is semantic, not syntactic:

> Corollarial deduction is where it is only necessary to imagine any case in which the premisses are true in order to perceive immediately that the conclusion holds in that case. Ordinary syllogisms and some deductions in the logic of relatives belong to this class. Theorematic deduction is deduction in which it is necessary to experiment in the imagination upon the image of the premiss in order from the result of such experiment to make corollarial deductions to the truth of the conclusion. [46, 35–39]

As an example, Figure 21 is Euclid's diagram for Proposition 1: "On a given finite straight line to construct an equilateral triangle." That statement mentions a straight line and a triangle that consists of three straight lines. But it does not mention circles. The insight to imagine the circles requires creativity. To make the proofs intelligible by students, Euclid added the results of imagination to the diagram for each theorem. For Figure 21, an experiment in the imagination added two circles.

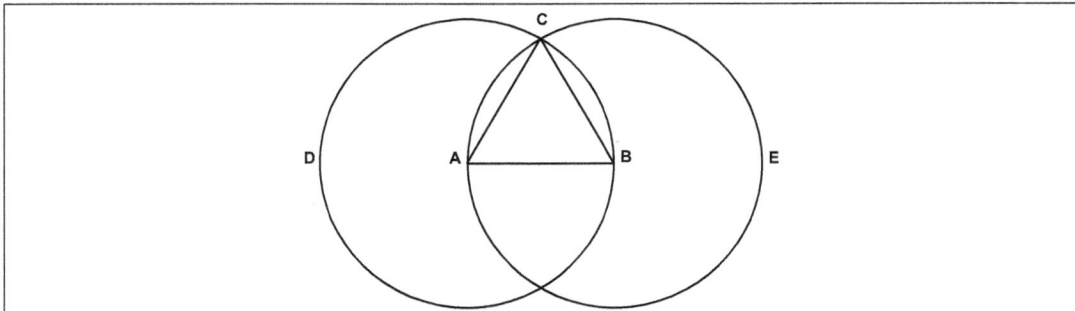

Figure 21: Diagram for Euclid's Proposition 1

Since Figure 21 already includes the result of imagination, the remainder of the proof uses only observations and applications of axioms (postulates and common notions): Apply Postulate 3 to draw two circles. Observe point C at the intersection of the two circles. Draw two more lines by Postulate 1. Observe that the two new lines are radii of one circle or the other. Finally, use Common Notion 1 to show that the three lines are congruent. Euclid's proof adds letters that relate each step to points on the diagram. In Heath's [6] translation,

> Let AB be the given finite straight line. Thus it is required to construct an equilateral triangle on the straight line AB. With center A and distance AB, let the circle BCD be described [Postulate 3]. Again with center B and distance BA, let the circle ACE be described [Post. 3]. And from the point C, in which the circles cut one another, to the points A, B let the straight lines CA, CB be joined [Post. 1]. Now, since the point A is the center of the circle CDB, AC is equal to AB [Definition 15]. Again, since the point B is the center of the circle CAE, BC is equal to BA [Def. 15]. But CA was also proved equal to AB. Therefore, each of the straight lines CA, CB is equal to AB. And things which are equal to the same thing are also equal to one another [Common Notion 1]. Therefore, CA is also equal to CB. Therefore, the three straight lines CA, AB, BC are equal to one another. Therefore, the triangle ABC is equilateral, and it has been constructed on the given finite straight line. QED [6]

Peirce maintained that continuity is the basis for generality. For Figure 21, there are no constraints on the diagram: nothing in the statement of Proposition 1 or its proof depends on the size, position, or orientation of the line AB. On any plane, there is an uncountable infinity of possible line segments, and Euclid's method could be used to construct an equilateral triangle on any of them.

Another criticism of Euclid's proofs is that they are not as precisely specified as modern algebraic versions. To address that criticism, every proof by Euclid can be translated, line by line, to operations on existential graphs, and every step of the graphic proof may be stated in the formally defined EGIF notation. Since there is a one-for-one correspondence, both notations can be used. Graphic EGs are more readable and teachable. They conform to the way people think, and they are language independent. For algebraic rigor, every step of an EGIF proof is sufficiently precise to be implemented in a computer program.

Finally, Euclid's proofs are stated as procedures that construct a figure, not as logical assertions that can be proved. Then the proof mixes imperative statements about what to draw with declarative statements about the result. That issue can be addressed by translating Euclid's imperatives to declaratives. For example, his Proposition 1, "On a given finite straight line to construct an equilateral triangle," may be expressed as a conditional: "If there is a finite straight line AB, then there is an equilateral triangle with AB as one of its sides." In the EG or EGIF proof, every imperative sentence beginning with *Let* may be used as a directive for the next step.

Since icons in an EG cannot occur in EGIF, icons shall be replaced by names for the figure types: Point, Line, Circle, Triangle. To name the instances of a figure type, Euclid concatenated the letters that name the points in the order they were used to construct that instance. For example, a line drawn from point A to point B would be named AB. But the same line, if drawn from B to A would be named BA. To guarantee unique names in EGIF, the EGIF names AB and BA shall be written in alphabetical order as AB. In case of name conflicts, the name of the figure type shall be concatenated at the end: `XYZ_circle` or `XYZ_triangle`.

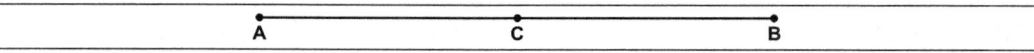

Figure 22: The line AB has a midpoint C

Any translation of English to logic requires some ontology of the relevant entities and relations. To define an ontology that describes Figure 22, begin with an English description of the details of interest: "There is a line AB from point A to point B. The point C is on AB. The lines AC and BC are congruent." The following list of relations constitutes a small ontology that may be used to express the English description in EGs or EGIF:

- (`Point X`): X is a point.
- (`Line XY X Y`): XY is a line with endpoints X and Y.
- (`On Z XY`): point Z is on line XY.
- (`"≅" XZ YZ`): XZ and YZ are congruent.

With this ontology, the English description of Figure 22 may be translated to EGIF:

(Point *A) (Point *B) (Line *AB A B) (Point *C) (On C AB)
(Line *AC A C) (Line *BC B C) ("≅" AC BC)

Names that begin with an asterisk, such as *A or *AB, are defining labels with an implicit existential quantifier. Names with the same spelling, but without an asterisk are bound labels that refer to the same entity as their defining label. The character ≅ represents the congruence relation. Since it is not an alphanumeric character, its EGIF name is enclosed in double quotes as "≅". The details of EGIF are specified in the appendix of this article.

Figure 22, by the way, raises an issue that has been debated since antiquity. Zeno and Cantor assumed that a line consists of an infinity of points. But Aristotle and Peirce maintained that points are markers on a line, not parts of it [26]. With Cantor's ontology, it's impossible to break the line AB in two congruent halves, because the midpoint C could not be a part of both halves. To avoid this issue, assume the Aristotle-Peirce ontology: the parts of a line are shorter line segments. Therefore, the midpoint is just a marker, and the two halves are congruent.

Given these preliminaries, write the EGIF proof by following Euclid's proof, line by line. The first sentence states the if-part of a conditional, and the second states the implication: "Let AB be the given finite straight line. Thus it is required to construct an equilateral triangle on the straight line AB." In English, if there is a finite straight line AB, then there is an equilateral triangle with AB as one of its sides. In EGIF, this statement would be written

```
[If (Point *A)(Point *B)(Line *AB A B)]
    [Then (Point *C) (Line *AC A C) (Line *BC B C)
    (Triangle *ABC AB AC BC) ("≅" AB AC BC)]]
```

Literally, this says that if there exist a point A, a point B, and a line AB from A to B, then there exist a point C, a line AC from A to C, a line BC from B to C, and a triangle ABC with congruent sides AB, AC, and BC. Common Logic allows relations to be *polyadic* (having any number of arguments). Therefore, the relation named "≅" may be used to assert that any number of figures are congruent.

All that detail is shown by the EG in Figure 23. To draw it, insert line AB in the shaded area (the hypothesis); insert, triangle ABC in the unshaded area (the conclusion); draw lines of identity (shown in red) from A and B in the shaded area to the unshaded area; insert the congruent symbol ≅ inside the triangle and draw lines of identity to each of the three sides.

Note that Figure 23 is language independent: all the details are shown, not said. If necessary, any detail may be observed and stated in any version of language or logic. As Peirce said, an icon plus an indexical can state a proposition. The letters A, B, and C are indexicals that correspond to pronouns in natural languages. In EGs, they serve the same purpose as lines of identity. Therefore, the two red lines from A to A and B to B are redundant. Either the lines or the letters could be deleted. Since the letters are the same as Euclid's, they are kept in Figure 24.

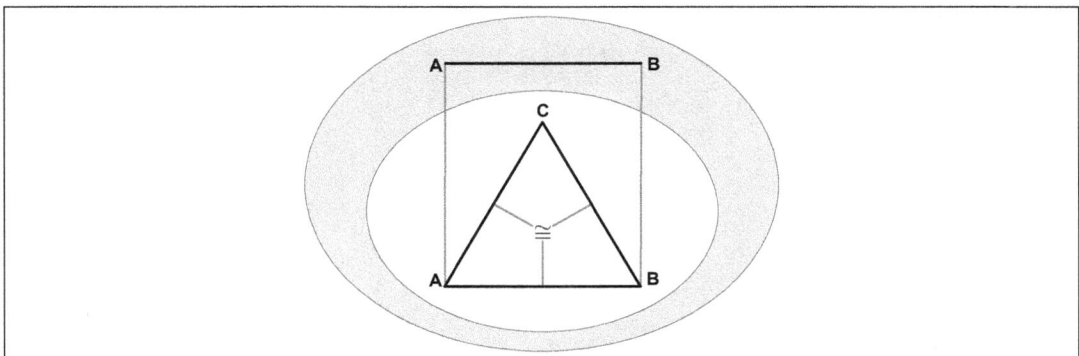

Figure 23: EG for Euclid's Proposition 1

But the three red lines attached to the congruent symbol are retained. The EGs in Figure 23 or 24 represent the conclusion to be proved.

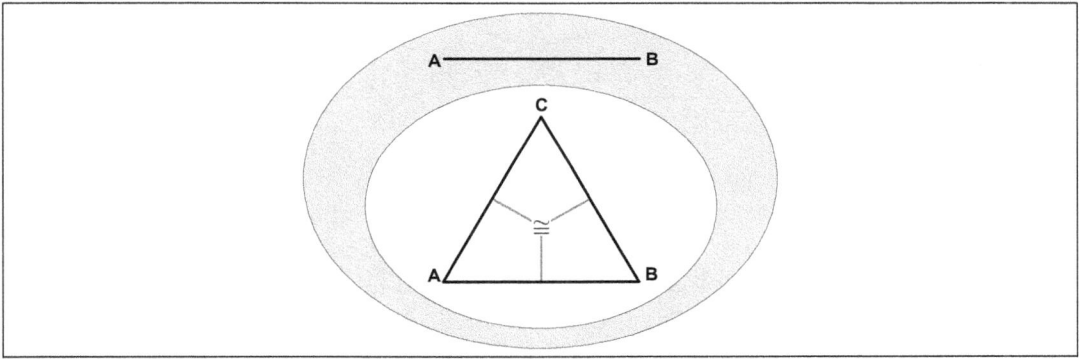

Figure 24: Keep copies of the letters A and B, but delete the connecting lines

Before starting the proof, begin with a blank sheet of assertion (SA), and assert EG diagrams or EGIF sentences for the postulates and common notions to be used in the proof. The first is Postulate 1. As Euclid stated it, "To draw a straight line from any point to any point." As a conditional, "If there are two points, then there exists a line from one point to the other point." Figure 25 shows the EG. The corresponding EGIF is [If (Point *X) (Point *Y) [Then (Line *XY X Y)]].

The next is Postulate 3: "To describe a circle with any center and distance." As a conditional, "If there are two points, then there exists a circle with one point at its center and a radius from the center to the other point." Figure 26 shows the EG. The EGIF is [If (Point *X) (Point *Y) [Then (Circle *XYZ) (Center XYZ X) (Radius XYZ XY)]].

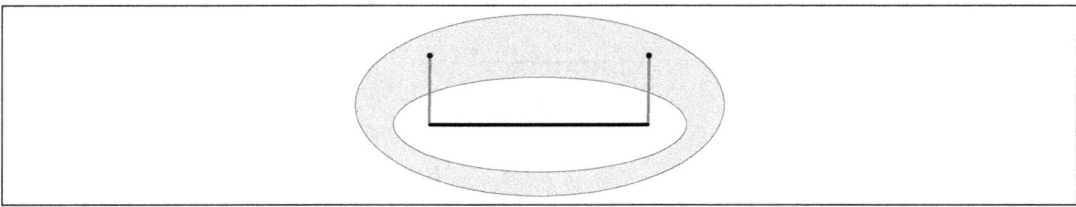

Figure 25: EG for Postulate 1

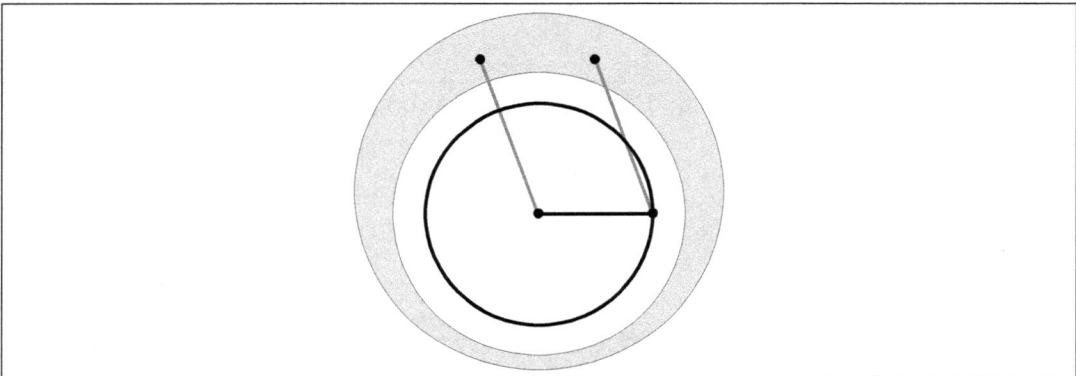

Figure 26: EG for Postulate 3

Finally, Common Notion 1 as Euclid stated it: "Things which are equal to the same thing are also equal to one another." As a conditional, "If X is congruent to Z and Y is congruent to Z, then X is congruent to Y." The EGIF for Figure 27: [If ("≅" *X *Z) ("≅" *Y Z) [Then ("≅" X Y)]].

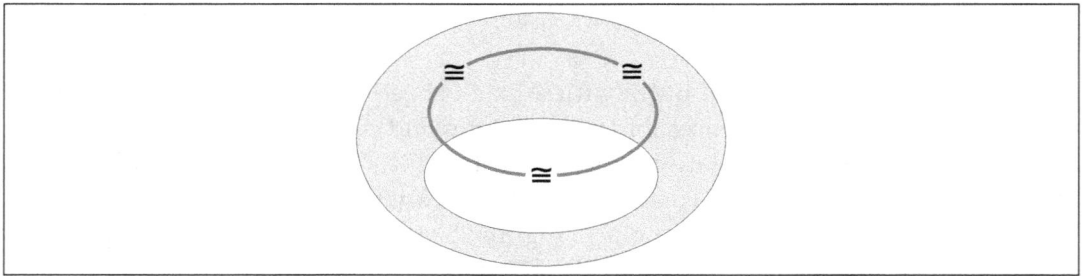

Figure 27: EG for Common Notion 1

Note that Aristotle's definition uses the plural word *things*. That implies two or more, and the three red lines in Figure 24 show the congruence relation as triadic. Since Common Logic allows polyadic relations with any number of arguments, that option may be specified by a recursive definition. For this proof, the triadic option is sufficient, and it may be specified with one addition to the EGIF:

[If ("≅" *X *Z) ("≅" *Y Z) [Then ("≅" X Y) ("≅" X Y Z)]].

But the word *things* implies an unordered set. More axioms would be needed to state that all possible permutations of the arguments of ≅ are equivalent. To avoid those details, just assume that ≅ is a dyadic or triadic relation and that the order of the arguments is irrelevant.

Given the two postulates and the common notion on the Sheet of Assertion, the first step of the proof uses rule 3i to draw a double negation around a blank area anywhere on SA. To make it more readable in EGIF, use the keywords If and Then:

1. [If [Then]]

By 1i, insert the hypothesis into the negative area:

2. [If (Point *A) (Point *B) (Line *AB A B) [Then]]

By 2i, copy the EG from the shaded area to the unshaded area; also copy the labels A and B instead of drawing lines of identity. The result is the EG in Figure 28. For EGIF, indicate the copies by inserting coreference nodes for A, B, and AB. The bound labels in the coreference nodes indicate the presence of the corresponding entities in that area.

3. [If (Point *A) (Point *B) (Line *AB A B) [Then [A] [B] [AB]]]

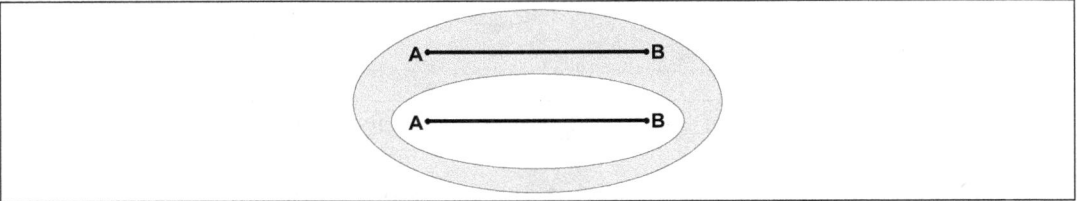

Figure 28: EG after step 3 of the proof

The next two sentences express Euclid's insight that circles are required: "With center A and distance AB, let the circle BCD be described [Postulate 3]. Again with center B and distance BA, let the circle ACE be described [Post. 3]." But his choice of names for the two circles indicate that has already made the observation that the two circles will intersect at point C. That observation is obvious from a construction of the diagram in an EG, but it is not obvious in EGIF. By 2i, insert Postulate 3 into the positive area:

4. [If (Point *A) (Point *B) (Line *AB A B)
   [Then [A] [B] [AB]
     [If (Point *X) (Point *Y)
     [Then (Circle *XYZ) (Center XYZ X) (Radius XYZ XY)]]]]

Since modus ponens is a derived rule of inference for EGs, it can be used to shorten EG proofs. By two applications of modus ponens, infer the circles BCD and ACE in Figure 29:

5. [If (Point *A) (Point *B) (Line *AB A B)
   [Then [A] [B] [AB]
     [If (Point *X) (Point *Y)
     [Then (Circle *XYZ) (Center XYZ X) (Radius XYZ X Y)]]
     (Circle *BCD) (Center BCD A) (Radius BCD AB)
     (Circle *ACE) (Center ACE B) (Radius ACE AB)]]

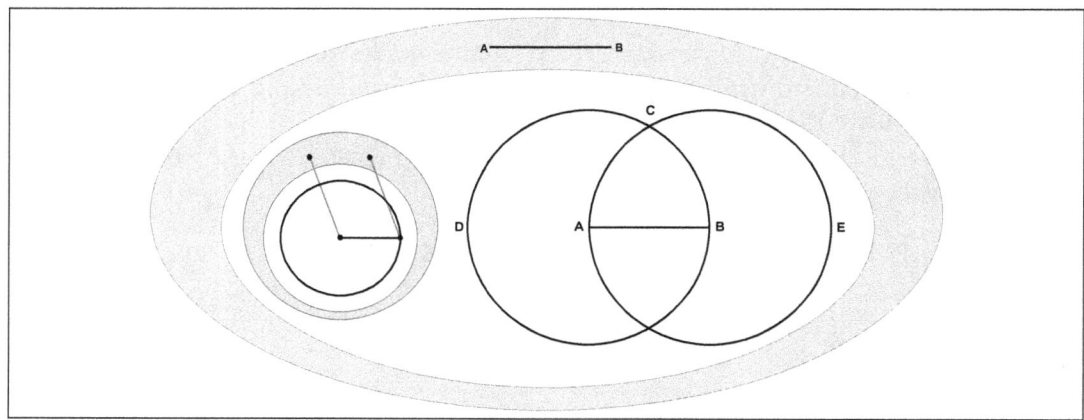

Figure 29: EG after step 5 of the proof

Since Postulate 3 is no longer needed, it may be erased by 1e:

6. [If (Point *A) (Point *B) (Line *AB A B)
   [Then [A] [B] [AB]
     (Circle *BCD) (Center BCD A) (Radius BCD AB)
     (Circle *ACE) (Center ACE B) (Radius ACE AB)]]

Euclid: "And from the point C, in which the circles cut one another, to the points A, B let the straight lines CA, CB be joined [Post. 1]." This sentence makes an observation that there exists a point C where the circles intersect. By 2i, insert Postulate 1 in the positive area:

7. [If (Point *A) (Point *B) (Line *AB A B)
   [Then [A] [B] [AB]
      (Circle *BCD] (Center BCD A) (Radius BCD AB)]
      (Circle *ACE] (Center ACE B) (Radius ACE AB)
      (Point *C) (On C BCD) (On C ACE)]
      [If (Point *X) (Point *Y) [Then (Line *XY) (Line XY X Y)]]]]

By two applications of modus ponens, infer lines AC and BC. These are the same lines that Euclid named CA and CB, but with the letters written in alphabetical order.

8. [If (Point *A) (Point *B) (Line *AB A B)
   [Then [A] [B] [AB]
      (Circle *BCD] (Center BCD A) (Radius BCD AB)
      (Circle *ACE] (Center ACE B) (Radius ACE AB)
      (Point *C) (On C BCD) (On C ACE)
      [If (Point *X) (Point *Y) [Then (Line *XY) (Line XY X Y)]]
      (Line *AC) (Line *BC)]]

By 1e, erase Postulate 1 in the positive area:

9. [If (Point *A) (Point *B) (Line *AB A B)
   [Then [A] [B] [AB]
      (Circle *BCD] (Center BCD A) (Radius BCD AB)
      (Circle *ACE] (Center ACE B) (Radius ACE AB)
      (Point *C) (On C BCD) (On C ACE)
      (Line *AC) (Line *BC)]]

"Now, since the point A is the center of the circle CDB, AC is equal to AB [Definition 15]. Again, since the point B is the center of the circle CAE, BC is equal to BA [Def. 15]." These two sentences depend on the observations that line AC is a radius of circle BCD and that BC is a radius of ACE. Therefore, AC is congruent to AB, and BC is congruent to AB:

10. [If (Point *A) (Point *B) (Line *AB A B)
    [Then [A] [B] [AB]
       (Circle *BCD] (Center BCD A) (Radius BCD AB)
       (Circle *ACE] (Center ACE B) (Radius ACE AB)
       (Point *C) (On C BCD) (On C ACE)
       (Line *AC) (Radius BCD AC) ("≅" AC AB)
       (Line *BC) (Radius ACE BC) ("≅" BC AB)]]

"But CA was also proved equal to AB. Therefore, each of the straight lines CA, CB is equal to AB. And things which are equal to the same thing are also equal to one another [Common Notion 1]. Therefore, CA is also equal to CB. Therefore, the three straight lines CA, AB, BC are equal to one another." By 2i, insert Common Notion 1 in the positive area and apply it to infer the dyadic and triadic congruences in the last line:

11. [If (Point *A) (Point *B) (Line *AB A B)
    [Then [A] [B] [AB]
      (Circle *BCD] (Center BCD A) (Radius BCD AB)
      (Circle *ACE] (Center ACE B) (Radius ACE AB)
      (Point *C) (On C BCD) (On C ACE)
      (Line *AC) (Radius BCD AC) ("≅" AC AB)
      (Line *BC) (Radius ACE BC) ("≅" BC AB)
      [If ("≅" *X *Z) ("≅" *Y Z)
      [Then ("≅" X Y) ("≅" X Y Z)]]
      ("≅" AB AC) ("≅" AB AC BC)]]

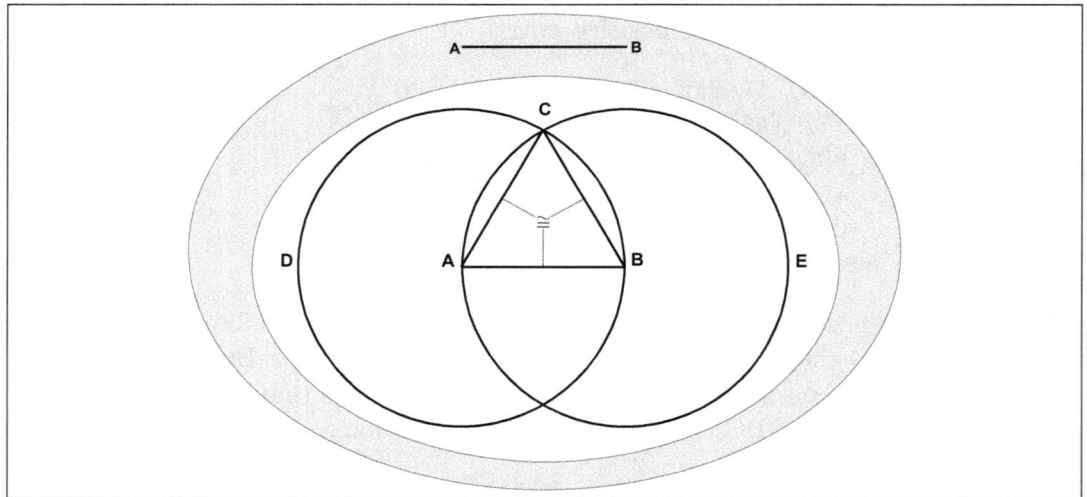

Figure 30: EG after step 1

"Therefore, the triangle ABC is equilateral, and it has been constructed on the given finite straight line. QED" By observation, there exists a triangle ABC with congruent sides AB, AC, BC. By rule 1e, erase all the nodes in the positive area of step 11 that are not required to represent the EG in Figure 22 or23. The result is the theorem to be proved:

```
12. [If (Point *A) (Point *B) (Line *AB A B)
    [Then (Point *C) (Line *AC A C) (Line *BC B C)
      (Triangle *ABC AB AC BC) ("≅" AB AC BC)]]
```

This example shows that a Euclidean proof can be translated, line by line, to a proof by EG rules of inference supplemented with the rules of observation and imagination. But it has implications that go far beyond techniques for analyzing and formalizing Euclid's proofs. Although the formalists were correct in discovering the lack of rigor in any step that requires observation or imagination, they were wrong in claiming that those steps were weaknesses or even flaws. For over two thousand years, Euclid's proofs were considered the ultimate standard of rigor. Many mathematicians found explanations and proofs that could be improved, but nobody found theorems that were false. Newton, for example, used Descartes' analytic geometry to discover his famous principles. But when he published them, he used a Euclidean style of definition and proof because that was the accepted standard of rigor in his day. Most readers did not know or trust the "new fangled" methods. They had more confidence in the methods they had learned and used in the universities.

For logic and mathematics, they show that diagrams have a legitimate role to play in formal reasoning. For artificial intelligence, they introduce computational techniques for combining and relating symbols and images. For psychology and neuroscience, mental models and image processing are intimately connected with all aspects of logic, language, and reasoning. Finally, they develop and extend Peirce's insights, hints, and conjectures about the continuum of semiotic processes in every aspect of cognition and life.

6 Research Issues

Peirce's voluminous manuscripts about logic are a gold mine of insights and innovations. Their relative neglect during the 20th century makes them a largely unexplored treasure for the 21st. His existential graphs and their rules of inference are still at the forefront of research in logic and cognitive science. Since each rule inserts or erases one semantic unit, the rules can be adapted to any notation just by defining its negative areas, positive areas, and semantic units. The option of inserting arbitrary icons in EGs opens new avenues of research in theory and applications.

The symmetry of the EG structure and rules of inference simplifies proof theory. For example, the EG rules imply the *reversibility theorem* and the *cut-and-paste theorem*. The proofs are simplest for the EG and EGIF notations, but they can be adapted to many other notations, including Peirce-Peano algebra.

- **Reversibility Theorem.** Any proof of $p \vdash q$ by Peirce's rules of inference can be converted to a proof of $\sim q \vdash \sim p$ by negating each step and reversing the order.

- **Proof.** Let $s_0, ..., s_n$ be the steps of the proof with s_0 as the EG that represents p, s_n as the EG that represents q, and each r_i from 1 to n as the rule that converts s_{i-1} to s_i. For the reverse sequence of negated EGs, note that $\sim[s_n]$ represents $\sim q$, and $\sim[s_0]$ represents $\sim p$. Furthermore, each $\sim[s_i]$ is converted to $\sim[s_{i-1}]$ by the inverse of rule r_i. The only question is whether the inverse conversions are permissible. Since the rules 2e, 2i, 3e, and 3i are equivalence operations, their inverses are permissible in any area, positive or negative. But 1e can only be performed in a negative area, and its inverse 1i can only be performed in a positive area. Since each step of the reverse proof is negated, the polarity (negative or positive) of every nested area in each $\sim[s_i]$ is reversed. Therefore, any application of 1i or 1e in the forward direction can be undone by its inverse (1e or 1i) in the reverse direction.

With traditional rules of inference, a proof of $p \vdash q$ can usually be converted to a proof of $\sim q \vdash \sim p$. But the conversions are more complex, because the traditional rules don't have exact inverses. For some kinds of proofs, Peirce's method is much shorter because of a property that is not shared by other common proof procedures: his rules can perform surgical operations (insertions or erasures) in an area of a graph or formula that is nested arbitrarily deep. Furthermore, the rules depend only on whether the area is positive or negative. Those properties imply the *cut-and-paste theorem* [56], which can be adapted to any notation for first-order logic:

- **Cut-and-Paste Theorem.** If a proof $p \vdash q$ is possible on a blank sheet sheet of assertion, then in any positive area of an EG where p occurs, q may be substituted for p.

- **Proof.** Since the nested area in which p occurs is positive, every step of a proof on a blank sheet can be carried out in that area. Therefore, it is permissible to "cut out" the steps of a proof from p to q in the outer area and "paste" them into the nested area. After q has been derived, Rule 1e can be used to erase the original p and any remaining steps of the proof other than q.

From the cut-and-paste theorem (CAPT), other important theorems can be proved as corollaries. One example is the deduction theorem:

- **Deduction Theorem.** If a proof $p \vdash q$ is possible, then $p \supset q$ is provable.

- **Proof.** To show that $p \supset q$ is provable, use CAPT to derive the corresponding EGIF from a blank:

    1. By 3i, draw a double negation around a blank: ~[~[ ]]
    2. By 1i, insert (p) in the negative area: ~[(p) ~[ ]]
    3. By 2i, iterate (p) to derive the equivalent of $p \supset p$: ~[(p) ~[(p)]]
    4. By CAPT, replace the inner (p) with (q): ~[(p) ~[(q)]]

The *constructive dilemma* is another another corollary of CAPT that is difficult to prove in many systems:

- **Constructive Dilemma.** If $p_1 \vdash q$ and $p_2 \vdash q$, then $p_1 \vee p_2 \vdash q$.

- **Proof:** Use two applications of CAPT.

    0. Starting EGIF: ~[~[(p1)] ~[(p2)]]
    1. By CAPT, replace (p1) with q: ~[~[(q)] ~[(p2)]]
    2. By CAPT, replace (p2) with q: ~[~[(q)] ~[(q)]]
    3. By 2e, deiterate one copy of ~[(q)]: ~[~[(q)]]
    4. By 3e, erase the double negation: (q)

In a positive area, erasing a graph or part of a graph makes the result more general because it is true in a broader range of cases. For example, (animal x) is more general than (cat x) because a cat is defined as an animal with certain feline characteristics. Replacing (cat x) with (animal x) is permissible because it has the effect of erasing the feline characteristics. Conversely, in a negative area, inserting a graph makes the result more specialized because it is true in a narrower range of cases. Therefore, replacing (animal x) with (cat x) in a negative area is permissible because it has the effect of inserting those feline characteristics.

The rules of generalization and specialization may be used as derived rules of inference. The proof of generalization uses CAPT in a one-step proof. But the proof of specialization takes six steps because it's harder to justify the erasure of graphs in a negative area. Such proofs usually require more steps with any proof procedure.

- **Generalization and Specialization.** For any propositions p and q stated in EG, EGIF, or semantically equivalent notations, if $p \vdash q$, then p is said to be more specialized than q, and q is said to be more generalized than p.

- **Rule of generalization.** In any positive area, p may be replaced by a generalization q.

- **Proof:** Assume that the area that contains p also contains some arbitrary EG (A), which may be blank.

    0. Starting EGIF in some positive area: (p) (A)
    1. By CAPT, replace (p) with (q): (q) (A)

- **Rule of specialization.** In any negative area, q may be replaced by a specialization p.

- **Proof:** Assume that the area that contains q also contains some arbitrary EG (A), which may be blank.

    0. Starting EGIF in some positive area: ~[(q) (A)]
    1. Since $p \supset q$, insert the corresponding EGIF in the same positive area: ~[(q) (A)] ~[(p) ~[(q)]]
    2. By rule 2i, iterate ~[(q) (A)] into the innermost area with (q): ~[(q) (A)] ~[(p) ~[(q) ~[(q) (A)]]]
    3. By 1e, erase the original copy of ~[(q) (A)]: ~[(p) ~[(q) ~[(q) (A)]]]
    4. By 2e, erase the innermost copy of (q): ~[(p) ~[(q) ~[(A)]]]
    5. By 1e, erase (q): ~[(p) ~[~[(A)]]]
    6. By 3e, erase the double negation: ~[(p) (A)]

Converting various proof procedures to EG form provides fundamental insights into the nature of the proofs and their interrelationships. Gentzen [11] developed two proof procedures called *sequent calculus* and *natural deduction*. A proof by either one can be systematically converted to a proof by Peirce's rules. The converse, however, does not hold because Gentzen's rules cannot operate on deeply nested expressions. For some proofs, many steps are needed to bring an expression to the surface of a formula before those rules can be applied. An example is the *cut-free* version of the sequent calculus, in which proofs can sometimes be exponentially longer than proofs in the usual version. Dau [3] showed that with Peirce's rules, the corresponding cut-free proofs are longer by just a polynomial factor.

Like Peirce's rules, Gentzen's rules for natural deduction also come in pairs, one of which inserts an operator, which the other eliminates. Gentzen, however, required six pairs of rules for each of his six operators: $\wedge, \vee, \supset, \sim, \forall$, and $\exists$. Peirce had only three operators and three pairs of rules. Even when limited to the same three operators, Gentzen's rules are highly irregular. They lack the symmetry of paired rules that are exact inverses of one another. Four of the twelve rules even require a provability test, expressed by the operator ⊢. Yet Gentzen's rules can all be proved as derived rules of inference in terms of Peirce's rules. In fact, two of them were just proved above: the rule for ⊃-introduction is the deduction theorem, and the rule for ∨-elimination is the constructive dilemma. Figure 31 shows Gentzen's six pairs of rules.

	Introduction Rules	Elimination Rules
∧	$\dfrac{A,\ B}{A \wedge B}$	$\dfrac{A \wedge B}{A} \qquad \dfrac{A \wedge B}{B}$
∨	$\dfrac{A}{A \vee B} \qquad \dfrac{B}{A \vee B}$	$\dfrac{A \vee B,\ A \vdash C,\ B \vdash C}{C}$
⊃	$\dfrac{A \vdash B}{A \supset B}$	$\dfrac{A,\ A \supset B}{B}$
~	$\dfrac{A \vdash \bot}{\sim A} \qquad \dfrac{\bot}{A}$	$\dfrac{A,\ \sim A}{\bot} \qquad \dfrac{\sim\sim A}{A}$
∀	$\dfrac{A(a)}{(\forall x)A(x)}$	$\dfrac{(\forall x)A(x)}{A(t)}$
∃	$\dfrac{A(t)}{(\exists x)A(x)}$	$\dfrac{(\exists x)A(x),\ A(a) \vdash B}{B}$

Figure 31: Rules of inference for Gentzen's natural deduction

Gentzen's rules can be proved as derived rules of inference in terms of Peirce's rules, but some of them require further explanation. The symbol ⊥ represents a proposition that is always false. In EGs, ⊥ is represented by an oval with nothing inside; in EGIF, it is ~[ ]. Peirce called ~[ ] the *pseudograph* because it is always false. In the resolution method for theorem proving, it is called the *empty clause*. From the pseudograph, any proposition (A) can be derived. Start with ~[ ]; by 1e, insert ~[(A)] to derive ~[~[(A)]]; by 3e, erase the double negation to derive (A). By the way, *intuitionistic logic* is a version of FOL that blocks the derivation of an arbitrary proposition p from a contradiction; that proof can be blocked by prohibiting rule 3e for erasing a double negation.

Two other special symbols are the arbitrary term t in the rules for ∃-introduction and ∀-elimination and the arbitrary value or free variable a in the rules for ∀-introduction and ∃-elimination. The arbitrary term t is an ordinary expression with no free variables, and it can be represented as an EG —t or EGIF (t *x), where t is any relation or graph with one peg designated to represent the value of the term. The rule for ∀-elimination, also called universal instantiation, was proved in Figure 17. The proof for ∃-introduction takes one step:

0. Starting EGIF: (t *x) (A x)

1. By 1e, erase part of the line that connects *x to x and introduce a new defining label for (A *y).

Since EGs don't have variables, they don't have free variables. But proof procedures that use arbitrary values or free variables have been a source of controversy for years. Kit Fine [8] interpreted Gentzen's rule of ∀-introduction as an assumption that some implicit proposition φ implies that for any individual a, the statement $A(a)$ is true.

0. Starting assumption: [*x] ⊢ (A x)

1. By the deduction theorem, conclude If [*x] [Then (A x)]]. This EGIF is equivalent to $(\forall x)A(x)$.

For Gentzen's rule of ∃-elimination, $A(a)$ is true of some arbitrary values, but not all. The conclusion B is provable from $A(a)$ only for values of a for which $A(a)$ is true.

0. Starting assumptions: [*x] (A x); (A *a) ⊢ (B)

1. Let *a be some *x for which (A x) true. Therefore: (B)

Although each of Gentzen's rules can be derived from Peirce's rules, proofs in the two systems take different paths. EG proofs proceed in a straight line from a blank sheet to the conclusion: each step inserts or erases one subgraph in the immediately preceding graph. No bookkeeping is required to keep track of information from any earlier step. But Gentzen's proofs interrupt the flow of a proof at every rule that contains the pattern $p \vdash q$. When deriving q from p, those rules use a method of bookkeeping called *discharging an assumption*:

1. Whenever the symbol ⊢ appears in one of Gentzen's rules, some proposition p is assumed.

2. The current proof is suspended, and the state of the current rule is recorded.

3. A side proof of q from p is initiated.

4. When the conclusion q is derived, the assumption p is discharged by resuming the previous proof and using q in the rule of inference that had been interrupted.

Such side proofs might be invoked recursively to an arbitrary depth. EG proofs avoid that bookkeeping by an option that most notations for logic can't represent: drawing a double negation around a blank. For example, in the proof of the Praeclarum Theorema (Figure 18), the first step is to draw a double negation on a blank sheet, and the second step is to insert the hypothesis into the shaded area. The result is a well-formed EG, and no bookkeeping is necessary to keep track of how that EG had been derived. In Gentzen's method, the first step is to assume the hypothesis, and a record of that assumption must be retained until the very end of the proof, when it is finally discharged to form the conclusion. This observation is the basis for converting a proof by Gentzen's method of natural deduction to a proof by Peirce's method:

- Replace each of Gentzen's rules that does not contain the symbol ⊢ with one or more of Peirce's rules that produce an equivalent result.

- Whenever one of Gentzen's rules containing ⊢ is invoked, apply rule 3i to insert a double negation around the blank, copy the hypothesis into the negative area by rule 1i, and continue.

An EG proof generated by this procedure will be longer than a direct proof that starts with Peirce's rules. As an exercise, use Gentzen's rules and his method of discharging assumptions to prove the Praeclarum Theorema. To avoid the bookkeeping, introduce a nest of two ovals when the symbol ⊢ appears in a rule. Insert the hypothesis into the shaded area, but continue to use the algebraic notation for the formulas. Finally, translate each formula to an EG, and add any intermediate steps needed to complete the proof according to Peirce's rules. Proofs by Gentzen's other method, the sequent calculus, can be converted to EG proofs more directly because they don't require extra steps to discharge assumptions.

Although Peirce understood the importance of functions in mathematics, he did not distinguish functions and relations in his versions of logic. In effect, he treated a function, such as $f(x) = y$, as a special case of a relation $f(x,y)$. What distinguishes a function from other relations is that for every value of x, there is exactly one value for y. Since EGIF has the semantics of Common Logic, which does distinguish functions from relations, EGIF uses the notation (f x | y) for $f(x) = y$. In general, a function may have any number of arguments on the left of the vertical bar, and exactly one value on the right: (g x y z | w). EGIF also allows functions with zero arguments, such as (h | x). For details, see the EGIF grammar in the appendix.

In computational systems, the widely used method of resolution applies a single rule of inference to a set of *clauses* [55]. Each clause is a disjunction of positive or negative *literals*, which are single atoms or their negations. In the following example, the two clauses on the left of ⊢ are combined by resolution to derive the clause on the right:

$$\sim p \vee \sim q \vee r, \ \sim r \vee u \vee v \ \vdash \ \sim p \vee \sim q \vee u \vee v.$$

The leftmost clause has a positive literal r, and the middle clause has a negated copy $\sim r$. Resolution is a cancellation method that erases both r and $\sim r$ and merges the remaining literals to form the clause on the right. Figure 32 shows the effect of resolution as expressed in EGs:

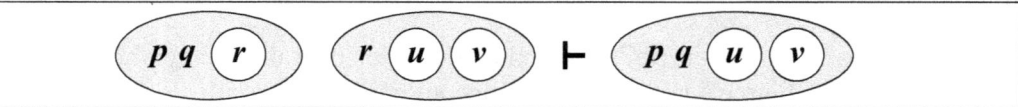

Figure 32: Resolution expressed in EGs

With EGs, resolution takes four steps: by 2i, iterate the middle EG into the doubly nested (unshaded) area of the EG on the left; by 2e, erase the innermost copy of r; by 1e, erase the remaining copy of r; by 3e, erase the double negation. The result is the EG on the right. Resolution is commonly used in a refutation procedure, which attempts to prove a statement s by deriving a contradiction from $\sim s$:

1. Negate the statement to be proved: $\sim s$.

2. Convert $\sim s$ to clause form.

3. Use repeated steps of resolution to derive the empty clause or pseudograph, [ ].

4. Since the empty clause is false, the original s must be true.

Many logicians have observed that proofs by resolution are "almost" the inverse of proofs by natural deduction. They aren't exact inverses, however, because the rule of resolution is simple and regular, but Gentzen's six pairs of insertion and elimination rules are highly irregular. Showing how one method can be derived from the other is a challenge that Larry Wos [65], a pioneer in automated reasoning methods, stated as problem 24 in his list of 33 unsolved problems:

> Is there a mapping between clause representation and natural deduction representation (and corresponding inference rules and strategies) that

causes reasoning programs based respectively on the two approaches or paradigms to attack a given assignment in an essentially identical fashion?

Problem 24 remained unsolved for years. But the solution in terms of Peirce's version of natural deduction follows from the reversibility theorem. For Gentzen's version, translate each step in Peirce's version to one or more steps in terms of Gentzen's rules:

1. Start by deriving an EG or EGIF proof by resolution.

2. Draw a negation around each step of the proof, and reverse the order.

3. Since the last step was the pseudograph, ~[ ], the first step of the reversed proof will be a double negation around a blank, ~[~[ ]]. By rule 3e, erase the double negation to derive the blank as the new starting point.

4. The final step of the reversed proof will be a double negation of the theorem to be proved. By 3e, erase the double negation.

For propositional logic, there are no quantifiers. Each application of the resolution rule corresponds to four EG rules, which can be reversed inside a negation. With quantifiers, two additional procedures must also be checked for reversibility: *skolemization* in step 2 and *unification* in step 3. In clause form, all variables are governed by universal quantifiers: $\forall x_1, ..., \forall x_n$. Skolemization replaces each existential quantifier with a term t, which is a constant or a function applied to one or more of the universally quantified variables. Unification involves universal instantiations followed by joining the pegs of functions or relations that are being unified. After the pegs are joined, rule 2e allows redundant copies of relation or function symbols to be erased.

To show that skolemization and unification are reversible with EG rules, note that the universal quantifiers of an algebraic formula correspond to lines of identity or defining nodes [*x1],..., [*xn] just inside the shaded area of the corresponding EG. To show that skolemization is reversible inside a negation, consider the following example, in which the skolem function $s(x)$ replaces the variable y:

$$\forall x \exists y R(x,y) \Rightarrow \exists f \forall x R(x, f(x)) \Rightarrow \forall x R(x, s(x))$$

If the left formula is true, the middle one must be true for at least one function f, and the right formula names such a function s. When negated, these implications are reversible: if the right formula is false, no function f exists for the middle one, and the left one is also false. This procedure can be generalized to any number of universal quantifiers, including zero for a skolem constant. In Common Logic and EGIF, a skolem constant may be represented by a function with zero arguments.

Unification is an application of ∀-elimination followed by merging identical functions or relations applied to identical values. In EGs, it is performed by joining a subgraph that represents a constant or functional term to a line of identity followed by joining pegs according to the rule of inference discussed in Appendix A.3. Its inverse inside a negation corresponds to Gentzen's ∃-introduction. In EGs, it is performed by cutting the line and erasing the subgraph. The operation of joining lines whose values are known to be identical is an equivalence that can be reversed in any area.

For issues of efficiency and computability, the humanly readable notation is almost irrelevant, since most theorem provers translate any input formats to their internal format before beginning a proof. The translation time is usually trivial compared to the time required to find a proof. Stewart [57] showed that a theorem prover based on EGs with Peirce's rules of inference was comparable in performance to a resolution theorem prover. But the world's fastest theorem provers use multiple methods that are more specialized for the kinds of problem than for the input notations. They also depend on efficient methods for storing intermediate results and recognizing which are the most relevant at each step.

7 Natural Logic

Since the 1970s, psychologists, philosophers, linguists, and logicians have been searching for some innate *natural logic or language of thought*. The goal was a theory and notation that could support the full range of human reasoning from the babbling of an infant to the most advanced science [30, 10, 60]. But Richard Montague [38]) went one step further. He claimed that language and logic have the same foundation:

> There is in my opinion no important theoretical difference between natural languages and the artificial languages of logicians; indeed, I consider it possible to comprehend the syntax and semantics of both kinds of languages within a single natural and mathematically precise theory.

Montague's claim stimulated 40 years of research on formal semantics for natural language (NL). But MacCartney and Manning [33]) noted that "truly natural language is fiendishly complex. The formal approach faces countless thorny problems: idioms, ellipsis, paraphrase, ambiguity, vagueness, lexical semantics, the impact of pragmatics, and so on." Instead of a complete translation to a logical form, they applied the term "natural logic" to a system of local features "which characterizes valid patterns of inference in terms of syntactic forms which are as close as possible to surface forms."

Montague did not approve of Chomsky's emphasis on syntax for NLs, and he would not accept a syntactic foundation for logic. Nor would Peirce, Polya, or Euclid. Peirce would treat Montague's claim as an empirical issue: What are the criteria for being humanly *natural*? Is there a unique logic for all languages? Or would the logic vary for different languages and cultures? How would variations in subject matter, levels of education, or businesses and professions affect the logical forms?

Wittgenstein [63] claimed that every natural language supports an open-ended variety of ways of talking and thinking, which he called *Sprachspiele* (language games). In 1939, he taught a course on the foundations of mathematics and logic [4]. Alan Turing attended all the lectures and debated the issues with Wittgenstein. Some logicians dismissed Wittgenstein's claims as eccentric, but the fact that Turing took them seriously is significant. Nobody knows exactly how Peirce would have replied to Montague or Wittgenstein, but the variety of logics that Peirce himself developed suggests that he might have had more sympathy with Wittgenstein [39].

Since all humans have similar bodies, sensory organs, and requirements for food, shelter, and social relations, they share a common core of concepts and ways of thinking. But cultures that have been isolated for centuries may have patterns of thought that don't have direct translations to one another. The Pirahã language of Brazil, for example, appears to be unusually difficult for outsiders to learn, even for linguists who know other languages of the region [7]. Cognitive linguists have explored these issues in depth [62, 29, 64, 31, 58].

In contrast with the linear, symbolic languages and logics, the psychologist Allan Paivio [40, 41] proposed a dual coding approach for relating language and imagery. The psycholinguist David McNeill [35] went one step further. He claimed that speech itself is multimodal: "Gestures orchestrate speech":

> Images and speech are equal and simultaneously present in the mind... Gestures look upward, into the discourse structure, as well as downward, into the thought structure. A gesture will occur only if one's current thought contrasts with the background discourse. If there is a contrast, how the thought is related to the discourse determines what kind of gesture it will be, how large it will be, how internally complex it will be, and so forth. [34, 2]

> Forced suppressions only shift the gesture to some other part of the body — the feet or an overactive torso. We once taped an individual who had

been highly recommended to us as an elaborate and vigorous gesturer. Somewhat maliciously, when asked to recount our cartoon stimulus, he sat on his hands yet unwittingly began to perform the gestures typical of the experiment with his feet, insofar as anatomically possible. [35, 3]

When Peirce's rules of inference are applied to icons in two, three, or even four dimensions (3D plus time), they can accommodate images, diagrams, and gestures as an integral part of the formalism. Instead of being heuristic aids, the diagrams become formal components of the propositions, axioms, arguments, and proofs. In Peirce's own words, they provide "a moving picture of the action of the mind in thought" [46, p. 298]. In the same letter in which he wrote the tutorial on EGs (quoted in Section 2), Peirce explained what he meant by moving pictures:

> Boole plainly thought in algebraic symbols; and so did I, until, at great pains, I learned to think in diagrams, which is a much superior method. I am convinced there is a far better one, capable of wonders; but the great cost of the apparatus forbids my learning it. It consists in thinking in stereoscopic moving pictures. Of course one might substitute the real objects moving in solid space; and that might not be so very unreasonably costly. [48, 3: 191]

As Peirce said, thinking in moving pictures is far better than thinking in diagrams, which is superior to thinking in algebraic symbols. All three ways of thinking use icons, and the methods of observation and imagination may be used for all of them. With the modern software for virtual reality (VR), Peirce's dream can become a reality: an integrated combination of EG and VR.

Although semantics is the foundation for any logic, a convenient syntax can simplify the mapping to and from natural languages (NLs). The *discourse representation structure* (DRS) by Hans Kamp [23] is widely used as a logical notation for NL semantics. Coincidentally, the DRS notation, as developed by Kamp and Reyle [24] happens to be isomorphic to EGs. In fact, EGIF has a more direct mapping to DRS than to EG. As an example, Figure 33 shows an EG and DRS for the sentence *If a farmer owns a donkey, then he beats it.*

Kamp's primitives are the same as Peirce's: the default *and* operator is omitted, and the default quantifier is existential. DRS negation is represented by a box marked with the ¬ symbol, and implication is represented by two boxes connected by an arrow. As Figure 33 illustrates, nested EG ovals allow lines in the *if* oval to extend into the *then* oval. For DRS, Kamp made an equivalent assumption: the quantifiers for x and y in the *if* box govern x and y in the *then* box. Since their structures are isomorphic and they use the same operators with the same scoping

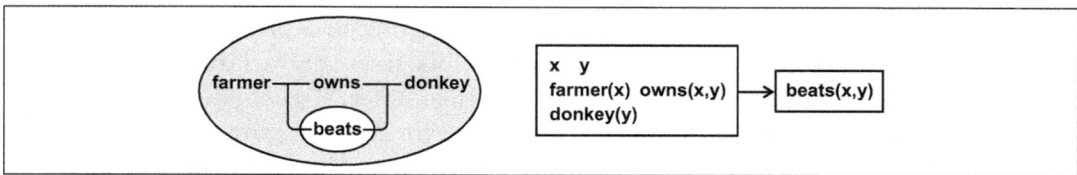

Figure 33: EG (left) and DRS (right) for *If a farmer owns a donkey, then he beats it.*

rules, the EG and DRS in Figure 33 can be translated to the same EGIF and to equivalent formulas in Peirce-Peano algebra:

[If (farmer *x) (donkey *y) (owns x y) [Then (beats x y)]]
$\sim (\exists x \exists y\ \text{farmer}(x) \wedge \text{donkey}(y) \wedge \text{owns}(x,y) \wedge \sim \text{beats}(x,y))$
$\forall x \forall y\ \text{farmer}(x) \wedge \text{donkey}(y) \wedge \text{owns}(x,y) \supset \text{beats}(x,y)$

As these translations show, the unlabeled EG lines are assigned labels x and y in both DRS and EGIF. The nested negations in the EG may be represented by the keywords If and Then in the EGIF, and those same keywords may be used to represent the DRS boxes. The first Peirce-Peano formula uses only existential quantifiers and negations. The second formula uses $\supset$ for implication, but the scoping rules require the $\exists$ quantifiers to be moved to the front and be replaced by $\forall$ quantifiers.

As another example of the similarity between EG and DRS, consider the following pair of sentences: *Pedro is a farmer. He owns a donkey.* Kamp and Reyle [24] observed that proper names like Pedro are not rigid identifiers. In DRS, proper names are represented by predicates rather than constants. That convention is similar to Peirce's practice with EGs. On the left of Figure 34 are the EGs for the two sentences; on the right are the DRSs. Following are the EGIF and the algebraic formulas. (Periods are added to separate the two formulas, but no periods are needed for EGIF.)

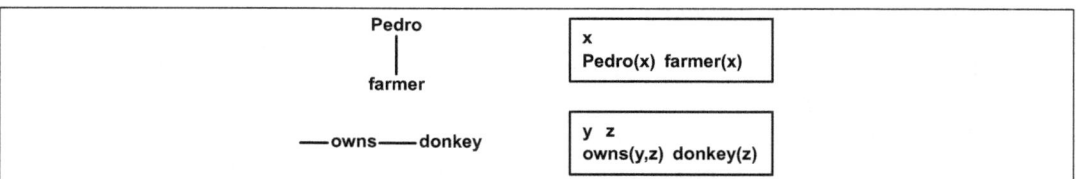

Figure 34: EGs and DRSs for *Pedro is a farmer. He owns a donkey.*

(Pedro *x) (farmer x) (owns *y *z) (donkey z)
$\exists x\ \text{Pedro}(x)\ \text{farmer}(x).\ \exists y \exists z\ \text{owns}(y,z) \wedge \text{donkey}(z).$

In English, a pronoun such as *he* can refer to something in a previous sentence, but the variable y in the second formula cannot be linked to the variable x in the first. For the EG, DRS, and EGIF, the linkage is much simpler. To combine the two EGs, connect the line of identity for *Pedro* to the line for *he* in Figure 34. To combine the two DRSs, transfer the contents of one DRS box to the other, move the list of variables to the top, and insert the equality $x = y$. Following is the EGIF for Figure 35 and its simplified form after deleting the coreferences node [x y]:

```
(Pedro *x) (farmer x) (owns *y *z) (donkey z) [x y]
(Pedro *x) (farmer x) (owns x *z) (donkey z)
```

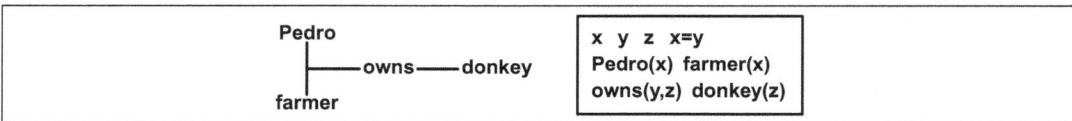

Figure 35: Combining the EGs and DRSs in Figure 34

Since the EG and DRS notations are isomorphic and they're represented by the same EGIF, Peirce's rules of inference can be applied to them in exactly equivalent steps. The proof in Figure 36 begins with the EGs for the sentences *Pedro is a farmer who owns a donkey* and *If a farmer owns a donkey, then he beats it*. Then the rules 2i, 1i, 2e, and 3e derive an EG for the conclusion *Pedro is a farmer who owns and beats a donkey*. The proof in Figure 36 takes four steps, but the first arrow combines two steps: extending lines by rule 2i and connecting lines by rule 1i. The following EGIF proof shows all four steps. For the DRS proof, translate each EGIF step to DRS.

0. Starting graphs: To avoid a name clash, the label **x** was replaced by **w**.
   ```
   (Pedro *w) (farmer w) (owns w *z) (donkey z)
   [If (farmer *x) (donkey *y) (owns x y) [Then (beats x y)]]
   ```

1. By rule 2i, extend the lines of identity by inserting nodes [w] and [z] in the area of the if-clause:
   ```
   (Pedro *w) (farmer w) (owns w *z) (donkey z)
   [If (farmer *x) (donkey *y) (owns x y) [w] [z]
   [Then (beats x y)]]
   ```

2. By 1i, connect the line for **w** to **x** and **z** to **y** and relabel:
   ```
   (Pedro *w) (farmer w) (owns w *z) (donkey z)
   [If (farmer w) (donkey z) (owns w z) [Then (beats w z)]]
   ```

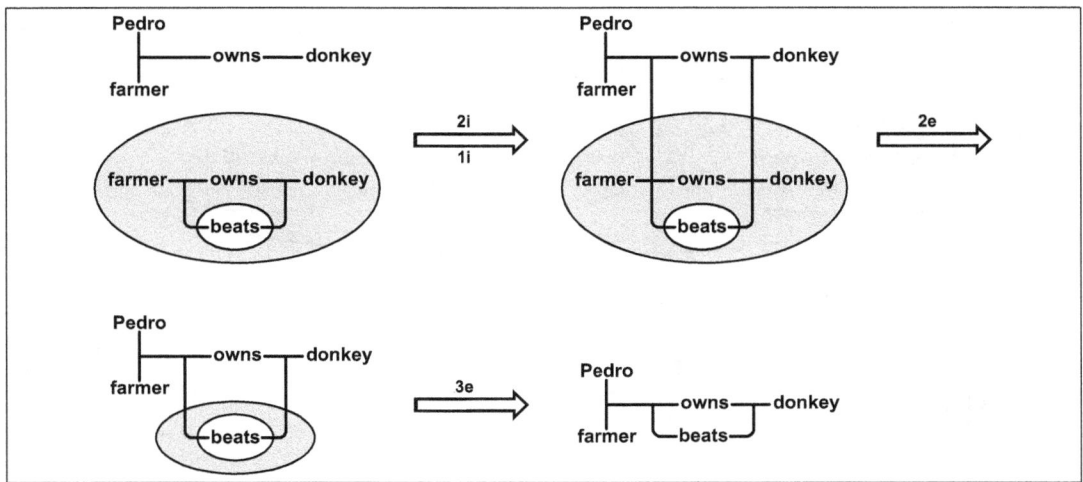

Figure 36: A proof according to Peirce's rules

3. By 2e, erase the copy of (farmer w) (donkey z) (owns w z) in the if-clause:
   (Pedro *w) (farmer w) (owns w *z) (donkey z)
   [If [Then (beats w z)]]

4. By 3e, erase the double negation (the if-then pair with an empty if-clause):
   (Pedro *w) (farmer w) (owns w *z) (donkey z) (beats w z)

For disjunctions, Figure 37 shows the EG and DRS for an example by Kamp and Reyle [24, p. 210]: *Either Jones owns a book on semantics, Smith owns a book on logic, or Cooper owns a book on unicorns.* The EG at the top of Figure 37 shows that the existential quantifiers for Jones, Smith, and Cooper are in the outer area, but the quantifiers for the three books are inside the alternatives. Both Peirce and Kamp allowed spaces inside relation names, but CGIF requires names with spaces or other special characters to be enclosed in quotes:

```
(Jones *x) (Smith *y) (Cooper *z)
~[ ~[ ("book on semantics" *u) (owns x u)]
   ~[ ("book on logic" *v) (owns y v)]
   ~[ ("book on unicorns" *w) (owns z w)] ]
```

For better readability, the Conceptual Graph Interchange Format allows the keywords Either and Or as synonyms for the negation symbol. That option could be added to EGIF:

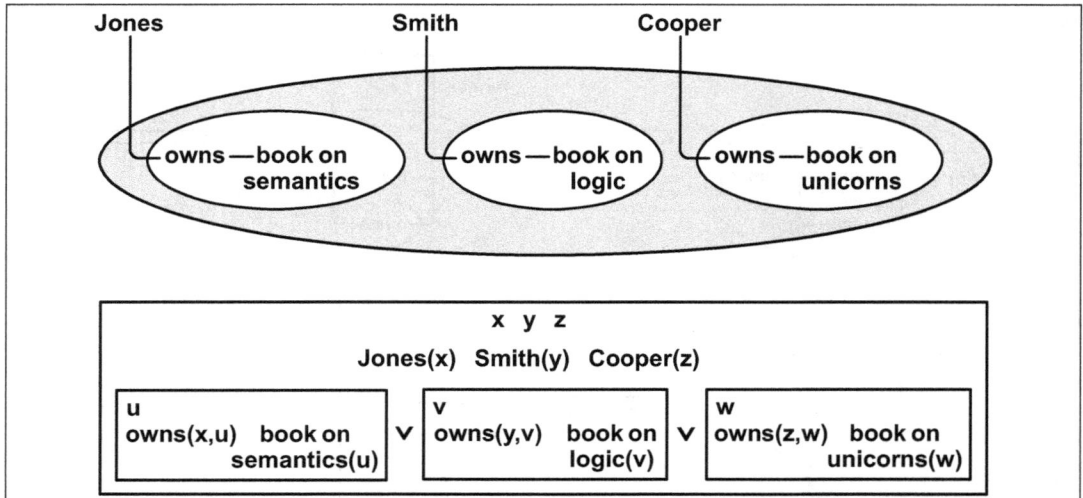

Figure 37: EG and DRS for a disjunction with three alternatives

```
(Jones *x) (Smith *y) (Cooper *z)
[Either [Or ("book on semantics" *u) (owns x u)]
        [Or ("book on logic" *v) (owns y v)]
        [Or ("book on unicorns" *w) (owns z w)] ]
```

The simplicity and generality of the EG structure, rules of inference, and methods for evaluating truth or falsity makes existential graphs a good candidate for a mental logic. But Figure 38 shows that Peirce's rules, as he stated them, aren't always applicable to the words and phrases of natural languages. Peirce's rules 1i and 1e, however, may be extended with the rules of specialization and generalization, which were derived in Section 6:

- **Specialization (Rule 1i):** In any negative area, any proposition q expressed in any notation may be replaced by a specialization p that implies q. In particular, p may be identical to $q \wedge A$ where A is an arbitrary proposition.

- **Generalization (Rule 1e):** In any positive area, any proposition p expressed in any notation may be replaced by a generalization q that is implied by p. In particular, p may be identical to $q \wedge A$ where A is an arbitrary proposition.

In Figure 38, the word *every*, which is on the boundary of an oval, creates the same pattern of nested negations as *if-then*. The word *cat*, by itself, expresses the proposition *There is a cat*. The phrase *cat in the house* expresses the conjunction *There is a cat and it is in the house*. By the rule of specialization (1i), the word cat

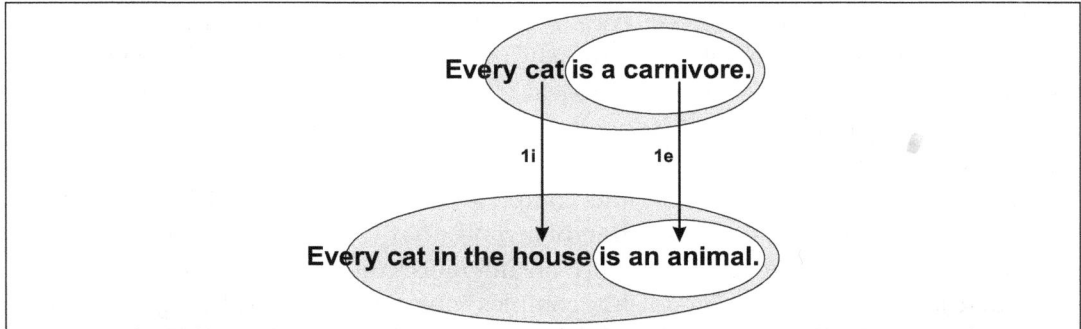

Figure 38: Specializing and generalizing English phrases

may be specialized to the phrase *cat in the house*. By the rule of generalization (1e), the word carnivore may be generalized to *animal*. In this example, the propositions are stated by printed words, but the mental models could represent them by icons with indexicals that link them to a real or imaginary world.

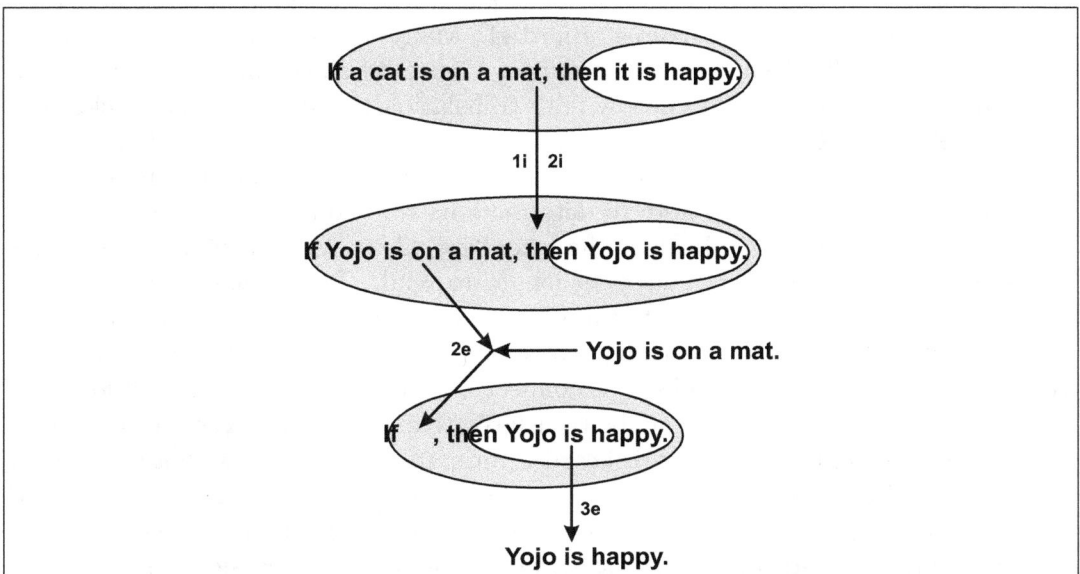

Figure 39: A four-step proof by Peirce's rules

Peirce's rules of iteration/deiteration and double negation may be used as he stated them, but some prior changes by the rules of generalization/specialization may be required. Figure 39 shows a four step proof. By specialization (1i), the clause *a cat is on a mat* is specialized to *Yojo is on a mat*. The pronoun *it* represents an EG line of identity or an EGIF coreference node. By iteration (2i), a copy of the name Yojo replaces the coreferent pronoun *it*. By deiteration (2e), the clause *Yojo is on a mat* is erased because it is identical to a sentence in the outer area. Finally, the shaded area and the words *if* and *then* on the boundaries of the ovals are erased by the rule of double negation (3e).

Since Peirce's rules are sound and complete for EG, EGIF, and DRS, they are also sound and complete for that subset of any NL that can be translated to and from DRS. They are also sufficient to support the inferences discussed by MacCartney and Manning [33], but not the "fiendishly complex" aspects of natural languages. Majumdar and Sowa [37], however, have shown that a full translation of NL documents to conceptual graphs is practical. Furthermore, many of those complex aspects can be handled by using background knowledge derived from simpler passages in the same documents or from related documents and textbooks. Finally, if all else fails, a computer system can do what any human would do: ask a question. People rarely understand everything they read or hear. A large part of any discourse. is devoted to questions and metalevel discussions aspects that may be unclear or debatable.

Although EGs can represent a moving picture of the mind in thought, they are not a perfect picture, as Peirce admitted. Many kinds of diagrams are better tailored to specialized subjects. For music, Figure 40 shows one measure in the usual notation. Below it is a note-by-note translation to the boxes and circles of a conceptual graph (CG). An experienced musician can read music notation at sight and play it at full speed. By contrast, the CG reflects the laborious analysis of a beginner who examines each note to determine its tone, duration, and relationship to other tones on the same beat or the next beat. A translation of the CG to EG would have more nodes and take even longer to read. Traditional music notation is more iconic than an EG, but the most iconic is the music in a musician's head. An EG supplemented with arbitrary icons may include the traditional notation, a recording of the music, and links that connect both of them to EG lines of identity.

Although music notations may be translated to a logic that states the equivalent information, the music as played or heard is intimately connected with a wide range of feelings and previous experience. The connection of those feelings to any kind of language, logic, or diagrams can only be expressed by informal metaphors. Different musicians and listeners, or even the same person on different occasions, may use use vastly different metaphors.

Unlike Frege and Russell, who focused on the foundations of mathematics, Peirce integrated logic with semiotics and applied it to every aspect of science, philosophy, language, and life. To represent modal logics in existential graphs, Peirce experimented with colors and dotted lines. Frege insisted on a single domain of quantification called *everything*, but Peirce used *tinctures* to distinguish different *universes*:

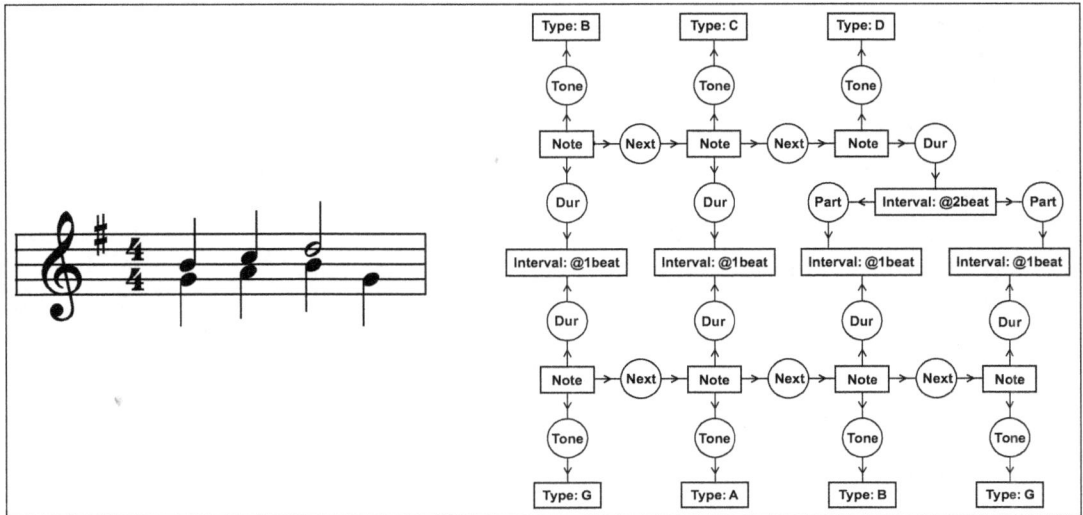

Figure 40: Two diagrams for representing the logic and ontology of music

> The nature of the universe or universes of discourse (for several may be referred to in a single assertion in the rather unusual cases in which such precision is required) is denoted either by using modifications of the heraldic tinctures, marked in something like the usual manner in pale ink upon the surface, or by scribing the graphs in colored inks. [46, 670:18]

Peirce considered three universes: actualities, possibilities, and the necessitated. He subdivided each universe in four ways to define 12 modes. In the universe of possibilities, for example, he distinguished objective possibility (an alethic mode), subjective possibility (epistemic), social possibility (deontic), and an interrogative mode, which corresponds to scientific inquiry by hypothesis and experiment. For the necessitated, he called the four subdivisions the rationally necessitated, the compelled, the commanded, and the determined. Most of his writings on these topics were unpublished, and he changed his terminology from one manuscript to the next. Peirce admitted that a complete analysis and classification would be "a labor for generations of analysts, not for one" [46, 478:165].

For the semantics of modal logic, Hintikka [18] proposed an infinite family of possible worlds, each described by the set of propositions that are true in that world. Kripke [28] showed how an *accessibility relation* among those worlds would imply the various axioms for different versions of modal logic. Dunn [5] combined Hintikka's method with Kripke's. He defined each world w by a set of laws and facts (L, F): the facts F of w are identical to Hintikka's propositions, and the laws

L are the subset of F that are necessarily true. Together, L and F imply Kripke's accessibility relation. Dunn's semantics has a direct mapping to Peirce's system. For each modality, the laws correspond to the necessitated; the facts correspond to the actualities; and the possibilities consist of all propositions consistent with the laws. For each world w, the possibilities include all the facts of w and of every world accessible from w.

Peirce's icons and his goals for 3D moving pictures correspond to the mental models by Johnson-Laird and his colleagues. But those models are rarely as well behaved as the theories that logicians specify. In fact, people typically accept statements that are false according to all versions of modal logic, including Peirce's: "Possibly it's raining and possibly it isn't" and "It isn't raining, but it might have been" [53]. The modal statements that people typically make also diverge from probability theory, and they're more consistent with the hypothesis that people are thinking in terms of some sort of mental model [17]. However, people who have some training in modal logic or in professions that require precision, such as science and the law, tend stay closer to the constraints of modal logics.

Peirce would not be surprised by the discrepancies between ordinary thinking and formal logics. In his classification of the sciences [45, 1.180-202], he placed formal logic under mathematics as pure theory, prior to any application. But he placed normative logic under philosophy as a standard for evaluating the way people actually think. Peirce realized that vagueness is inevitable, and that a vague statement, such as "Maybe it will rain, maybe not," can be preferable to an irrelevant attempt at precision:

> The vague might be defined as that to which the principle of contradiction does not apply... But wherever degree or any other possibility of continuous variation subsists, absolute precision is impossible. Much else must be vague, because no man's interpretation of words is based on exactly the same experience as any other man's. Even in our most intellectual conceptions, the more we strive to be precise, the more unattainable precision seems. It should never be forgotten that our own thinking is carried on as a dialogue, and though mostly in a lesser degree, is subject to almost every imperfection of language. [45, 5:505-506]

In summary, Peirce's writings, Wittgenstein's language games, and research in anthropology suggest that no single language, logic, or system of diagrams can include everything that might be called natural. As guidelines for analyzing all the possibilities, Peirce stated two fundamental principles:

> Do not block the way of inquiry. [45, 1.135]

> The elements of every concept enter into logical thought at the gate of perception and make their exit at the gate of purposive action; and whatever cannot show its passports at both those two gates is to be arrested as unauthorized by reason. [47, 2.241]

The first principle implies that the search must be open ended, not limited to any particular language, culture, or profession. The second requires a grounding in perception and action. Since anything perceptible can be represented by an icon, the option of including arbitrary icons in EGs suggests a way to support everything: use EGs as a metalanguage that may include, describe, and relate icons of any representation, linear or diagrammatic.

References

[1] Barwise, Jon, and John Etchemendy (1993). *Tarski's World*. Stanford, CA: CSLI Publications.

[2] Damasio, Antonio R. (2010) *Self Comes to Mind: Constructing the Conscious Brain*, New York: Pantheon Books.

[3] Dau, Frithjof (2006). Some notes on proofs with Alpha graphs. In *Conceptual Structures: Inspiration and Application*, (LNAI 4068), H. Schärfe, P. Hitzler, and P. Øhrstrom (eds.), 172-188. Berlin: Springer.

[4] Diamond, Cora (ed) (1975) *Wittgenstein's Lectures on the Foundations of Mathematics*, Cambridge 1939, Chicago: University of Chicago Press.

[5] Dunn, J. Michael (1973) A truth value semantics for modal logic, in H. Leblanc, ed., *Truth, Syntax and Modality*, Amsterdam: North-Holland, pp. 87-100.

[6] Euclid, *The Thirteen Books of the Elements*, translated by Thomas L. Heath, Second Edition, Dover: New York.

[7] Everett, Daniel L. (2008) *Don't Sleep, There Are Snakes: Life and Language in the Amazon Jungle*, London: Profile Books.

[8] Fine, Kit (1985). *Reasoning with Arbitrary Objects*. Oxford: Basil Blackwood.

[9] Frege, Gottlob (1879). Begriffsschrift. In *From Frege to Gödel*, J. van Heijenoort (ed.) (1967), 1-82. Cambridge, MA: Harvard University Press.

[10] Fodor, Jerry A. (1975) *The Language of Thought*, Cambridge, MA: Harvard University Press.

[11] Gentzen, Gerhard (1934). Untersuchungen über das logische Schließen [Investigations into logical deduction]. In *The Collected Papers of Gerhard Gentzen*, M. E. Szabo, editor and translator (1969), pp. 68-131. Amsterdam: North-Holland Publishing Co.

[12] Hadamard, Jacques (1945) *The Psychology of Invention in the Mathematical Field*, Princeton University Press, Princeton.

[13] Halmos, Paul R. (1968) Mathematics as a creative art, American Scientist 56, 375-389.

[14] Hayes, Patrick, and Chris Menzel (2006) IKL Specification Document, http://www.ihmc.us/users/phayes/IKL/SPEC/SPEC.html

[15] Henkin, Leon (1961). Some remarks on infinitely long formulas. In *Infinitistic Methods, (Proceedings of symposium on foundations of mathematics)*, 176-183. London: Pergamon Press.

[16] Hilpinen, Risto (1982). On C. S. Peirce's theory of the proposition: Peirce as a precursor of game-theoretical semantics. *The Monist* 65: 182-188.

[17] Hinterecker, Thomas, Markos Knauff, & Philip Johnson-Laird (2016) Modality, probability, and mental models, *Journal of Experimental Psychology: Learning, Memory, and Cognition* 42, 1606-1620.

[18] Hintikka, Jaakko (1961) Modality and quantification, *Theoria* 27, 110-128.

[19] Hintikka, Jaakko (1973). *Logic, Language Games, and Information*. Oxford: Clarendon Press.

[20] ISO/IEC (2007). Common Logic (CL) — A Framework for a family of Logic-Based Languages, (IS 24707). Geneva: International Organisation for Standardisation.

[21] ISO/IEC (1996) Extended BNF, IS 14977, International Organisation for Standardisation, Geneva.

[22] Johnson-Laird, Philip N. (2002) Peirce, logic diagrams, and the elementary processes of reasoning, *Thinking and Reasoning* **8:2**, 69-95. http://mentalmodels.princeton.edu/papers/2002peirce.pdf

[23] Kamp, Hans (1981) Events, discourse representations, and temporal references, *Langages* 64, 39-64.

[24] Kamp, Hans, & Uwe Reyle (1993) *From Discourse to Logic*, Dordrecht: Kluwer.

[25] Kamp, Hans, Josef van Genabith, & Uwe Reyle (2011) *Discourse Representation Theory: An Updated Survey*, in D. Gabbay (ed.), Handbook of Philosophical Logic, 2nd ed., Vol XV, pp. 125-394. http://www.ims.uni-stuttgart.de/institut/mitarbeiter/uwe/Papers/DRT.pdf

[26] Ketner, Kenneth Laine, & Hilary Putnam (1992) *Introduction to Peirce* (RLT).

[27] Knight, Will (2016) Can this man make AI more human? Interview with Gary Marcus, *MIT Technology Review*.

[28] Kripke, Saul A. (1963) Semantical considerations on modal logic, Acta Philosophica Fennica, Modal and Many-valued Logics, pp. 83-94.

[29] Lakoff, George (1987) *Women, Fire, and Dangerous Things*, University of Chicago Press, Chicago.

[30] Lakoff, George (1970) Linguistics and natural logic, *Synthese* 22, 151-271.

[31] Langacker, Ronald W. (1999) A view from cognitive linguistics, *Behavioral and Brain Sciences* 22, 625.

[32] Limber, John (1973) The genesis of complex sentences. In , T. Moore, ed., Cognitive *Development and the Acquisition of Language*, Academic Press, New York, 169-186.

[33] MacCartney, Bill, & Christopher D. Manning (2009) An extended model of natural

logic, *Proc. 8th International Conference on Computational Semantics*, Tilburg, pp 140–156.

[34] McNeill, David (1992) *Hand and Mind: What Gestures Reveal about Thought*, Chicago: University of Chicago Press.

[35] McNeill, David (2016) Why We Gesture: The surprising role of hand movements in communication, Cambridge: University Press.

[36] Mossakowski, Till, Mihai Codescu, Oliver Kutz, Christophe Lange, & Michael Grüninger (2014) Proof support for Common Logic, ARONL IJCAR, http://www.iltp.de/ARQNL-2014/download/arqnl2014_paper5.pdf

[37] Majumdar, Arun K., and John F. Sowa (2009). Two paradigms are better than one, and multiple paradigms are even better. In *Proceedings of ICCS 2009, (LNAI 5662)*, S. Rudolph, F. Dau, and S.O. Kuznetsov (eds.), 32-47. Berlin: Springer.

[38] Montague, Richard (1970) English as a formal language, reprinted in Montague (1974), pp. 188-221.

[39] Nubiola, Jaime (1996) Scholarship on the relations between Ludwig Wittgenstein and Charles S. Peirce, in I. Angelelli & M. Cerezo, eds., *Proceedings of the III Symposium on the History of Logic*, Gruyter, Berlin.

[40] Paivio, Allan (1971) *Imagery and Verbal Processes*, New York: Holt, Rinehart and Winston.

[41] Paivio, Allan (2007) *Mind and Its Evolution: A Dual Coding Approach*, New York: Psychology Press.

[42] Peirce, Charles Sanders (1869). Grounds of validity of the laws of logic. *Journal of Speculative Philosophy* 2: 193-208. http://www.peirce.org/writings/p41.html (accessed 15 November 2009).

[43] Peirce, Charles Sanders (1885). On the algebra of logic. American Journal of Mathematics 7:180-202.

[44] Peirce, Charles Sanders (1898) *Reasoning and the Logic of Things*, The Cambridge Conferences Lectures of 1898, ed. by K. L. Ketner, Cambridge, MA: Harvard University Press, 1992.

[45] Peirce, Charles Sanders (1931-1958 CP). Collected Papers of C. S. Peirce, C. Hartshorne, P. Weiss, and A. Burks (eds.), 8 vols., 1931-1958. Cambridge, MA: Harvard University Press.

[46] Peirce, Charles Sanders (1902) *Logic, Considered as Semeiotic*, MS L75, edited by Joseph Ransdell, http://www.iupui.edu/~arisbe/menu/library/bycsp/L75/175.htm

[47] Peirce, Charles Sanders (EP) The Essential Peirce, ed. by N. Houser, C. Kloesel, and members of the Peirce Edition Project, 2 vols., Indiana University Press, Bloomington, 1991-1998.

[48] Peirce, Charles Sanders (NEM) The New Elements of Mathematics, ed. by Carolyn Eisele, 4 vols., The Hague: Mouton, 1976. Peirce, Charles Sanders (RLT). Reasoning and the Logic of Things, (The Cambridge Conferences Lectures of 1898), K. L. Ketner (ed) (1992). Cambridge, MA: Harvard University Press.

[49] Petitto, Laura-Ann (2005) How the brain begets language: On the neural tissue underlying human language acquisition. In J. McGilvray, ed., *The Cambridge Companion to Chomsky*, Cambridge: University Press, pp 84-101.

[50] Pietarinen, Ahti-Veikko (2006) *Signs of Logic: Peircean Themes on the Philosophy of Language, Games, and Communication*, Synthese Library, vol. 329, Berlin: Springer.

[51] Pólya, George (1954) *Mathematics and Plausible Reasoning*, Volume I: *Induction and Analogy in Mathematics*, Volume II: *Patterns of Plausible Inference*, Princeton: University Press.

[52] Putnam, Hilary (1982) Peirce the Logician, *Historia Mathematica* 9:290-301, reprinted in Putnam (1990) pp. 252-260.

[53] Ragni, Marco, & P. N. Johnson-Laird (2018) Reasoning about possibilities: human reasoning violates all normal modal logics, PNAS, http://mentalmodels.princeton.edu/papers/2018humans-vs-modal.pdf

[54] Roberts, Don D. (1973). *The Existential Graphs of Charles S. Peirce*. The Hague: Mouton.

[55] Robinson, J. Alan (1965). A machine oriented logic based on the resolution principle. *Journal of the ACM* 12: 23-41.

[56] Sowa, John F. (2000) *Knowledge Representation: Logical, Philosophical, and Computational Foundations*, Brooks/Cole Publishing Co., Pacific Grove, CA.

[57] Stewart, John (1996). *Theorem Proving Using Existential Graphs*, MS Thesis, Computer and Information Science. Santa Cruz: University of California.

[58] Talmy, Leonard (2000) *Toward a Cognitive Semantics, Volume I: Concept Structuring Systems, Volume II: Typology and Process in Concept Structure*, Cambridge, MA: MIT Press.

[59] Tarski, Alfred (1936). Über den Begriff der logischen Folgerung [On the concept of logical consequence]. In *Logic, Semantics, Metamathematics*, A. Tarski (1982), 2nd ed, 409-420. Indianapolis: Hackett.

[60] van Benthem, Johan (2008) A brief history of natural logic, Technical Report PP-2008-05, Institute for Logic, Language, and Information. http://www.illc.uva.nl/Publications/ResearchReports/PP-2008-05.text.pdf

[61] Whitehead, Alfred North, & Bertrand Russell (1910) *Principia Mathematica*, 2nd ed. 1925, Cambridge: Cambridge University Press.

[62] Whorf, Benjamin Lee (1956) *Language, Thought, and Reality*, Cambridge, MA: MIT Press.

[63] Wittgenstein, Ludwig (1953) *Philosophische Untersuchungen*, translated as *Philosophical Investigations* by G. E. M. Anscombe, P. M. S. Hacker, & Joachim Schulte, revised 4th edition, Oxford: Wiley-Blackwell, 2009.

[64] Wierzbicka, Anna (1996) *Semantics: Primes and Universals*, Oxford: Oxford University Press.

[65] Wos, Larry (1988) *Automated Reasoning: 33 Basic Research Problems*, Englewood Cliffs, NJ: Prentice Hall.

A Appendix: EGIF Grammar

Existential Graph Interchange Format (EGIF) is a linear notation that serves as a bridge between EGs and other notations for logic. Over the years, Peirce had written manuscripts with variant notations, terminology, and explanations [54]. Those variations have raised some still unresolved questions about possible differences in their semantics. With its formally defined semantics, EGIF provides one precise interpretation of each graph translated to it. Whether that interpretation is the one Peirce had intended is not always clear. But the EGIF interpretation serves as a fixed reference point against which other interpretations may be compared and analyzed.

Every EGIF statement has a formally defined mapping to CGIF (Conceptual Graph Interchange Format), whose semantics is specified in ISO/IEC standard 24707 for Common Logic (CL). The semantics of the corresponding CGIF statement shall be called the *default semantics* of EGIF. To express the full CL semantics, the grammar rules in Section 2.3 specify two extensions to EGIF: (1) functions and (2) bound labels as names of relations or functions. EG relations, by themselves, can represent functions. But treating functions as a special case can simplify the inferences and shorten the proofs.

The option of bound labels as names of relations and functions goes beyond first-order logic. It supports some features of Peirce's Gamma graphs, which add extensions for higher-order logic, modal logic, and metalanguage. Section A.4 adds grammar rules for metalanguage in a way that is consistent with the proposed IKL extensions to Common Logic. Whether IKL is compatible with Peirce's intentions is another question for further research.

Section 2.1 presents lexical rules for the symbols and character strings of EGIF. These rules support the features that Peirce used and add new features to support the data types of modern programming languages. Section 2.2 presents the phrase-structure rules. Section 2.3 states context-sensitive constraints. Section 2.4 discusses extensions beyond Common Logic. But Peirce also speculated about variations that are not supported by these rules. Anyone who wishes to extend EGIF to represent them should state where they go beyond the version of EGIF specified here.

A.1 Lexical Rules

All EGIF grammar rules are stated as Extended Backus-Naur Form (EBNF) rules, as defined by [21]. The lexical rules specify names and identifiers, which exclude white space except as noted. The phrase-structure rules in Section 2.2 specify larger combinations that may have zero or more characters of white space between constituents. Each EBNF rule is preceded by an English sentence that serves as an informative description of the syntactic category. If any question arises, the EBNF rule shall be normative.

The following four lexical categories are defined formally in Section A.2 of the ISO standard for Common Logic [20]. The brief definitions here are informative summaries. White space, which is any sequence of one or more white characters, is permitted only in quoted strings and enclosed names.

- A *digit* is any of the ten decimal digits: 0, 1, 2, 3, 4, 5, 6, 7, 8, 9.

- An *integer* is an optional sign (+ or −) followed by a sequence of one or more digits.

- An *enclosed name* is any sequence of Unicode characters, except control characters, preceded and followed by a double quote ". Any double quote internal to an enclosed name shall be represented by the string \". Any backslash internal to an enclosed name shall be represented by the string \\.

- A *letter* is any of the 26 upper case letters from A to Z or the 26 lower case letters from a to z.

- A *white character is* a space, a tab, a new line, a page feed, or a carriage return.

In addition to the above lexical categories, EGIF uses the following lexical category, which is specified in Section B.2 of ISO/IEC 24707:

- An *identifier* is a letter followed by zero or more letters, digits, or underscores.
  `identifier = letter, {letter | digit | '_'};`

Identifiers and enclosed names are case sensitive: the identifier `Apple` is distinct from `apple`. But an identifier is considered identical to the enclosed name with the same case and spelling: Apple is identical to `"Apple"`, and apple is identical to `"apple"`. An integer is also considered identical to the enclosed name that contains the same sign and string of digits.

A.2 Phrase-Structure Rules

Unlike the lexical rules, the phrase-structure rules for EGIF permit an arbitrary amount of white space between constituents. White space is only necessary to separate adjacent identifiers. For example, the following two EGIF statements are semantically identical:

~[(mother*x)~[(woman x)]]
~ [(mother * x) ~ [(woman x)]]

In English, either statement may be read "If some x is a mother, then x is a woman." For better readability, a nest of two negations, which Peirce called a *scroll* may be written with the first negation ~[replaced by [If and the second negation replaced by [Then:

[If (mother *x) [Then (woman x)]]

Each phrase-structure rule is described by an informal English comment and by a formal EBNF rule. For any question, the EBNF rule shall be normative. For any derivation, the rule that defines EG shall be the starting rule. The next paragraph defines informal terms and conventions that relate EGIF to the graphic EGs.

An *area* is a space where existential graphs may be drawn or asserted. The outermost area on which EGs may be asserted is called the *sheet of assertion* (SA). The SA may be cut or fenced off by oval enclosures, which are areas that contain *nested* EGs. For propositional and first-order logic (Peirce's Alpha and Beta graphs), the only oval enclosures are used to represent negations. In EGIF, the area of a negation is represented as ~[EG], where EG represents an existential graph as a set of *nodes* (possibly empty). Some of the nodes in the EG may be negations whose *nested areas* may contain more deeply nested EGs. There is no limit to the depth of nesting. In an informal definition, a term is printed in italics, as in *bound label*. But the same term in EBNF is printed, as in Bound Label. EBNF permits spaces in an identifier, but EGIF does not.

1. A *bound label* is an identifier. The bound label shall be *bound* to a defining label with the same identifier as the bound label.
   Bound Label = identifier;

2. A *coreference node* consists of a left bracket [, one or more names, and a right bracket]. A *defining node* is a coreference node that contains exactly one defining label. An *extension node* is a coreference node that contains exactly one name that is not a defining label.
   Coreference Node = '[', Name, {Name}, ']';

3. A *defining label* consists of an asterisk * and an identifier.
   Defining Label = '*', identifier;

4. A *double negation* is either a negation whose enclosed EG is a negation or a scroll whose first EG is a blank.

```
Double Negation = ' ', ' [', Negation, ']' | '[', 'If',
'[' 'Then', EG, ']' ']';
```

5. An *existential graph* (EG) is a set of nodes. An existential graph with zero nodes is called a blank or an empty existential graph. The order of nodes in the set is semantically irrelevant; but any node that contains a bound label shall follow (occur to the right of) the node that contains its defining label.
   ```
   EG = {Node};
   ```

6. A *function* consists of a left parenthesis (, a type label, zero or more names, a vertical bar |, a name, and a right parenthesis). The names to the left of the bar may be called the inputs, and the one to the right may be called the output.
   ```
   Function = '(', Type Label, {Name}, '|', Name, ')';
   ```

7. A *name* is one of a defining label, a bound label, an identifier, an enclosed name, or an integer.
   ```
   Name = Defining Label | Bound Label | identifier | enclosed name
   | integer;
   ```

8. A *negation* consists of a tilde ~, a left bracket [, an existential graph, and a right bracket].
   ```
   Negation = ' ', '[', EG, ']';
   ```

9. A *node* is a coreference node, a relation, a function, a negation, or a scroll.
   ```
   Node = Coreference Node | Relation | Function | Negation
   | Scroll;
   ```

10. A *relation* consists of a left parenthesis (, a type label, zero or more names, and a right parenthesis).
    ```
    Relation = '(', Type Label, {Name}, ')';
    ```

11. A *scroll* consists of a left bracket [, the letters **If**, an EG, a left bracket [, the letters **Then**, an EG, and two right brackets]]. Syntactically, a scroll is an optional notation for replacing the tilde in two negations with the keywords **If** and **Then**. Both notations have identical semantics.
    ```
    Scroll = '[', 'If', EG, '[', 'Then', EG, ']', ']';
    ```

12. A *type label* is any name except a defining label. If the name is a bound label, the value associated with its defining label determines the type of some relation or function.
    ```
    Type Label = Name - Defining Label;
    ```

A.3 Constraints and Examples

As the grammar rules show, an EG is represented by zero or more EGIF nodes. The only constraints on the nodes or their ordering are determined by the location of defining labels and their bound labels. The following six rules are equivalent to the rules for scope of quantifiers in predicate calculus. These rules are not needed for the graphic EGs, because the lines of identity show the scope by direct connections, not by labels.

1. No area may contain two or more defining labels with the same identifier.

2. The *scope* of a defining label shall include the area in which it occurs and any area nested directly or indirectly in this area, unless it is blocked by rule 5 below.

3. Every bound label shall be in the scope of exactly one defining label, which shall have exactly the same identifier. It is said to be *bound* to that defining label.

4. Every defining label shall precede (occur to the left of) every one of its bound labels.

5. If a defining label with some identifier x occurs in an area nested within the scope of a defining label in an outer area with the same identifier x, then the scope of the outer defining label shall be blocked from that area: every bound label with the identifier x that occurs in this inner area shall be bound to the defining label in this area.

6. In any area, all permutations of the nodes that preserve the above constraints shall be semantically equivalent.

A name enclosed in double quotes, such as "John Q. Public", may contain spaces and punctuation. To avoid quotes, other stylistic options may be used, such as John_Q_Public or JohnQPublic, but these three variants are distinct. To specify multiple names as synonyms, put them in a coreference node:

["John Q. Public" John_Q_Public JohnQPublic JQPublic JQP]

Peirce treated functions as special special cases of relations. He didn't have a notation that distinguished them from ordinary relations. In EGIF, a function may be considered a kind of relation for which the value represented by the argument (or peg) after the vertical bar is uniquely determined by the values of the pegs that precede the bar. The pegs to the left of the vertical bar may be called the *inputs*, and the one to the right of the bar may be called the *output*.

- If f is a function with n inputs and if for every i from 1 to n, the ith input of one instance of f is coreferent with the ith input of another instance of f in the same area, then a line of identity may be drawn to connect the output pegs of both instances.

The act of drawing a line of identity between the output pegs in the graphic notation corresponds to inserting a coreference node in EGIF or an equality in other notations for logic. This rule of inference is the basis for the method of unification in theorem proving systems. The EGIF grammar allows functions with zero inputs; every instance of such a function would have exactly the same output value. Therefore, the output pegs of all instances of that function may be joined. These rules imply that a function with zero input pegs may be used to represent a *Skolem constant*.

Peirce did not introduce an EG notation for proper names or constants. Instead, he used monadic relations, such as -Alexander or -is Alexander. This relation would be true of anyone named Alexander. In EGIF, a name that is true of exactly one individual may be used as the name of a function with zero inputs. In the following function, the defining label *p would represent a line of identity for the unique person with that name:

```
("Philippus Aureolus Theophrastus Bombastus von Hohenheim"
 | *p)
```

A name such as Alexander, which may be unique in a specific context, could be written as a function with an input peg for the context and an output peg for the person of that name: (Alexander c | *p). Although Peirce did not define a notation for functions in EGs, the output peg of a function could be distinguished by an arrowhead in the graphic form.

The EBNF rules for relation and function allow a bound label to be the type label. That option supports the feature of Common Logic that allows quantified variables to refer to functions and relations. As an example, consider the sentence "There is family relation between any two members of the same family." To translate that sentence to any version logic, restate it with explicit variables F, R, x, and y: "For any family F and any two members x and y of F, there is a family relation R that is true of x and y." That sentence may be translated to the following EGIF:

```
[If (family *F) (memberOf *x F) (memberOf *y F)
   [Then (familyRelation *R) (R x y) ] ]
```

The expression (familyRelation *R) asserts that there exists a family relation R, and the next expression uses R as a type label. For his Gamma graphs, which are discussed in the next section, Peirce experimented with graphical ways of representing such options.

A.4 Extensions for Gamma Graphs

Peirce developed various notations for logic, algebraic and graphic. He also experimented with semantic extensions that go beyond first-order logic. In 1885, he introduced the algebraic notation that became predicate calculus [52]. He used the terms *first-intentional logic* for quantifiers that range over simple individuals and *second-intentional logic* for quantifiers that range over relations. In that article, he used second intentional logic to define equality $x = y$ by a statement that for every relation R, $R(x)$ if and only if $R(y)$. Ernst Schröder translated Peirce's terms as *erste Ordnung* and *zweite Ordnung*, which Bertrand Russell translated back to English as *first order* and *second order*. Peirce also introduced notations for three-valued logic, modal logic, and metalanguage about logic. Roberts [54] summarized the various graphical and algebraic notations and cited the publications and manuscripts in which Peirce discussed them.

Peirce used the term *Gamma graphs* for the versions of EGs that went beyond first-order (or first-intentional) logic. As early as 1898, he used the following example of a metalevel statement in EGs:

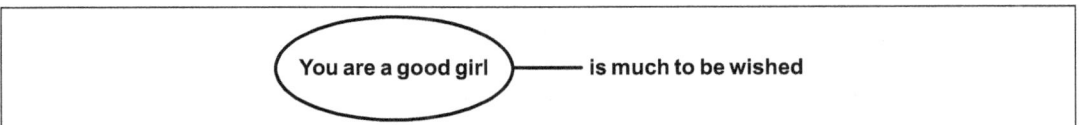

Figure 41: A metalevel existential graph by Peirce [44])

In describing that graph, Peirce wrote "When we wish to assert something about a proposition without asserting the proposition itself, we will enclose it in a lightly drawn oval, which is supposed to fence it off from the field of assertions." When translated to EGIF, the enclosed proposition would be a medad, a relation with no pegs: ("You are a good girl"). The line of identity, which refers to that proposition could be represented by a defining label *p. To show that p refers to the proposition expressed by the EG, another syntactic option is necessary:

> A proposition node is a coreference node that contains a name and an EG that expresses the proposition that the name refers to.
> Proposition Node = '[', Name, EG, ']';

With this option, the EGIF for Figure 41 would be

> [*p ("You are a good girl")] ("is much to be wished" p)

For the semantics of that option, the IKL extension to Common Logic could be used. Hayes and Menzel [14] introduced an operator named *that* as a kind of function that maps an IKL sentence to a name that refers to the proposition stated by that sentence. The IKL semantics for the that-operator could be adopted for proposition nodes in EGIF. These issues are discussed further in Section 6.

In some writings, Peirce used ovals with colors or dotted boundaries to represent modality. EGIF can use identifiers such as `Possible` or `Necessary` for relations applied to propositions. The following EGIF says "That you are a good girl is possible but not necessary."

[*p ("You are a good girl")] (possible p) ~[(necessary p)]

Examples like this may be generalized to any kind of metalevel propositions and methods for reasoning about them: modal, multivalued, intuitionistic, fuzzy, or statistical. In his published works and much more voluminous unpublished manuscripts, Peirce analyzed and proposed many semantic features. For Alpha and Beta graphs, the semantics of Common Logic or its FOL subset is consistent with Peirce's examples. But determining exactly what he had in mind for his many variations of Gamma graphs is still a research project. In some late manuscripts, he also mentioned a version of Delta graphs. No one knows what Peirce intended, but EGIF can be a useful tool for exploring the options.

A.5 Computer tools for processing EGIF

EGIF was originally defined as a subset of the Conceptual Graph Interchange Format (CGIF), one of the dialects of Common Logic (CL). But after some simplifications in the notation, the version of EGIF defined in this appendix still has a direct mapping to CGIF, and its semantics is still defined by its mapping to CGIF. But its syntax is simpler. As an example, "A cat is on a mat" would be translated to the following EGIF:

(Cat *x) (On x *y) (Mat y)

To translate any EGIF sentence to an equivalent CGIF sentence, put every defining label in a defining node at the beginning of the area in which it occurs. Then replace each defining label inside a relation node with a bound label with the same identifier. The result is an EGIF sentence that is logically equivalent to the original:

[*x] [*y] (Cat x) (On x y) (Mat y)

To derive a logically equivalent CGIF sentence, place a question mark ? in front of every bound label:

[*x] [*y] (Cat ?x) (On ?x ?y) (Mat ?y)

The equivalent sentence in the Common Logic Interchange Format (CLIF) requires the keyword exists for the quantifier and an explicit Boolean operator for the implicit conjunction:

```
(exists (x y) (and (Cat x) (On x y) (Mat y)))
```

The Heterogeneous Tool Set (HeTS) provides theorem provers for Common Logic and translators to and from CL and other logics, including the logics for the Semantic Web [36]. Any EGIF sentence may be translated to CGIF or CLIF by the above steps and be processed by any of the HeTS tools. For natural languages, discourse representation theory [25] addresses the linguistic analysis that is prior to any representation in logic. But the base logic (DRS) is isomorphic to EGs and CGIF. For examples, see Figures 33 to 37.

Playing with Anticipations as Abductions
Strategic Reasoning in an Eco-Cognitive Perspective

Lorenzo Magnani
University of Pavia, Italy
lmagnani@unipv.it

Abstract

In my opinion, it is only in the framework of a study concerning abductive inference that we can correctly and usefully grasp the cognitive status of reasoning strategies and related heuristics. To this aim, taking advantage of my *eco-cognitive model* (EC-model) of abduction, I will analyze strategic reasoning in the perspective of the so-called fill-up and cutdown problems, which characterize abductive cognition. I will also illustrate the abductive character of the concept of anticipation linking it to the cognitive problem of spatiality and of the genesis of space in the description of the abductive role of the *Abschattungen* (adumbrations), as they are described in the framework of the philosophical tradition of phenomenology. Anticipations share various features with visual and manipulative abduction and prove to be a useful tool to favor the characterization of the two kinds of strategic reasoning I am introducing in this article: locked and unlocked abductive strategies. The role of these different cognitive strategies, active both in humans and in computational machines, is also intertwined with the production of different kinds of cognitive hypothetical results, which range from poor to rich level of knowledge creativity. Differences that also have important consequences when we have to deal with computational AI programs devoted to perform the various types of abductive reasoning.

Keywords: Abduction, Anticipations, Creativity, Eco-Cognitive Model, Eco-Cognitive Openness, Locked and Unlocked Strategies, Go game, AlphaGo, Deep Learning

1 Unlocked Cognitive Strategies Normally Characterize Knowledge-Enhancing Abduction

The adjective strategic, which is relatively vague, is adopted – especially in the AI research – to refer to a smart composition in reasoning of several heuristic cognitive devices: the strategy pertains the smart consecutive choice of the next state of a cognitive process (for example the nearest one, according to some distance measure) and a heuristic is one of the tools a strategy can adopt to reach the desired state quickly. In the framework of game theory the word strategy immediately involves the consideration of other agents and the related adversarial, intertwined, or collective cognitive acts. Further, the tradition of the so-called ecological thinking (or ecological rationality) [12, 11, 43, 10] attributes to strategies a broader meaning: they also involve thinking processes which employ a huge quantity of information and high computational costs, instead heuristics are in general committed to be much more simple and efficient, even if less accurate. Finally, cognitive heuristics are often just considered as cognitive strategies in themselves. In this article I am inclined to follow the AI tradition, so thinking at strategies as a smart successive choice of appropriate heuristics.

In my opinion, it is only in the framework of abductive inference, I will consider in the present first section, that we can correctly and usefully grasp the cognitive problem of strategic reasoning. More precisely, the keystone concept of *knowledge-enhancing abduction* will make us able to deeply understand the logical and cognitive status of those kinds of cognitive strategies I am introducing here and that I call *unlocked abductive strategies* and the nature of what I define their eco-cognitive facets. It will be in the second section, exemplifying unlocked model-based and manipulative abductive strategies, that I will deal with the abductive character of the concept of anticipation in the light of the cognitive problem of the genesis of space in the description of the role of the *Abschattungen* (adumbrations), as they are described in the framework of the philosophical tradition of phenomenology. Anticipations prove to be a very useful tool to further deepening the concept of strategy: indeed I will devote the third and last section to illustrate the other of reasoning I am introducing in this article: the *locked abductive strategies*. I will also explain the importance of the distinction between locked and unlocked strategies in the perspective of computational AI programs devoted to perform the various types of abductive reasoning.

1.1 Strategies and Cutdown and Fill-Up Problems in the EC-Model of Abduction

As I have anticipated, it is useful to see cognitive strategies in the perspective of the so-called fill-up and cutdown problems of abduction, clearly defined by [48, p. 242]. Since for many abduction problems there are – usually – many guessed hypotheses, the abducer needs reduce this space to one: this means that the abducer has to produce the best choice among the members of the available group: "It is extremely difficult to see how this is done, both formally and empirically. [...] There is the problem of finding criteria for hypothesis *selection*. But there is the prior problem of specifying the conditions for *thinking up* possible candidates for selection. The first is a 'cutdown' problem. The second is a 'fill-up problem'; and with the latter comes the received view that it is not a problem for logic" ([48, p. 243] emphasis added). Obviously consistency and minimality constraints were emphasized in the "received view" on abduction established by many classical logical accounts, more oriented to illustrate selective abduction [26] – for example in diagnostic reasoning, where abduction is merely seen as an activity of "selecting" from an encyclopedia of pre-stored hypotheses – than to analyze *creative* abduction (abduction that generates new hypotheses).[1] It is at the fill-up level that strategic reasoning and related heuristics play a fundamental role, for example in the various cases of creative cognition.

When we are dealing with strong cases of creative cognition, such as scientific discovery, the consistency requirement results puzzling: it is here sufficient to note that Paul Feyerabend, in *Against Method* [8], correctly attributes a great importance to the role of contradiction in generating hypotheses, also against the role of similarity, and so implicity celebrates the value of creative abductive cognition. Speaking of induction and not of abduction (this concept was relatively unknown at the level of the international philosophical community at that time), he establishes a new "counterrule". This is the opposite of the neopositivistic one that it is "experience" (or "experimental results") which constitutes the most important part of our scientific empirical theories, a rule that formed the core of the so-called "received view" in philosophy of science (where inductive generalization, confirmation, and corroboration play a central role). The counterrule "[...] advises us to introduce and elaborate hypotheses which are inconsistent with well-established theories and/or well-established facts. It advises us to proceed counterinductively" [8, p. 20]. Counterinduction is seen more reasonable than induction, because appropriate to the needs of creative reasoning in science: "[...] we need a dream-world in order to discover the features of the real world we think we inhabit" (p. 29). We know that counterinduction, that is the act of introducing, inventing, and generating new inconsistencies and anomalies, together with new points of view incommensurable with the old ones, is congruous with the aim of inventing "alternatives" (Feyerabend contends that "proliferation of theories is beneficial for science"), and very important in all kinds of creative reasoning.

[9], by proposing their GW-Schema, contend that abduction presents an *ignorance-preserving* or (ignorance-mitigating) character. From this perspective abductive reasoning is a *response* to an ignorance-problem; through abduction the basic ignorance – that does not have to be considered a total "ignorance" – is neither solved nor left intact. Abductive reasoning is an ignorance-preserving accommodation of the problem at hand. Following this perspective (later on modified by Gabbay and Woods themselves [49]) knowledge can be adequately enhanced through abdduction only thanks to a necessary empirical evaluation phase, or an inductive phase, as Peirce called it.

However, Feyerabend's observations I have just resumed lead us to touch the core of the

---

[1] I have proposed the dichotomic distinction between selective and creative abduction in [26]. A recent and clear analysis of this dichotomy and of other classifications emphasizing different aspects of abduction is given in [36].

ambiguity of the ignorance-preserving character of abduction. Why?

- Because the cognitive processes of generation (fill-up) and of selection (cutdown) can both be sufficient – even in absence of the standard inductive evaluation phase – to *activate* and accept an abductive hypothesis, and so to reach cognitive results relevant to the context (often endowed with a knowledge-enhancing outcome, as I have illustrated in [28]. In these cases instrumental aspects (which simply enable one's target to be hit) often favor both abductive generation and abductive choice, and they are not necessarily intertwined with plausibilistic concerns, such as consistency and minimality.

In these special cases the best choice – often thanks to the exploitation of strategic reasoning and related appropriate heuristics – is immediately reached without the help of an experimental trial (which fundamentally characterizes the received view of abduction in terms of the so-called "inference to the best explanation").

Let us recall some basic information concerning the received view on strategic rules. Hintikka thinks that "strategic rules" (contrasted with definitory rules) are smart rules, even if they fail in individual cases, and show a propensity for cognitive success. I would add that they are tune with Peirce's consideration of abduction "as akin to truth".[2] Even if inclined to cognitive success, strategic rules, when exploited in abductive hypothetical reasoning, tacitly fulfil the ignorance condition I have illustrated above, thus abduction would aim at neither truth-preservation not probability-enhancement, as Peirce maintained. Moreover, Hintikka's definitory rules are recursive but in several important cases strategic rules are not: for example, playing a game strategically requires some kind of creativity.

I contended – few lines above – that special cases of cognitive processes of generation (fill-up) and of selection (cutdown) are sufficient to reach the acceptation of a hypothesis: even in absence of the standard inductive evaluation phase the best choice is immediately reached without the help of an experimental trial (which fundamentally characterizes the received view of abduction in terms of the so-called "inference to the best explanation"). The best choice is often reached through strategic reasoning and heuristics, that are not, in these cases, ignorance preserving, but instead knowledge-enhancing.

Furthermore, we have to strongly note that the generation process alone can be still seen sufficient considering the case of human visual *perception*, where the hypothesis generated is immediate and unique, even if no strategic reasoning appears to be involved.[3] Indeed, per-

---

[2]"It is a primary *hypothesis* underlying all abduction that the human mind is akin to the truth in the sense that in a finite number of guesses it will light upon the correct hypothesis" [37, 7.220].

[3]Woods observes that "[...] for wide ranges of cases knowledge will require more information than the conscious mind can hold at the time the knowledge is acquired and retained. The moral to draw is that most of that indispensable knowledge is held unconsciously. Unconscious information-processing has all or most of the following properties, often in varying degrees and harmonies" [50].

ception is considered by Peirce, as an "abductive" fast and uncontrolled (and so automatic) knowledge-production procedure. Perception, in this philosophical perspective, is a vehicle for the instantaneous retrieval of knowledge that was previously structured in our mind through more structured inferential processes. Peirce says: "Abductive inference shades into perceptual judgment without any sharp line of demarcation between them" [38, p. 304]. By perception, knowledge constructions are so instantly reorganized that they become habitual and diffuse and do not need any further testing: "[...] a fully accepted, simple, and interesting inference tends to obliterate all recognition of the uninteresting and complex premises from which it was derived" [37, 7.37].[4] Can we avoid to attribute a strategic role to perception or to other kinds of model-based/manipulatory non-propositional cognition? I do not think so: in the second section I will illustrate the strategic character of visual, kinesthetic, and motor sensations.

My abrupt reference to perception as a case of abduction (in this case I strictly follow Peirce) does not have to surprise the reader. Indeed, at the of center of my perspective on cognition is the emphasis on the "practical agent", of the individual agent operating "on the ground", that is, in the circumstances of real life. In all its contexts, from the most abstractly logical and mathematical to the most roughly empirical, I always emphasize the cognitive nature of abduction. Reasoning is something performed by cognitive systems. At a certain level of abstraction and as a first approximation, a cognitive system is a triple (A, T, R), in which A is an *agent*, T is a *cognitive target* of the agent, and R relates to the *cognitive resources* on which the agent can count in the course of trying to meet the target-information, time and computational capacity, to name the three most important. My agents are also *embodied distributed cognitive systems*: cognition is embodied and the interactions between brains, bodies, and external environment are its central aspects. Cognition is occurring taking advantage of a constant exchange of information in a complex distributed system that crosses the boundary between humans, artifacts, and the surrounding environment, where also instinctual and unconscious abilities play an important role. This interplay is especially manifest and clear in various aspects of abductive cognition.

It is in this perspective that we can appropriately consider perceptual abduction as a fast and uncontrolled knowledge production, that operates for the most part automatically and out of sight, so to speak. This means that – at least in this light – we cannot say that abduction is constitutively ignorance preserving an basically propositional, that is rendered by symbols carrying propositional content. My perspective adopts the wide Peircean philosophical framework, which approaches "inference" *semiotically* (and not simply "*logically*"): Peirce distinctly says that all inference is a form of sign activity, where the word sign includes "feeling, image, conception, and other representation" [37, 5.283]. It is clear

---

[4]An interesting research related to artificial intelligence (AI) presents a formal theory of robot perception as a form of abduction, so reclaiming the rational relevance of the speculative anticipation furnished by Peirce, cf. [45].

that this semiotic view is considerably compatible with my perspective on cognitive systems as embodied and distributed systems. It is in this perspective that we can fully appreciate the role of strategic cognition, which not only refers to propositional aspects but it is also performed in a framework of distributed cognition, in which also models, artifacts, internal and external representations, manipulations play an important role.

In a wide eco-cognitive perspective the cutdown and fill-up problems in abductive cognition appear to be spectacularly *contextual*.[5] I lack the space to give this issue appropriate explanation but it suffices for the purpose of this study – which instead aims at revisiting the concept of strategy and heuristic – to remember that, for example, one thing is to abduce a model or a concept at the various levels of scientific cognitive activities, where the aim of reaching rational knowledge dominates, another thing is to abduce a hypothesis in literature (a fictional character for example), or in moral reasoning (the adoption/acceptance of a hypothetical judgment as a trigger for moral actions), or in an adversarial board game such as Go or Chess (in which the hypothesis is an anticipation of long-term results that favors each move to the strategic aim of winning). However, in all these cases abductive hypotheses which are evidentially inert are accepted and activated as a basis for action, often highly creative or at least successful, even if of different kind.

The backbone of this approach can be found in the manifesto of my eco-cognitive model (EC-model) of abduction in [27].[6] It might seem awkward to speak of "abduction of a hypothesis in literature," but one of the fascinating aspects of abduction is that not only it can warrant for scientific discovery, but for other kinds of creativity as well. We must not necessarily see abduction as a *problem solving device* that sets off in response to a cognitive irritation/doubt: conversely, it could be supposed that esthetic abductions (referring to creativity in art, literature, music, games, etc.) arise in response to some kind of esthetic irritation that the author (sometimes a *genius*) perceives in herself or in the public. Furthermore, not only esthetic abductions are free from empirical constraints in order to become the "best" choice: as I am showing throughout this article, many forms of abductive hypotheses in traditionally-perceived-as-rational domains (such as the setting of initial conditions, or axioms, in physics or mathematics) are relatively free from the need of an empirical assessment. The same could be said of moral judgements: they are eco-cognitive abductions, inferred upon a range of internal and external cues and, as soon as the judgment hypothesis has been abduced, it immediately becomes prescriptive and "true," informing the agent's behavior as such. Assessing that there is a common ground in all of these works of what could be broadly defined as "creativity" does not imply that all of these forms of selective or creative abduction with their related cognitive strategies are the same, contrarily it should

---

[5]Some acknowledgment of the general contextual character of these kinds of criteria, and a good illustration of the role of coherence, unification, explanatory depth, simplicity, and empirical adequacy in the current literature on scientific abductive best explanation, is given in [25].

[6]Further details concerning the EC-model of abduction can be found in [29, 31].

spark the need for firm and sensible categorization: otherwise it would be like saying that to construct a doll, a machine-gun and a nuclear reactor are all the same thing because we use our hands in order to do so!

1.2 Is Abduction Knowledge-Enhancing?

In the previous subsection I have contended that abduction does not have to be considered a constitutively ignorance-preserving (or ignorance mitigating) reasoning, truth can easily emerge: we have to remember that Peirce sometimes contended that abduction "come to us as a flash. It is an act of insight" [37, 5.181] but nevertheless possesses a mysterious power of "guessing right" [37, 6.530]. Consequently abduction, preserves ignorance, in the logical sense I have illustrated above, but also can provide truth because has the power of guessing *right*.

In the light of the ignorance-preserving perspective we can also say that the inference to the best explanation – if considered as a truth conferring achievement justified by empirical approval – cannot be a case of abduction, exactly because abductive inference is instead and always constitutively ignorance-preserving. If we say that truth can be reached through a "simple" abduction (not intended as involving an evaluation phase, that is coinciding with the whole inference to the best explanation, fortified by an empirical evaluation), it seems we confront a manifest incoherence. Indeed, in this new perspective it is contended that even simple abduction can provide truth, even if it is epistemically "inert" from the empirical perspective. Why? We can solve the incoherence by observing that we should be compelled to consider abduction as ignorance-preserving only if we consider the empirical test *the only way* of conferring truth to a hypothetical knowledge content. If we admit that there are ways to accept a hypothetical knowledge content different from the empirical test, for example taking advantage of special knowledge-enhancing reasoning strategies – simple abduction is not necessarily constitutively ignorance-preserving: in the end we are dealing with a disagreement about the nature of *knowledge*, as Woods himself contends. As I have indicated at the beginning of this paragraph, those who consider abduction as an inference to the best explanation – that is as a truth conferring achievement involving empirical evaluation – obviously cannot consider abductive inference as ignorance-preserving. Those who consider abduction as a mere activity of guessing are more inclined to accept its ignorance-preserving character.

However, *we are objecting that abduction – and so the possible cognitive strategies and heuristics that substantiate it – is in this last case still knowledge-enhancing*.

At this point two important consequences concerning the meaning of the word *ignorance* in this context have to be illustrated:

1. abduction, also when intended as an inference to the best explanation in the "classi-

cal" sense I have indicated above, is always *ignorance-preserving* because abduction represents a kind of reasoning that is constitutively provisional, and you can withdraw previous abductive results (even if empirically confirmed, that is appropriately considered "best explanations"), in presence of new information. From the logical point of view this means that abduction represents a kind of nonmonotonic reasoning, and in this perspective we can even say that abduction interprets the "spirit" of modern science, where truths are never stable and absolute. Peirce also emphasized the "marvelous self-correcting property of reason" in general [37, 5.579]. So to say, abduction incarnates the human perennial search of new truths and the human Socratic awareness of a basic ignorance which can only be attenuated/mitigated. In sum, in this perspective abduction always preserves ignorance because it reminds us we can reach truths that can always be withdrawn; ignorance removal is at the same time constitutively related to ignorance regaining;

2. even if ignorance is preserved in the sense I have just indicated, which coincides with the spirit of modern science, abduction is also knowledge-enhancing because new truths can be and "are" discovered which *are not necessarily best explanations intended as hypotheses which are empirically tested.*

I have just said that knowledge can be attained in the absence of evidence; there are propositions about the world which turn to be true by virtue of considerations that lend them no evidential/empirical weight. They are true beliefs that are not justified on the basis of evidence. Is abduction related to the generation of knowledge contents of this kind? Yes it is.

Abduction is guessing reliable hypotheses, and humans are very good at it; abduction is akin to truth: it is especially in the case of empirical scientific cognition that abduction reveals its more representative epistemic virtues, because it provides hypotheses, models, ideas, thoughts experiments, etc., which, even if *devoid of initial* evidential support, constitute the fundamental rational building blocks for the generation of new laws and theories which only later on will be solidly empirically tested.

In the following subsections of this study I aim at illustrating this intrinsic character of abduction, which shows why we certainly can logically consider it a kind of ignorance-preserving cognition, but at the same time a cognitive – strategic – process that can enhance knowledge at various level of human cognitive activities, even if the empirical evaluation lacks. Consequently, strategic abductive processes and related heuristics are – occasionally – knowledge-enhancing in themselves.[7]

---

[7]In a previous article [28] I have shown that Peirce, to substantiate the truth-reliability of abduction – which coincides with its "ampliative" character, as illustrated in standard literature – provides philosophical and evolutionary justifications; furthermore, I have also illustrated some actual examples of knowledge-enhancing

1.3 Do Reasoning Strategies Justify Abduction?

I have just illustrated the main theoretical tools we need to reframe strategic reasoning and heuristics in an eco-cognitive perspective on abduction. Indeed, from an eco-cognitive point of view, in more hybrid and multimodal (cf. [27, chapter four]) (not merely inner) abductive processes, such as in the case of manipulative abduction,[8] the *assessment/acceptation* of a hypothesis is reached – and constrained – taking advantage of the gradual – strategic – acquisition of consecutive external information with respect to future interrogation and control, and not necessarily thanks to a final and actual experimental test, in the classical sense of empirical science.

Hintikka implicitly acknowledges the multimodality and hybridity of what I call *selective abduction* when, taking advantage of the intellectual atmosphere of his Socratic interrogative epistemology, observes that "[...] abduction as a method of guessing is based on the variety of different possible sources of answers. Such 'informants' must include not only testimony, observation, and experiments, but the inquirer's memory and background knowledge" [17, p. 56].[9] Moreover, Hintikka further notes that also "creative abduction", generated by a kind of *oracle*, is often needed: "But what can an inquirer do when all such sources fail to provide an answer to a question? Obviously the best the inquirer can do is make an informed guess. For the purposes of a general theory of inquiry, what Peirce calls 'intelligent guessing' must therefore be recognized as one of the many possible 'oracles', alias sources of answers. Peirce may very well have been more realistic than I have so far been in emphasizing the importance of this particular 'oracle' in actual human inquiry" (*ibid.*). In sum, the sources of answers(and information) have to be very wide and rich in the case of selective abduction, but this requirement obviously does not change in the case of creative abduction, in which a further cognitive "leap" has to be performed.

In summary, at least four kinds of actions can be involved in the manipulative abductive strategic processes (and we would have to also take into account the motoric aspect (*i*) of inner "thoughts" too). In the eco-cognitive interplay of abduction the cognitive agent

---

abductions active in science, that nevertheless are evidentially inert, such as in the case of guessing the so-called "conventions". They are extremely important in physics, evidentially inert fruits of abduction – at least from the point of view of their impossible falsification – but nevertheless knowledge-enhancing.

[8]The concept of *manipulative abduction* – which also takes into account the external dimension of abductive reasoning in an eco-cognitive perspective – captures a large part of scientific thinking where the role of action and of external models (for example diagrams) and devices is central, and where the features of this action are implicit and hard to be elicited. Action can provide otherwise unavailable information that enables the agent to solve problems by starting and by performing a suitable abductive process of generation and/or selection of hypotheses. Manipulative abduction happens when we are thinking through doing and not only, in a pragmatic sense, about doing (cf. [27, chapter one]).

[9]We will see below that in my perspective this means that good abductions need be performed in a situation characterized by what I call (cf. below section 3) *optimization of eco-cognitive situatedness* in which eco-cognitive openness is fundamental.

further triggers internal *thoughts* "while" modifying the environment and so (*ii*) acting on it (thinking through doing). In this case the "motor actions" directed to the environment have to be intended as part and parcel of the whole embodied abductive inference, and so have to be distinguished from the *final* (*iii*) "actions" as a possible consequence of the reached abductive result.

In this perspective the proper experimental test involved in the Peircean evaluation phase, which for many researchers reflects in the most acceptable way the idea of abduction as inference to the best explanation, just constitutes a *special* subclass of the process of the adoption of the abductive hypothesis – the one which involves a terminal kind (*iv*) of actions (experimental tests), and should be considered ancillary to the nature of abductive cognition, and inductive in its essence. We have indeed to remark again that in Peirce's mature perspective on abduction as embedded in a cycle of reasoning, induction just plays an evaluative role. Hintikka usefully notes, and I agree with him, that Peirce was right in denying the role of "naked" induction in forming new hypotheses:

> Many philosophers would probably bracket abductive inference with inductive inference. Some would even think of all ampliative inference as being, at bottom, inductive. In this matter, however, Peirce is one hundred percent right in denying the role of naked induction in forming new hypotheses. [...] It might seem that the critical and evaluative aspect of inquiry that Peirce called inductive still remains essentially different from the deductive and abductive aspects. A common way of thinking equates all ampliative inferences with inductive ones. Peirce was right in challenging this dichotomy. Rightly understood, the ampliative versus non-ampliative contrast becomes a distinction between interrogative (ampliative) and deductive steps of argument. As in Peirce, we also need over and above these two also the kind of reasoning that is involved in testing the propositions obtained as answers to questions. I do not think that it is instructive to call such reasoning inductive, but this is a merely terminological matter [17, pp. 52 and 55].

In absence of empirical evaluation, can we attribute the *pure* abductive inclination to produce right guesses indicated by Peirce, conductive to the acquisition of truth, to the *reliability* of the process? Yes, we can, but only if we take into account the following warning, still illustrated by Hintikka, who stresses the importance of *strategic/heuristic aspects*, arguing for their fundamental role as the warrant and justification of abductive inference: "Many contemporary philosophers will assimilate this kind of justification to what is called a reliabilist one. Such reliabilist views are said to go back to Frank Ramsey, who said that 'a belief was knowledge if it is (1) true, (2) certain, (3) *obtained by a reliable process*'. Unfortunately for reliabilists, such characterizations are subject to the ambiguity that was pointed out earlier. By a reliable process one can mean either a process in which *each step* is conducive to acquiring and/or maintaining truth or closeness to truth, or one that *as a whole* is apt to lead the inquirer to truth. Unfortunately, most reliabilists unerringly choose the wrong interpretation – namely, the first one. As was pointed out earlier, the true justification of a rule of abductive inference is a strategic one" (emphases added) [17, p. 57]. The important thing is to stress that this strategic justification *does not warrant* any "specific step" of the whole process. Let us remember that abduction certainly provides new information into an argument, but this is not necessarily a true information, because it is not implied by what it is already known or accepted but it is constitutively hypothetical – that is, ignorance-preservation is constitutive, from the general logico-philosophical point of view, and Hintikka is in tune with this assumption when observing that specific steps do not warrant abduction.

2 Playing with Anticipations as Abductions in Natural and Artificial Games

2.1 Adumbrations and the Generation of the Three-Dimensional Space: Strategies in Embodiment and in Distributed Cognition Environment

As I promised at the beginning of this article this second section is devoted to study – in the light of abductive cognition – the so-called "anticipations": they will help us to delineate both the role of strategies in distributed hypothetical reasoning and to further illustrate the main features of the so-called "unlocked" strategies. Indeed, as I have already indicated, when describing manipulative abduction, strategic cognition not only refers to propositional aspects but it is also performed in a distributed cognition framework, in which models, artifacts, internal and external representations, sensations, and manipulations play an important role: indeed the phenomenological example illustrated in this section also shows that strategic cognition can involve, when clearly seen in embodied and distributed systems, visual, kinesthetic, and motor sensations. In this case we deal with a "natural game" between humans and their surroundings, in which "unlocked" strategies are at play: in the light of the analysis of the distinction between unlocked and locked strategies it will be useful to compare this case with the human made case of "artificial games", to show instructive analogies and differences (see below subsection 2.2 and section 3).

Looking at the philosophical explanations of the ways humans perform to build "idealities", geometrical idealities, and the objective space, Husserl contends that "facticities of every type [...] have a root in the essential structure of what is generally human", and that "human surrounding world is the same today and always" [23, p. 180].[10] However, the

---

[10]Of course this should not hold when we consider the possible evolutionary character of this surrounding

horizon of the rough surrounding prepredicative world of appearances and primordial and immediately given experiences – which is at the basis of the constructive cognitive activity – is a source of potentially infinite data,[11] which cannot "lock" cognitive strategies related to the multiple strategic abductive generation of idealities, geometrical ideal forms, and spatiality. Indeed, step by step, ideal objects in Husserlian sense are constructed and become *traditional* objects, and so they possess historicity as one of their multiple eidetic components. They become, Husserl says, "sedimentations of a truth meaning", which describe the cumulative character of human experience (not every "abiding possession" of mine is traceable to a self-evidence of my own). The research which takes advantage of the already available sedimented idealities (sedimentations of someone else's already accomplished experience) is at the basis of further abductive work to the aim, for example, of discovering new mathematical knowledge in the field of geometry.

Let us follow some Husserlian speculations that lead us to consider the important strategic role of anticipations as abductions. In subsection 1.1 above I have already illustrated the constitutive abductive character of perception in the light of Peircean philosophy. Now we will see the strategic abductive role of both perception and kinesthetic data in the Husserlian philosophical framework, integrating it with a reference to some of the current results of neuroscience. Indeed, the philosophical tradition of phenomenology fully recognizes the protogeometrical role of kinesthetic data in the generation of the so-called "idealities" (and of geometrical idealities). The objective space we usually subjectively experience has to be put in brackets by means of the transcendental reduction, so that pure lived experiences can be examined without the compromising intervention of any psychological perspective, any "doxa". By means of this transcendental reduction, we will be able to recognize perception as a structured "intentional constitution" of the external objects, established by the rule-governed activity of consciousness (similarly, space and geometrical idealities, like the Euclidean ones, are "constituted" objective properties of these transcendental objects).

The modality of appearing in perception is already markedly structured: it is not that of concrete material things immediately given, but it is mediated by sensible schemata constituted in the temporal continual mutation of adumbrations. So at the level of "presentational perception" of pure lived experiences, only partial aspects (*adumbrations* [*Abschattungen*]) of the objects are provided. Therefore, an activity of unification of the different adumbra-

---

world and so of organic beings. A similar kind of possibility was advanced by Helmholtz and Poincaré, when they hypothesized the famous "fantastic worlds" in which there are beings educated in an environment quite different from ours [42, pp. 64–68]. Their different "experience" will lead these beings to classify phenomena in a different way than we would, that is a non-Euclidean way, because it is more convenient, even though the same phenomena could be described in a Euclidean way. In fact, Poincaré says that these worlds can be described "without forsaking the use of ordinary geometrical language" [42, p. 71]. To the aim of my considerations in this article, which regards "beings like us" this objections can be disregarded.

[11] The prepredicative world is not yet characterized by predications, values, empirical manipulations and techniques of measurement as instead the Husserl's prescientific world is.

tions to establish they belong to a particular and single object (noema) is further needed.[12]

The analysis of the generation of idealities (and geometrical idealities) is constructed in a very thoughtful philosophical scenario. The noematic appearances are the objects as they are intuitively and immediately given (by direct acquaintance) in the constituting multiplicity of the so-called adumbrations, endowed with a morphological character. The noematic meaning consists of a syntactically structured categorical content associated with judgment. Its ideality is "logical". The noema consists of the object as deriving from a constitutive rule or synthetic unity of the appearances, in the transcendental sense [40]. To further use the complex Husserlian philosophical terminology – which surely motivates an interpretation in terms of abduction – we can say: hyletic data (that is immediate given data) are vivified by an intentional synthesis (a noetic apprehension) that transforms them into noematic appearances that adumbrate objects, etc.

As illustrated by Husserl in *Ding und Raum* [1907] [22] the geometrical concepts of point, line, surface, plane, figure, size, etc., used in eidetic descriptions are not spatial "in the thing-like sense": rather, in this case, we deal with the problem of the generation of the objective space itself. Husserl observes: it is "senseless" to believe that "the visual field is [...] in any way a surface on objective space" (§48, p. 166), that is, to act "as if the oculomotor field were located, as a surface, in the space of things" (§67, p. 236).[13] What about the phenomenological genesis of geometrical global three-dimensional space?

The process of making adumbrations represents a strategy which is distributed in visual, kinesthetic, and motor activities usually involving the manipulations of some parts of the external world. The adumbrative aspects of things are part of the visual field. To manage them a first requirement is related to the need of gluing different fillings-in of the visual field to construct the temporal continuum of perceptive adumbrations in a global space: the visual field is considered not translation-invariant, because the images situated at its periphery are less differentiated than those situated at its center (and so resolution is weaker at the periphery than at the center), as subsequently proved by the pyramidal algorithms in neurophysiology of vision research.

Perceptual intentionality basically depends on the ability to realize kinesthetic situations and sequences. In order for the subject to have visual sensations of the world, he/she must be able not only to possess kinesthetic sensations but also to freely initiate kinesthetic strategic

---

[12]On the role of adumbrations in the genesis of ideal space and on their abductive and nonmonotonic character cf. below subsection 2.2. An interesting article [34] deals with the relationship between perceptual intentionality, agency, and bodily movement and acknowledges the abductive role of adumbrations. In the remaining part of this section I will try to clarify their meaning.

[13]Moreover, Husserl thinks that space is endowed with a double function: it is able to constitute a phenomenal extension at the level of sensible data and also furnishes an intentional moment. Petitot says: "Space possesses, therefore, a noetic face (format of passive synthesis) and a noematic one (pure intuition in Kant's sense)" [40, p. 336].

"sequences": this involves a bodily sense of agency and awareness on the part of the doer [34, p. 20]. The kinesthetic control of perception is related to the problem of generating the objective notion of three-dimensional space, that is, to the phenomenological constitution of a "thing",[14] as a single body unified through the multiplicity of its appearances. The "meaning identity" of a thing is of course related to the continuous flow of adumbrations: given the fact that the incompleteness of adumbrations implies their synthetic consideration in a temporal way, the synthesis in this case, *kinetic*, involves eyes, body, and objects.

Visual sensations are not sufficient to constitute objective spatiality. Kinesthetic sensations[15] (relative to the movements of the perceiver's own body)[16] are required. Petitot observes, de facto illustrating the abductive role of kinesthetic sensations:

> Besides their "objectivizing" function, kinesthetic sensations share a "subjectivizing" function that lets the lived body appear as a proprioceptive embodiment of pure experiences, and the adumbrations as subjective events. [...] There exists an obvious equivalence between a situation where the eyes move and the objects in the visual field remain at rest, and the reciprocal situation where the eyes remain at rest and the objects move. But this trivial aspect of the relativity principle is by no means phenomenologically trivial, at least if one does not confuse what is constituting and what is constituted. Relativity presupposes an *already* constituted space. At the preempirical constituting level, one must be able to discriminate the two equivalent situations. The kinesthetic control paths are essential for achieving such a task [40, pp. 354–355].

Multidimensional and hierarchically organized, the space of kinesthetic controls includes several degrees of freedom for movements of eyes, head, and body. Kinesthetic controls are kinds of *spatial* gluing operators. They are able to compose, in the case of visual field, different partial aspects – identifying them as belonging to the same object, (cf. Figure 1), that is constituting an ideal and transcendent "object". They are realized in the pure consciousness and are characterized by an intentionality that demands a temporal lapse of time.

With the help of very complex eidetic descriptions, that further develop the strategic operations we sketched, Husserl is able to explain the constitution of the objective parametrized time and of space, dealing with stereopsis, three-dimensional space and three-dimensional things inside it. Of course, when the three-dimensional space (still inexact)

---

[14] Cf. also [20, §40, p. 129] [originally published in 1913].

[15] Husserl uses the terms "kinestetic sensations" and "kinesthetic sequences" to denote the subjective awareness of position and movement in order to distinguish it from the position and movement of perceived objects in space. On some results of neuroscience that corroborate and improve several phenomenological intuitions cf. [35, pp. 211–216] and [2, 39].

[16] The ego itself is only constituted thanks to the capabilities of movement and action.

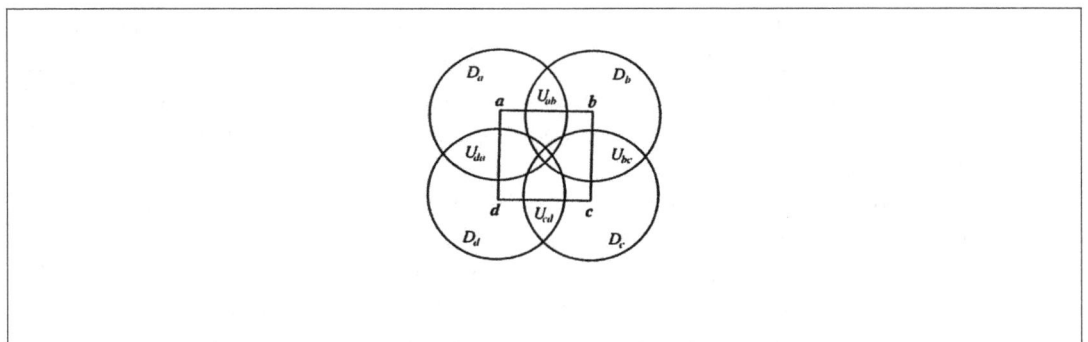

Figure 1: Scanning square S with corners a, b, c, d. To each position p corresponds a token Dp of the visual field D centered on p (focalization on p). The neighboring Dp overlap. (From [41], ©1999 by the Board of Trustees of the Leland Stanford Junior University, Stanford University Press, Stanford, reprinted by permission).

is generated (by means of two-dimensional gluing and stereopsis) it is possible to invert the phenomenological order: the visual field is so viewed as a portion of surface in $\mathbf{R}^3$, and the objective constituted space comes first, instead of the objects as they are intuitively and immediately given by direct acquaintance. So the space is in this case an objective datum informing the cognitive agent about the external world where she can find objects from the point of view of their referentiality and denotation. The kinesthetic system "makes the oculomotor field (eventually enlarged to infinity) the mere projection of a three spatial thingness" [22, section 63, p. 227]. Adumbrations now also appear to be consequences of the objective three-dimensional space, as continuous transformations of two-dimensional images as if the body were embedded in the space $\mathbf{R}^3$.[17]

2.2 Anticipations as Abductions

Of course adumbrations, the substrate of gluing operations that give rise to the two-dimensional space, are multiple and infinite, and there is a potential co-givenness of some of them (those potentially related to single objects). They are incomplete and partial so for the complete givenness of an object a temporal process is necessary. Adumbrations, not

---

[17]The role of adumbrations in objectifying entities can be hypothesized in many cases of nonlinguistic animal cognition dealing with the problem of reification and the formation of a kind of "concept", cf. chapter five of [27]. In human adults objects are further individuated and reidentified by using both spatial aspects, such as place and trajectory information and static-property information (in this last case exploiting what was gained through previous adumbration activity); adults use this property information to explain and predict appearances and disappearances: "If the same large, distinctive white rabbit appears in the box and later on in the hat, I assume it's the same rabbit" [13].

only intuitively presented, can be also represented at the level of *imagination*. Just because incomplete, *anticipations* instead correspond to a kind of non-intuitive intentional expectation: when we see a spherical form from one perspective (as an adumbration), we will assume that it is effectively a sphere, but it could be also a hemisphere (an example already employed by Locke).

Anticipations share with visual and manipulative abduction (cf. subsection 1.3 above) various features: they are highly conjectural and nonmonotonic, so wrong anticipations have to be replaced by other plausible ones. Moreover, they constitute an activity of "generate and test" as a kind of action-based cognition: the finding of adumbrations involves kinesthetic controls, sometimes in turn involving manipulations of objects; but the activity of testing anticipations also implies kinesthetic controls and manipulations. Finally, not all the anticipations are informationally equivalent and work like attractors for privileged individuations of objects. In this sense the whole activity is toward "the best anticipation", the one that can display the object in an optimal way. Prototypical adumbrations work like structural-stable systems, in the sense that they can "vary inside some limits" without altering the apprehension of the object.

As in the case of selective abduction, anticipations are able to select possible paths for constituting objects, actualizing them among the many that remain completely tacit. As in the case of creative abduction, they can construct new ways of aggregating adumbrations, by delineating the constitution of new objects/things. In this case they originate interesting "attractors" that give rise to *new* "conceptual" generalizations.

Some of the Husserl's wonderful philosophical speculations are being further developed scientifically from the neurological and cognitive perspective in current cognitive science research. [14, 16] has built an emulation theory based on control theory where forward models as emulators (shared by humans and many other animals) are used to illustrate, in the case of humans, various cognitive processes like perception, imagery, reasoning, and language. He contends that simulation circuits are able to hypothesize forward mapping from control signals to the anticipated – and so abduced – consequences of executing the control command. In other words, they mimic the body and its interaction with the environment, enhancing motor control through sensorimotor abductive hypotheticals: "For example, in goal-directed hand movements the brain has to plan parts of the movement before it starts. To achieve a smooth and accurate movement proprioceptive/kinesthetic (and sometimes visual) feedback is necessary, but sensory feedback per se is too slow to affect control appropriately" [7]. The "solution" is an emulator/forward model that can predict the sensory feedback resulting from executing a particular motor command" [47, p. 1310]. The control theory framework is also useful to describe the emergence of implicit and explicit agency [16]. The humans' understanding of themselves as explicit agents is accomplished through an interplay between the standard egocentric point of view and the so-called "simulated alter-egocentric" point of view, which represents the agent itself as an entity in the environment.

Given the fact that motor imagery can be seen as the off-line driving force of the emulator via efference copies, it is noteworthy that the emulation theory can be usefully extended to account for visual imagery as the off-line operator behind an emulator of the motor-visual loop. In these systems a kind of *amodal* spatial imagery can be hypothesized: "Modal imagery [...] is imagery based on the operation of an emulator of the sensory system itself, whereas amodal imagery is based on the operation of an emulator of the organism and its environment: something like arrangements of solid objects and surfaces in egocentric space. I show how the two forms of emulation can work in tandem" [14, p. 386].[18]

The Husserlian phenomenological explanation of the generation of "idealities" leads to a moment in which, once the space as an objective datum is settled, it informs the cognitive agent about the external world where she can find objects from the point of view of their referentiality and denotation, like it is happening – currently – to beings like us.

2.3 Anticipations and the Activity of "Reading Ahead"

Let us abandon the phenomenological speculative story and come back to the current external world where we simply find objects from the point of view of their referentiality and denotation, as already created and then sedimented thanks to our ancestors. Indeed we can now turn our attention to the cognitive abductive strategies that are involved in the non phenomenological case of the moves in the adversarial board game Go with two players, which analogously to the phenomenological process, still concerns visual, kinesthetic, and motor sensations and actions, but also involves a strong role of visual, iconic, and propositional representations (both internal and external).

In the case of game Go we are no more dealing with the *natural* game between humans and their prepredicative surroundings of the Husserlian case but with the *artificial* game that concerns the interplay between two players and their human made surroundings.[19] We

---

[18]It is important to note that amodal imagery is neither sentential nor pictorial because the amodal environment space/objects emulators are closely tied to the organism's sensorimotor engagement with the environment. An interesting example of amodal abduction, in our terms, "where an object cannot currently be sensed by any sensory modality (because it is behind an occluder, is silent and odorless, etc.) yet it is represented as being at a location. I think it is safe to say that our representation of our own behavioral (egocentric) space allows for this, and it is not clear how a multisensory system, in which tags for specific modalities were always present, could accomplish this" [15, p. 434]. On Grush's approach cf. the detailed discussion illustrated in [4, chapter seven] in the framework of the theory of the extended mind; a treatment of current cognitive theories, such as the sensorimotor theory of perception, which implicitly furnish a scientific account of the phenomenological concept of anticipation, is given in chapter eight of the same book. A detailed treatment of recent neuroscience achievements which confirm the abductive character of perception is given in the article "Vision, thinking, and model-based inferences" [44], recently published in the *Handbook of Model-Based Science* [33].

[19]Cf. Wikipedia, entry Go (game) https://en.wikipedia.org/wiki/Go_(game), cf. also [6]. Of course many other books are available, which introduce to the richness of Go strategies.

will see that also in this case there are processes that remind the ones of adumbration and anticipation and that play a central strategic role in the reasoning performed during the game in between the two players and their respective changing surroundings. Surroundings that in this case are basically formed by board, stones, and possible artifactual assisting accessories.

One of the most important strategies required for efficient tactical play is the ability to *read ahead*, as the Go players commonly say. Reading ahead is a rich and complicated (either thoughtful or intuitive) group of various kinds of anticipations and involves considering

1. available moves to play and their potential consequences. By exploiting the Husserlian lexicon we can say that the observed scenario at time t_1, offered by the board, constitutes an adumbration of a further possible more advantageous scenario at time t_2, which indeed is abductively plausibly hypothesized: in turn another abduction is selected and activated, which – coherently and plausibly – triggers a certain move that can favor the reaching of the envisaged more advantageous scenario;

2. possible responses to each move;

3. subsequent chances after each of those responses. Some of the more skilled players of the game can read up to 40 moves ahead even in extremely complicated positions.

In a book concerning the illustration of various strategies that can be adopted Davies says:

> The problems in this book are almost all reading problems. [...] they are going to ask you to work out sequences of moves that capture, cut, link up, make good shape, or accomplish some other clear tactical objective. A good player tries to read out such tactical problems in his head before he puts the stones on the board. He looks before he leaps. Frequently he does not leap at all; many of the sequences his reading uncovers are stored away for future reference, and in the end never carried out. This is especially true in a professional game, where the two hundred or so moves played are only the visible part of an iceberg of implied threats and possibilities, most of which stays submerged. You may try to approach the game at that level, or you may, like most of us, think your way from one move to the next as you play along, but in either case it is your reading ability more than anything else that determines your rank [6, p. 6]

Other strategies used by human players in the game Go deal for example "with global influence, interaction between distant stones, keeping the whole board in mind during local fights, and other issues that involve the overall game. It is therefore possible to allow a tactical loss when it confers a strategic advantage". [20]

In these kinds of scenarios regarding artificial games, the sensible objects (stones and board) of the external scenario, in their consecutive configurations, are the effects of cognition sedimented[21] in their embodiment. Cognition which derives from the application of both the game permitted rules and the personal cognitive endowments possessed by the two players, strategies, tactics, heuristics, etc. is sedimented in those sensible objects (artifacts, in this case) that become *cognitive mediators*:[22] for example they constrain players' reasoning, communicate information, and mediate reasoning chances. Mental manipulations of each of the subsequent scenarios, suitable represented internally, are further made, to the aim of favoring the next successful move. The strategies which are activated are multiple but all are "locked" because the components of each scenario are always the same (just the number of present stones and their configurations change), in a finite and unchanging framework (no new rules, no new objects, no new boards, etc.) These strategies lack the part which could refer to the possibility of resorting to sources of information *different* from the ones available in the rigid given scenario.[23]

---

[20] Cf. Wikipedia, entry Go (game) https://en.wikipedia.org/wiki/Go_(game).

[21] An expressive adjective used by Husserl.

[22] This expression, I have introduced in [26], is derived from the cognitive anthropologist Hutchins, who coined the expression "mediating structure" to refer to various external tools that can be built to cognitively help the activity of navigating in modern but also in "primitive" settings. Any written procedure is a simple example of a cognitive "mediating structure" with possible cognitive aims, so mathematical symbols and diagrams: "Language, cultural knowledge, mental models, arithmetic procedures, and rules of logic are all mediating structures too. So are traffic lights, supermarkets layouts, and the contexts we arrange for one another's behavior. Mediating structures can be embodied in artifacts, in ideas, in systems of social interactions [...]" [24, pp. 290–291] that function as an enormous new source of information and knowledge. [46, p. 249] maintains that "epistemic tools support open-ended and counterfactually robust dispositions to succeed" and further stresses their social character.

[23] The concept of locked strategy refers to a cognitive classification that is not related to other more technical ones coming from game theory. Basically, in combinatorial game theory Go is considered a zero-sum (player choices do not increase resources available-colloquially), perfect-information, partisan, deterministic strategy game, belonging to the same class as chess, checkers (draughts) and Reversi (Othello). Also, Go is bounded (every game must finish with a victor within a finite number of moves), strategies are of course associative (function of board position), format is obviously non-cooperative (no teams are present), positions are extensible (that is they can be represented by board position trees). Cf. Wikipedia entry Go (game) https://en.wikipedia.org/wiki/Go_(game).

3 Locked Abductive Strategies Undermine the Maximization of Eco-Cognitive Openness

In the previous sections I have repeatedly emphasized the knowledge enhancing character of abduction and the fact that reasoning strategies can grant successful results. When I say that abduction can be knowledge-enhancing I am referring to various types of new produced knowledge of various novelty level, from that new piece of knowledge about an individual patient we have abductively reached (a case of selective abduction, no new biomedical knowledge is produced) to the new knowledge produced in scientific discovery, which Paul Feyerabend emphasized in *Against Method* [8], as I have illustrated in subsection 1.1. However, also knowledge produced in an artificial game thanks to a smart application of strategies or to the invention of new strategies and/or heuristics has to be seen as the fruit of knowledge enhancing abduction.

I contend that to reach rich selective or creative good abductive results efficient strategies have to be exploited, but it is also necessary to count on an environment characterized by what I have called *optimization of eco-cognitive situatedness*, in which eco-cognitive openness is fundamental [31]. Below in the subsection 3.1 I will illustrate in detail that to favor good creative and selective abduction reasoning strategies must not be "locked" in an external restricted eco-cognitive environment, such as in a scenario characterized by fixed definitory rules and finite material aspects, which would function as cognitive mediators able to constrain agents' reasoning.

At this point it is useful to provide a short introduction to the concept of eco-cognitive openness. The new perspective inaugurated by the so-called naturalization of logic [30] contends that the normative authority claimed by formal models of ideal reasoners to regulate human practice on the ground is, to date, unfounded. It is necessary to propose a "naturalization" of the logic of human inference. Woods holds a naturalized logic to an adequacy condition of "empirical sensitivity" [49]. A naturalized logic is open to study many ways of reasoning that are typical of actual human knowers, such as for example fallacies, which, even if not truth preserving inferences, nonetheless can provide truths and productive results. Of course one of the best examples is the logic of abduction, where the naturalization of the well-known fallacy "affirming the consequent" is at play. Gabbay and Woods [9, p. 81] clearly maintain that Peirce's abduction, depicted as both a) a surrender to an idea, and b) a method for testing its consequences, perfectly resembles central aspects of practical reasoning but also of creative scientific reasoning.

It is useful to refer to my recent research on abduction [31], which stresses the importance in good abductive cognition of what has been called *optimization of situatedness*: abductive cognition is for example very important in scientific reasoning because it refers to that activity of creative hypothesis generation which characterizes one of the more valued aspects of rational knowledge. The study above teaches us that situatedness is related to the so-called eco-cognitive aspects, referred to various contexts in which knowledge is "traveling": to favor the solution of an inferential problem – not only in science but also in other abductive problems, such as diagnosis – the richness of the flux of information has to be maximized.

It is interesting to further illustrate this problem of optimization of eco-cognitive situatedness taking advantage of simple logical considerations. Let $\Theta = \{\Gamma_1, ..., \Gamma_m\}$ be a theory, $P = \{\Delta_1, ..., \Delta_n\}$ a set of true sentences corresponding – for example – to phenomena to be explained and $\Vdash$ a consequence relation, usually – but not necessarily – the classical one. In this perspective an abductive problem concerns the finding of a suitable improvement of $A_1, ..., A_k$ such that $\Gamma_1, ... \Gamma_m, A_1, ..., A_k \Vdash_L \Delta_1, ..., \Delta_n$ is *L-valid*. It is obvious that an improvement of the inputs can be reached both by additions of new inputs but also by the modification of inputs already available in the given inferential problem. In [31] I contend that to get good abductions, such as for examples the creative ones that are typical of scientific innovation, the input and output of the formula $\Lambda_1, ..., \Lambda_i, ?_I \Vdash_L^X \Upsilon_1, ..., \Upsilon_j$, (in which $\Vdash_L^X$ indicates that inputs and outputs do not stand each other in an expected relation and that the modification of the inputs $?_I$ can provide the solution) have to be thought as *optimally positioned*. Not only, this optimality is made possible by a *maximization of changeability* of both input and output; again, not only inputs have to be enriched with the possible solution but, to do that, other inputs have usually to be changed and/or modified.[24]

Indeed, in our eco-cognitive perspective, an "inferential problem" can be enriched by the appearance of new outputs to be accounted for and the inferential process has to restart. This is exactly the case of abduction and the cycle of reasoning reflects the well-known non-monotonic character of abductive reasoning. Abductive consequence is ruptured by new and newly disclosed information, and so defeasible. In this perspective abductive inference is *not only* the result of the modification of the inputs, but, in general, actually involves the intertwined modification of both input and outputs. Consequently, abductive inferential processes are highly *information-sensitive*, that is the flux of information which interferes with them is continuous and systematically human(or machine)-promoted and enhanced when needed. This is not true of traditional inferential settings, for example proofs in classical logic, in which the modifications of the inputs are *minimized*, proofs are usually taken with "given" inputs, and the burden of proofs is dominant and charged on rules of inferences, and on the smart choice of them together with the choice of their appropriate sequentiality. This changeability first of all refers to a wide psychological/epistemological openness in which knowledge transfer has to be maximized.

In sum, considering an abductive "inferential problem" as symbolized in the above formula, a suitably anthropomorphized logic of abduction has to take into account a continuous

---

[24] More details are illustrated in [31, section three].

flux of information from the eco-cognitive environment and so the constant modification of both inputs and outputs on the basis of both

1. the *new information available*,

2. the *new information inferentially generated*, for example new inferentially generated inputs aiming at solving the inferential problem.

To conclude, optimization of situatedness is the main general property of logical abductive inference, which – from a general perspective – defeats the other properties such as minimality, consistency, relevance, plausibility, etc. These are special subcases of optimization, which characterize the kind of situatedness required, at least at the level of the appropriate abductive inference to generate the new inputs of the above formula.

3.1 Locking Strategies Affects Creativity

I have said above that to favor good creative and selective abduction reasoning strategies must not be "locked" in an external restricted eco-cognitive environment (that is a scenario determined by fixed definitory rules and finite material aspects which would serve as cognitive mediators). To make an example of a poor scenario from the point of view of the lack of eco-cognitive openness we have already considered in subsection 2.3 the game Go (but also Chess and other games could be exploited). We have seen that in the game Go stones, board, and rules are fixed and so completely expected; what "can be" unexpected are the strategies and related heuristics that are learned seeing at the way the adversary is playing the game, and the ones later on produced to play the game[25].

As I have already said the available strategies and the adversary's ones are always *locked* in the fixed scenario I have indicated above: it is impossible, to make an imaginary example, that when you are playing Go, you can play for five minutes Chess or another game or you perform another external cognitive process, claiming that that strange part of the game is still legitimate as a part of the Go game you are playing. You cannot decide to adopt a different scenario so *unlocking* your strategic reasoning, because for example you think this will improve your performance against your adversary. You cannot activate at your discretion a process of eco-cognitive opening in that artificial game, such as it is instead occurring, for example, in the case of scientific discovery, in which it is common to recur to disparate external models[26] to make analogies or to favor other cognitive strategies (prediction, simplification, confirmation, etc.) to support the abductive creative process.

---

[25]Of course if you are an expert player your mind is also full of rich strategies you have learned in previous games and when attending other games involving other people.

[26]A myriad of examples can be found in the recent [33].

This example exactly illustrates a scenario which is poor from the point of view of its eco-cognitive openness. Indeed, the reasoning strategies you can adopt, even if multiple and potentially infinite, are *locked* in a finite perspective in which the elements do not change (the stones can just diminish). We can say that the fixed scenario establishes a kind of *autoimmunization* [32, 1] that prevents the players from applying strategies to not pre-established knowledge contents, extraneous to the ones embedded in the elements of the fixed scenario. I have already said these elements play the role of *cognitive mediators*, which constrain a great part of the entire cognitive process of the game.

An analysis of what is occurring in these cognitive cases in the light of creative and selective abduction has to be provided:

1. contrarily to the case of high level "human" creative processes (creative abductions) such as for example the ones regarding scientific discovery or other cases of exceptional intellectual achievements, the situation of artificial games is very poor from the point of view of the *non-strategic knowledge* involved. We are facing with stones, few rules, and a board. Step by step, during the game, the configurations of the scenario strongly change but no new cognitive mediators (objects) are presented: for example we cannot expect the appearing of multiply colored stones or the adoption of a new pentagonal board. In scientific discovery (for example in empirical science) first of all the empirical source (evidence) can be very rich and full of novel aspects (not only amenable to the change of configurations of the usual objects, as in the case of artificial games). Secondly, the knowledge involved is hyper-rich, and involves analogies, thought experiments, modeling activities, imageries, mathematical schemas, etc. that can derive from very disparate disciplines. In sum in this last case we are dealing with a situation of optimal eco-cognitive situatedness (further details on this kind of creative abduction are illustrated in [27, 29, 31]);

2. what is occurring in the case of selective abduction? Let us consider the case of medical diagnosis: first of all information (evidence) freely flows from multiple empirical sources regarding somatic symptoms and data mediated by complicated artifacts (which also change thanks to new discoveries and/or technological improvements). The hypothetical knowledge in which selective abduction can operate is instead *locked*,[27] but this does not impede that also at this level new knowledge can be adopted (of course not "created") so enriching the diagnostic process thanks to scientific advancements. Thirdly, new reasoning strategies and related heuristics can be invented and old ones exploited in new unexpected ways but, what is important, strategies are not locked. In sum, the creativity involved is of a lower level with re-

---

[27]For example in medical diagnosis the task is to "select" from an encyclopedia of pre-stored diagnostic entities.

spect to the one present in scientific discovery, but richer that the one involved in the locked reasoning strategies of the games I have considered;

3. in the artificial games in which heuristics are "locked"[28] heuristics are exactly the only part of the game cognitive process that can be improved: strategies and related heuristics can be used in a novel way and new ones can be invented. Anticipations as abductions (which refer to the activities of "reading ahead") just concern the reconfigurations and re-aggregations of the same components. No other kinds of knowledge will grow, everything else remains stable.[29] Of course this concentration on the strategies is the beauty of Go, Chess, and other games, and also reflects the spectacularity of the skilled performances of the human champions players. Regrettably, this concentration also explains the fact the creativity involved is nevertheless even lower than the one involved in the previous case of selective abduction (diagnosis). We will see soon that this is the reason why the skilled performances of Go or Chess games can be relatively more easily reproduced, with respect to the processes of scientific discovery, by the artificial intelligence programs.[30]

All the three cases I have just illustrated are occurring in a distributed cognition framework typical of human knowers, in which model-based and manipulative aspects are crucial, but the second and the third show that the optimization of situatedness lowers the eco-cognitive openness. Some parts of the process are locked and cannot take advantage of new, fresh, and disparate pieces of information and knowledge as in the case, for example, of scientific discovery.

I do not mean to downplay the importance of creative heuristics in Go and other board games. As John Holland extensively studied [18, 19], board games such as checkers, but also Go, are impressive examples of "emerging" processes, where virtually infinite possibilities open up for the performance of the system from the most simple set of rules regulating the moves of its pieces – and they cannot be predicted from the initial status. It is the issue of the emergence of complexity out of simplicity. While other domains incorporate what could be seen as "vertical" creativity (unlocked), board games can be examples of "horizontal" creativity: albeit being locked by the constraints of the game, "horizontal" creativity can reach amazing levels within the rules. While it has been satisfactorily captured

---

[28] Already Aristotle presented a seminal perspective on abduction, which emphasizes the important of – so to speak – non locked, but extremely open, reasoning, in the famous passage of the chapter B25 of *Prior Analytics* concerning ἀπαγωγή ("leading away"), also studied by Peirce. I contend that some of the current well-known distinctive characters of abductive cognition are already expressed, which are in tune with the EC-Model I have introduced above (more details are illustrated in cf. [29]).

[29] Of course, for example, new rules and new boards can be adopted, proposing new kinds of game, but this possibility does not affect my argumentation.

[30] I have discussed the problem of automated scientific discovery with AI programs in [27, chapter two, section 2.7 "Automatic Abductive Scientists"].

by artificial intelligence software (see the following paragraph), it has been an undisputed human achievement for many decades: furthermore it was tackled by artificial intelligence heuristics that were able to *learn from* human games. What are the important consequences when we have to deal with computational AI programs devoted to perform cognitive abductive processes characterized by "locked" strategic reasoning?

It is well known that in 2015 Google DeepMind's program AlphaGo beat Fan Hui, the European Go champion and a 2 dan (out of 9 dan possible) professional, five times out of five with no handicap on a full size 19x19 board. In March 2016, Google also challenged Lee Sedol, a 9 dan considered the top world player, to a five-game match. The program shot down Lee in four of the five games. It seems the looser acknowledged the fact the program adopted one unconventional move – never played by humans – leading to a new strategy, so performing a very "human" capacity, and I have to say, better than the one of the more skilled humans. AlphaGo learned to play the game by checking data of thousands of games, and may be also those played by Lee Sedol, exploiting the so-called "reinforcement learning", which means the machine plays against itself to further enrich and adjusts its own neural networks based on trial and error. Of course the program also implicitly performs what we call "reasoning strategies" to reduce the search space for the next best move from something almost infinite to a more calculable quantity.

Cohleo and Thompsen Primo de facto testify in the below passage that for an AI program as AlphaGo is relatively easy to reproduce at the computational level what I have called in this article *locked reasoning strategies*. In summary, a kind of general reason of this simplicity would be that this kind of human reasoning is less creative than others, even if it is so spectacular and performed in an optimal way only by very skilled and intelligent subjects.

> Let us compare the key ideas behind Deep Blue (Chess) and AlphaGo (Go). The first program used values to assess potential moves, a function that incorporated lots of detailed chess knowledge to evaluate any given board position and immense computing power (brute force) to calculate lots of possible positions, selecting the move that would drive the best possible final possible position. Such ideas were not suitable for Go. A good program may capture elements of human intuition to evaluate board positions with good shape, an idea able to attain far-reaching consequences. After essays with Monte Carlo tree search algorithms, the bright idea was to find patterns in a high quantity of games (150,000) with deep learning based upon neural networks. The program kept making adjustments to the parameters in the model, trying to find a way to do tiny improvements in its play. And, this shift was a way out to create a policy network through billions of settings, i.e., a valuation system that captures intuition about the value of different board position. Such search-and-

optimization idea was cleverer about how search is done, but the replication of intuitive pattern recognition was a big deal. The program learned to recognize good patterns of play leading to higher scores, and when that happened it reinforces the creative behavior (it acquired an ability to recognize images with similar style) [5].

We humans with our organic brains do not have to feel humiliated by these bad news... Human portentous performances with the game Go and other human ways of reasoning, even more creative than the ones involved in a locked strategic reasoning, cannot reach the global echo AlphaGo gained. The reason is simple, human-more-skillful-abductive creative performances – still cognitively gorgeous – are not sponsored by Google, which is a powerful corporation that can easily obtain a huge attention by aggressive media, a lot of internet web sites, and social networks enthusiast ignorant followers, more easily impressionable by the "miracles" of AI, robotics, and in general, information technologies, than by exceptional human knowledge achievements, always out of their material and intellectual reach.

Google managers also believe AI programs similar to AlphaGo could be used to help scientists solve tough real-world problems in healthcare and other areas. This is more than welcome. Of course I guess Google will also expect to implement some business thanks to a commercialization of new AI capacities to gather information and making abductions on it. Marketing aims are always important in these cases.[31] Academic epistemologists and logicians have to monitor the exploitation of these AI tools (the uses that can be less transparent than the simple and clear – and so astonishing – performance of AlphaGo in games against humans). Good software, which represents a great opportunity for science and data analytics, can be transformed in a tool that does not respect epistemological rigor. For example, in the different case concerning the management of big data, results can lead to unsubstantial computer-discovered correlations, (may be instead interesting from a commercial point of view), but they are presented as aiming at substituting human centered scientific understanding as a guide to prediction and action. Calude and Longo say: "Con-

---

[31] The Wikipedia entry DeepMind (https://en.wikipedia.org/wiki/DeepMind) [DeepMind is a British artificial intelligence company founded in September 2010 and acquired by Google in 2014, the company realized the AlphaGo program] reports the following non contested passage: "In April 2016 New Scientist obtained a copy of a data-sharing agreement between DeepMind and the Royal Free London NHS Foundation Trust, which operates the three London hospitals which an estimated 1.6 million patients are treated annually. The revelation has exposed the ease with which private companies can obtain highly sensitive medical information without patient consent. The agreement shows DeepMind Health is gaining access to admissions, discharge and transfer data, accident and emergency, pathology and radiology, and critical care at these hospitals. This included personal details such as whether patients had been diagnosed with HIV, suffered from depression or had ever undergone an abortion. This led to some public outcry and officials from Google have yet to make a statement but many regard this move as controversial and question the legality of the acquisition generally. The concerns were widely reported and have led to a complaint to the Information Commissioner's Office (ICO), arguing that the data should be pseudonymised and encrypted".

sequently, there will be no need to give scientific meaning to phenomena, by proposing, say, causal relations, since regularities in very large databases are enough: 'with enough data, the numbers speak for themselves' ". Unfortunately, some "correlations appear only due to the size, not the nature, of data. In 'randomly' generated, large enough databases too much information tends to behave like very little information". Certainly we cannot consider some correlations examples of pregnant scientific creative abduction, but just uninteresting generalizations, even if made thanks to sophisticated artifacts.[32] This is another new problem regarding sad issues linked to the relationship between ethics and technology I cannot afford in this article, limited to cognitive, logical, and epistemological problems.

4 Conclusion

In this article, with the help of the concepts of knowledge enhancing abduction, adumbration, anticipation, optimization of eco-cognitive openness, I have illustrated some basic aspects of the cognitive status of reasoning strategies and related heuristics. Taking advantage of my *eco-cognitive model* (EC-model) of abduction, I have illustrated the abductive character of the concept of anticipation linking it to the cognitive problem of spatiality and of the genesis of space in the description of the abductive role of the *Abschattungen* (adumbrations), as described in the framework of the philosophical tradition of phenomenology. I have stressed that anticipations share various features with visual and manipulative abduction and are useful conceptual tools to favor the analysis of the two new kinds of strategic reasoning I have introduced in this article: *locked* and *unlocked abductive strategies*, in turn related to different kinds of exploitation of heuristics. I have illustrated that this distinction is very important for delineating crucial aspects in the light of both philosophy of creativity and computational models of abduction. Locked abductive reasoning strategies are much easier to be reproduced at the computational level but show a kind of autoimmunity with respect to their possible productive role in strong human creative reasoning, because of the poorness of their related eco-cognitive environment.

Acknowledgements

This paper derives from the keynote lecture to the conference on the Logical Foundations of Strategic Reasoning, at the Korean Advanced Institute of Science and Technology, Daejeon, November 3, 2016. For the instructive criticisms and precedent discussions and correspondence that helped me to develop my analysis of strategic reasoning in an eco-cognitive perspective, I am indebted and grateful to John Woods, Woosuk Park, Atocha Aliseda,

---

[32]On this analysis and related warnings regarding recent computational tools cf. [3].

Luís Moniz Pereira, Paul Thagard, Athanassios Raftopoulos, Michael Hoffmann, Gerhard Schurz, Walter Carnielli, Akinori Abe, Yukio Ohsawa, Cameron Shelley, Oliver Ray, John Josephson, Ferdinand D. Rivera, to the two reviewers, and to my collaborators Tommaso Bertolotti and Selene Arfini.

References

[1] S. Arfini and L. Magnani. An eco-cognitive model of ignorance immunization. In P. Li L. Magnani and W. Park, editors, *Philosophy and Cognitive Science II Western & Eastern Studies*, volume 20, pages 59–75. Springer, Switzerland, 2015.

[2] R. Barbaras. The movement of the living as the originary foundation of perceptual intentionality. In J. Petitot, F. J. Varela, B. Pachoud, and J.-M. Roy, editors, *Naturalizing Phenomenology*, pages 525–538. Stanford University Press, Stanford, CA, 1999.

[3] C. S. Calude and G. Longo. The deluge of spurious correlations in big data. *Foundations of Science*, pages 1–18, 2016. forthcoming.

[4] A. Clark. *Supersizing the Mind. Embodiment, Action, and Cognitive Extension*. Oxford University Press, Oxford/New York, 2008.

[5] H. Coelho and T. Thompsen Primo. Exploratory apprenticeship in the digital age with ai tools. *Progress in Artificial Intelligence*, 2016. Onlime first: DOI 10.1007/s13748-016-0100-6.

[6] J. Davies. *Tesuji. Elementary Go Series. 3.* Kiseido Publishing Company, Tokyo, 1995.

[7] M. Desmurget and S. Grafton. Forward modeling allows feedback control for fast reaching movements. *Trends in Cognitive Sciences*, 4:423–431, 2002.

[8] P. Feyerabend. *Against Method*. Verso, London-New York, 1975.

[9] D. M. Gabbay and J. Woods. *The Reach of Abduction*. North-Holland, Amsterdam, 2005.

[10] G. Gigerenzer and H. Brighton. Homo heuristicus: why biased minds make better inferences. *Topics in Cognitive Science*, 1:107–143, 2009.

[11] G. Gigerenzer and R. Selten. *Bounded Rationality. The Adaptive Toolbox*. The MIT Press, Cambridge, MA, 2002.

[12] G. Gigerenzer and P. Todd. *Simple Heuristics that Make Us Smart*. Oxford University Press, Oxford/New York, 1999.

[13] A. Gopnik and A. Meltzoff. *Words, Thoughts and Theories (Learning, Development, and Conceptual Change)*. The MIT Press, Cambridge, MA, 1997.

[14] R. Grush. The emulation theory of representation: Motor control, imagery, and perception. *Behavioral and Brain Sciences*, 27:377–442, 2004.

[15] R. Grush. Further explorations of the empirical and theoretical aspects of the emulation theory. *Behavioral and Brain Sciences*, 27:425–435, 2004. Author's Response to Open Peer Commentary to R. Grush, The emulation theory of representation: Motor control, imagery, and perception.

[16] R. Grush. Agency, emulation and other minds. *Cognitive Semiotics*, 0:49–67, 2007.

[17] J. Hintikka. Socratic Epistemology. Explorations of Knowledge-Seeking by Questioning. Cambridge University Press, Cambridge, 2007.

[18] J. H. Holland. *Hidden Order*. Addison-Wesley, Reading, MA, 1995.

[19] J. H. Holland. *Emergence: From Chaos to Order*. Oxford University Press, Oxford, 1997.

[20] E. Husserl. *Ideas. General Introduction to Pure Phenomenology* [First book, 1913]. Northwestern University Press, London and New York, 1931. Translated by W. R. Boyce Gibson.

[21] E. Husserl. *The Crisis of European Sciences and Transcendental Phenomenology* [1954]. George Allen & Unwin and Humanities Press, London and New York, 1970. Translated by. D. Carr.

[22] E. Husserl. *Ding und Raum: Vorlesungen (1907)*. Nijhoff, The Hague, 1973. Husserliana 16, edited by U. Claesges.

[23] E. Husserl. The Origin of Geometry (1939). In J. Derrida, editor, *Edmund Husserl's "The Origin of Geometry"*, pages 157–180. Nicolas Hays, Stony Brooks, NY, 1978. Translated by D. Carr. Originally published in [21], pp. 353-378.

[24] E. Hutchins. *Cognition in the Wild*. The MIT Press, Cambridge, MA, 1995.

[25] A. Mackonis. Inference to the best explanation, coherence and other explanatory virtues. *Synthese*, 190:975–995, 2013.

[26] L. Magnani. *Abduction, Reason, and Science. Processes of Discovery and Explanation*. Kluwer Academic/Plenum Publishers, New York, 2001.

[27] L. Magnani. *Abductive Cognition. The Epistemological and Eco-Cognitive Dimensions of Hypothetical Reasoning*. Springer, Heidelberg/Berlin, 2009.

[28] L. Magnani. Is abduction ignorance-preserving? Conventions, models, and fictions in science. *Logic Journal of the IGPL*, 21(6):882–914, 2013.

[29] L. Magnani. The eco-cognitive model of abduction. Ἀπαγωγή now: Naturalizing the logic of abduction. *Journal of Applied Logic*, 13:285–315, 2015.

[30] L. Magnani. Naturalizing logic. Errors of reasoning vindicated: Logic reapproaches cognitive science. *Journal of Applied Logic*, 13:13–36, 2015.

[31] L. Magnani. The eco-cognitive model of abduction. Irrelevance and implausibility exculpated. *Journal of Applied Logic*, 15:94–129, 2016.

[32] L. Magnani and T. Bertolotti. Cognitive bubbles and firewalls: Epistemic immunizations in human reasoning. In L. Carlson, C. Hölscher, and T. Shipley, editors, *CogSci 2011, XXXIII Annual Conference of the Cognitive Science Society*. Cognitive Science Society, Boston MA, 2011.

[33] L. Magnani and T. Bertolotti, editors. *Handbook of Model-Based Science*. Springer, Heidelberg/Berlin, 2016. Forthcoming.

[34] S. Overgaard and T. Grünbaum. What do weather watchers see? Perceptual intentionality and agency. *Cognitive Semiotics*, 0:8–31, 2007.

[35] B. Pachoud. The teleological dimension of perceptual and motor intentionality. In J. Petitot, F. J. Varela, B. Pachoud, and J.-M. Roy, editors, *Naturalizing Phenomenology*, pages 196–219. Stanford University Press, Stanford, CA, 1999.

[36] W. Park. On classifying abduction. *Journal of Applied Logic*, 13:215–238, 2015.

[37] C. S. Peirce. *Collected Papers of Charles Sanders Peirce*. Harvard University Press, Cambridge, MA, 1931-1958. vols. 1-6, Hartshorne, C. and Weiss, P., eds.; vols. 7-8, Burks, A. W., ed.

[38] C. S. Peirce. Perceptual judgments. In *Philosophical Writings of Peirce*, pages 302–305. Dover, New York, 1955. Edited by J. Buchler.

[39] J.-L. Petit. Constitution by movement: Husserl in the light of recent neurobiological findings. In J. Petitot, F. J. Varela, B. Pachoud, and J.-M. Roy, editors, *Naturalizing Phenomenology*, pages 220–244. Stanford University Press, Stanford, CA, 1999.

[40] J. Petitot. Morphological eidetics for a phenomenology of perception. In J. Petitot, F. J. Varela, B. Pachoud, and J.-M. Roy, editors, *Naturalizing Phenomenology*, pages 330–371. Stanford University Press, Stanford, CA, 1999.

[41] J. Petitot, F. J. Varela, B. Pachoud, and J.-M. Roy, editors. *Naturalizing Phenomenology*. Stanford University Press, Stanford, CA, 1999.

[42] H. Poincaré. *La science et l'hypothèse*. Flammarion, Paris, 1902. English translation by W. J. G. [only initials indicated], 1958, *Science and Hypothesis*, with a Preface by J. Larmor, The Walter Scott Publishing Co., New York, 1905. Also reprinted in *Essential Writings of Henri Poincaré*, Random House, New York, 2001.

[43] M. Raab and G. Gigerenzer. Intelligence as smart heuristics. In R. J. Sternberg and J. E. Prets, editors, *Cognition and Intelligence. Identifying the Mechanisms of the Mind*, pages 188–207. Cambridge University Press, Cambridge, MA, 2005.

[44] A. Raftopoulos. Vision, thinking, and model-based inferences. In L. Magnani and T. Bertolotti, editors, *Handbook of Model-Based Science*. Springer, Heidelberg/Berlin, 2016. Forthcoming.

[45] M. Shanahan. Perception as abduction: Turning sensory data into meaningful representation. *Cognitive Science*, 29:103–134, 2005.

[46] K. Sterelny. Externalism, epistemic artefacts and the extended mind. In R. Schantz, editor, *The Externalist Challenge*, pages 239–254. De Gruyter, Berlin–New York, 2004.

[47] H. Svensson and T. Ziemke. Making sense of embodiment: simulation theories and the sharing of neural circuitry between sensorimotor and cognitive processes. In K. D. Forbus, D. Gentner, and T. Regier, editors, *CogSci 2004, XXVI Annual Conference of the Cognitive Science Society*, Chicago, IL, 2004. CD-Rom.

[48] J. Woods. Recent developments in abductive logic. *Studies in History and Philosophy of Science*, 42(1):240–244, 2011. Essay Review of L. Magnani, *Abductive Cognition. The Epistemologic and Eco-Cognitive Dimensions of Hypothetical Reasoning*, Springer, Heidelberg/Berlin, 2009.

[49] J. Woods. *Errors of Reasoning. Naturalizing the Logic of Inference*. College Publications, London, 2013.

[50] J. Woods. Inconsistency-management in big information systems: Tactical and strategic challenges to logic, November 3-4, 2016. Abstract of the Lecture presented at the Workshop "Logical Foundations of Strategic Reasoning", KAIST and Korean Society for Baduk Studies, Daejeon, S. Korea.

Abductive Cognition: Affordance, Curation, and Chance

Akinori Abe
Faculty of Letters, Chiba University, Chiba, JAPAN
Dwango Artificial Intelligence Laboratory, Tokyo, JAPAN
ave@chiba-u.jp

Abstract

In this paper I will discuss human cognition as an abductive cognition. Thus I briefly show the mechanism of abduction. Then I will combine several concepts such as affordance and curation with abduction. Especially this type of cognition is achieved in the form of chance discovery.

1 Introduction

In the title I used the phrase "abductive cognition." Cognition is a very important activity in our life. All living things have a certain cognitive system for their lives. Without cognition, they cannot survive. From several viewpoints, a cognitive system of human beings is discussed. For instance, affordance [16, 17] proposed by Gibson will be the most famous one.

In this paper, I will discuss human cognition from abduction, affordance, and curation's viewpoints. I will combine these concepts and explain the human cognition system especially from the viewpoint of chance discovery [30]. For the chance discovery, I will explain in the following section.

In this paper many concepts and strategies are combined to explain the human cognition system.

2 Abduction

2.1 Abduction and induction

In this section, as an incomplete knowledge reasoning (reasoning dealing with incomplete knowledge), I briefly introduce logical reasoning system —induction, and abduction.

Peirce classified *abduction* from a philosophical point of view as the operation of adopting an explanatory hypothesis and characterized its form.

(1) The surprising fact, C, is observed;

(2) But if A were true, C would be a matter of course,

(3) Hence, there is reason to suspect that A is true.

Where 'reason (hypothesis)' can not be easily assumed from A and C. In addition, he characterized *induction* as the operation of dealing and then testing a hypothesis by experiments.

(1) Suppose that I have been led to surmise that among our coloured population there is greater tendency toward female birth than among our whites.

(2) I say, if that be so, the last census must show it.

(3) I examine the last census report and find that, sure enough, there was a somewhat greater proportion of female births among coloured births than white births in that census year.

Thus Peirce characterized abduction and induction as follows [32]:

- Abduction is an operation for adopting an explanatory hypothesis, which is subject to certain conditions, and that in pure abduction, there can never be justification for accepting the hypothesis other than through interrogation.

  Inference for (novel) discovery

- Induction is an operation for testing a hypothesis by experiment, and if it is true, an observation made under certain conditions ought to have certain results.

  Inference for classification and learning, which are (generalized) discovery

Thus although abduction and induction are categorized to an incomplete knowledge reasoning and discover something "new," those which abduction discovers are rather different from those which induction discovers. If we want to discover general tendencies or classification induction will be better. On the other hand, if we want to discover something rare or novel, abduction will be better.

2.2 Abductive discovery

Abduction can be applied to applications for new discovery. Very typical application of abduction will be discoveries or solutions in affairs. For instance, the following is a scene from a detective novel "A Study In Scarlet" by Arthur Conan Doyle.

"Dr. Watson, Mr. Sherlock Holmes," said Stamford, introducing us. "How are you?" he (= Holmes) said cordially, gripping my hand with a strength for which I (= Dr. Watson) should hardly have given him credit. "You have been in Afghanistan, I perceive."...

Of course, for the sudden utterance from a stranger which was astonishingly correct, Dr. Watson asked that "How on earth did you know that?" in astonishment. In fact, during several minutes when Holmes shook hands with Dr. Watson, Holmes concluded (=abduced) Dr. Watson had been in Afghanistan. He did not have any previous information of Dr. Watson, but with several observations he had such a conclusion. He illustrated his abduction procedure as below;

Nothing of the sort. I (= Holmes) knew you (= Dr. Watson) came from Afghanistan. From long habit the train of thoughts ran so swiftly through my mind, that I arrived at the conclusion without being conscious of intermediate steps. There were such steps, however. The train of reasoning ran, 'Here is a gentleman of a medical type, but with the air of a military man. Clearly an army doctor, then. He has just come from the tropics, for his face is dark, and that is not the natural tint of his skin, for his wrists are fair. He has undergone hardship and sickness, as his haggard face says clearly. His left arm has been injured. He holds it in a stiff and unnatural manner. Where in the tropics could an English army doctor have seen much hardship and got his arm wounded? Clearly in Afghanistan.' The whole train of thought did not occupy a second. I then remarked that you came from Afghanistan, and you were astonished.

In the above scene, Sherlock Holmes determined Dr. Watson's vocation from the observation from Dr. Watson. Then Holmes guessed Dr. Watson's situation. The process of the guesswork was not based on a "chance" but a very formal and logical inference. Of course, this process can be explained by abduction. Half of the above procedure are deduction to obtain (infer) observations for abduction and can be logically described as follows:

1) Dr. Watson is an army doctor ← medical type & with the air of a military man.

2) Dr. Watson is not colored ← wrists are fair.

3) Dr. Watson has just come back from the tropics ← face is dark & not_colored.

4) Dr. Watson has undergone hardship and sickness ← haggard face & left arm has been injured.

5) Afghanistan ← English army doctor have much hardship and sickness & tropics.

That is, we can conduct deduction as follows:

- Observations: $medical\_type$, $wrists(fair)$, $face(dark)$, $haggard\_face$, $injured$, $air\_of\_a\_military\_man$

- deduction phase
    - $medical\_type \vee air\_of\_a\_military\_man \models army\_doctor$.
    - $wrists(colored) \models colored$.
    - $wrists(fair) \models not\_colored$.
    - $face(dark) \vee not\_colored \models tropics$.
    - $haggard\_face \vee injured \models hardship\_and\_sickness$

Then the rest of the inference process was logically performed based on observations (abduction). That is, Holmes generated Afghanistan as a hypothesis to explain various observations from Dr. Watson. In addition he used knowledge such as world situation in those days. The above inference process can be logically described as follows.

- abduction phase
    - Observations O: $hardship\_and\_sickness$, $tropics$, $army\_doctor$, $Englishman$
    - Facts F: knowledge sets in Holmes's brain
    - $\{Afghanistan, Malaysia, Russia, Japan, \ldots\} \in H$

Actually, Holmes knew another feature of Afghanistan that Afghanistan is a harder place to live in than other countries in the tropics etc. Accordingly he could conclude (abduce) that Dr. Watson had been in Afghanistan. Thus hypothesis (h) which is for "$in\ Afghanistan$" will be generated (selected) from H. The above is an inference by Sherlock Holmes (human inference). A computational inference will be illustrated in the following sections.

2.3 Computational abduction

Abduction in the Artificial Intelligence field is generally understood as reasoning from observation to explanations, and induction as the generation of general rules from specific data. Sometimes, both types of inferences are regarded as the same because they can be viewed as being an inverse of deduction. For computation, Pople mechanized abduction as an inverse of deduction [34], although he seemed to distinguish abduction from induction. Muggleton and Buntine have formalised induction as an inverted resolution [27]. Both formalizations are realized as an inverse of deduction. In this paper, I will not discuss a relationship between abduction and induction. It was discussed in [1]. I will focus on a discussion on abduction.

Thus, abduction is usually used to find the reason (set of hypotheses) in a logical way to explain an observation. For instance, the inference mechanism of **Theorist** [33] that explains an observation (O) by a consistent and minimal hypotheses set (h) selected from a set of hypotheses (H) is shown as followings.

$$F \nvdash O. \qquad (O \text{ can not be explained by only } F.) \qquad (1)$$

$$F \cup h \vdash O. \qquad (O \text{ can be explained by } F \text{ and } h.) \qquad (2)$$

$$F \cup h \nvdash \Box. \qquad (F \text{ and } h \text{ is consistent.}) \qquad (3)$$

Where F is a fact (background knowledge) and $\Box$ is an empty clause. A hypothesis set (h) is selected from a hypothesis base ($h \in H$).

Thus, "reason" is usually selected from the knowledge (hypotheses) base. For instance, when Theorist is used for an LSI circuit design, F includes knowledge about the devices' function and their connections, and the knowledge of other rules. In addition, H includes candidate devices and their candidate connections. If the relation between input and output of the circuit is given as an observation O, Theorist computes the name of devices and their connections as hypotheses h. Therefore, usual abduction requires a perfect hypotheses base from which a consistent hypotheses set is selected to explain an observation. Here, "perfect hypotheses base" means the hypotheses base that contains all the necessary hypotheses.

Clause Management System (CMS) was proposed by Reiter and de Kleer [37] and it was a database management system. Its mechanism is illustrated as follows:

When $\Sigma \nvDash C$, if propositional clause C (observation) is given, CMS returns a set of minimal clauses S to clause set Σ such that

$$\Sigma \models S \vee C. \tag{4}$$

$$\Sigma \not\models S. \tag{5}$$

A clause S is called a minimal support clause, and $\neg S$ is a clause set that is missing from clause set Σ that can explain C. Therefore, although CMS was not proposed as abduction, since from the abductive point of view $\neg S$ can be thought of as an abductive hypothesis, CMS can be used for abduction.

In addition, I proposed **Abductive Analogical Reasoning (AAR)** which combines CMS-like abduction and analogical mapping [2]. As shown above, CMS generates only the minimal hypothesis set. Thus it is not always the case we can obtain the sufficient hypothesis set. Accordingly I proposed Abductive Analogical Reasoning (AAR) that logically and analogically generates missing hypotheses. Its generation mechanism is similar to CMS's. Structures of generated knowledge sets are analogous to the known knowledge sets. In the framework of AAR, not completely unknown but rather unknown hypotheses can be generated. In addition, by the introduction of analogical mapping, we can adopt new hypothesis evaluation criteria other than Occam's Razor (for instance, criteria such as explanatory coherence [41]). The inference mechanism is briefly illustrated as follows (for notations, see [2]):

When

$$\Sigma \not\models O, \quad (O \text{ cannot only be explained by } \Sigma.) \tag{6}$$

Σ (background knowledge) lacks a certain set of clauses to explain O. Consequently, AAR returns a set of minimal clauses S such that

$$\Sigma \models S \vee O, \tag{7}$$

$$\neg S \notin \Sigma. \tag{8}$$

The result is the same as CMS's. This is not always a guaranteed hypothesis set. To guarantee the hypothesis set, we introduced analogical mapping from known knowledge sets.

$$S \mapsto S', \quad (S' \text{ is analogically transformed from } S.) \tag{9}$$

$$\neg S' \in \Sigma, \tag{10}$$

$$S' \mid\!\sim S'', \tag{11}$$

$$\Sigma \models S'' \vee O, \tag{12}$$

$$\neg S'' \notin \Sigma. \tag{13}$$

O is then explained by $\neg S''$ as an hypotheses set. Thus we can generate a new hypothesis set that is logically abduced whose structure is similar to authorized (well-known) knowledge sets.

3 Affordance and curation

3.1 Affordance

Gibson ecologically introduced the concept of affordance for perceptional phenomena [16, 17]. It emphasizes the environmental information available in extended spatial and temporal pattern in optic arrays, for guiding the behaviors of animals, and for specifying ecological events. Thus he defined the affordance of something as "a specific combination of the properties of its substance and its surfaces taken with reference to an animal." For instance, the affordance of climbing a stair step in a bipedal fashion has been described in terms of the height of a stair riser taken with reference to a person's leg length [43]. That is, if a stair riser is less than 88% of a person's leg length, then that means that the person can climb that stair. On the other hand, if a stair riser is greater than 88% of the person's leg length, then that means that the person cannot climb that stair, at least not in a bipedal fashion. For that Jones pointed out that "it should be noted also that this is true regardless of whether the person is aware of the relation between his or her leg length and the stair riser's height, which suggests further that the meaning is not internally constructed and stored but rather is inherent in the person's environment system" [19].

In the context of human-machine interaction Norman extended the concept of affordance from Gibson's definition. He pointed our that "...the term affordance refers to the perceived and actual properties of the thing, primarily those fundamental properties that determine just how the thing could possibly be used. [...] Affordances provide strong clues to the operations of things. Plates are for pushing. Knobs are for turning. Slots are for inserting things into. Balls are for throwing or bouncing. When affordances are taken advantage of, the user knows what to do just by looking: no picture, label, or instruction needed" [29]. Thus Norman defined affordance as something of both actual and perceivable properties. Accordingly his interpretation has effectively been introduced to interaction designs.

Zhang categorized several types of affordance into the following categories [45]:

- Biological Affordance

  For instance, a healthy mushroom affords nutrition, while a toxic mushroom affords dying.

- Physical Affordance

  For instance, the flat horizontal panel on a door can only be pushed. Many of this type of affordances can be found in Norman [29].

- Perceptual Affordance

  In this category, affordances are mainly provided by spatial mappings. For instance, if the switches of the stove top burners have the same spatial layout as the burners themselves, the switches provide affordances for controlling the burners. Examples of this type include the pictorial signs for ladies' and men's restrooms.

- Cognitive Affordance

  Affordances of this type are provided by cultural conventions. For instance, for traffic lights, red means "stop," yellow means "prepare to stop," and green means "go."

- Mixed Affordance

  For instance, a mailbox, which is one of the examples used by Gibson, does not provide the affordance of mailing letters at all for a person who has no knowledge about postal systems. In this case, internal knowledge is involved in constructing the affordance in a great degree.

Thus since Gibson's introduction, affordance has been widely discussed, and the other perspective and extensions have been added. Especially, it has been effectively introduced to interface designs after several extensions. I also introduced the concept of affordance in several applications. I will show some of them in the following.

3.2 Curation

Recently, in several situations, the word "curation" has been used. For instance, in the marketing strategies, for fashion shows, and for a DJ etc. This section reviews various types of curation. Actually curatorial task is usually used for tasks in (art) museum. Many cases introduced in this section are curatorial works in gallery and (art) museum. In addition a new type of curation, digital data curation, is also reviewed.

3.2.1 (General) curation

There is at least a person who is responsible as "curator" in (special) exhibitions, galleries, archive, or (art) museums. Their main task of curator is a curatorial task, which is multifaceted. Curator comes from a Latin word "cura" which means cure. Then originally it used for a person who take care of a cultural heritage.

In the report by American Association of Museums Curators Committee (AAMCC) [12], they pointed out "curators are highly knowledgeable, experienced, or educated in a discipline relevant to the museum's purpose or mission. Curatorial roles and responsibilities vary widely within the museum community and within the museum itself, and may also be fulfilled by staff members with other titles." Then they showed the definition of curator as follows;

- Remain current in the scholarly developments within their field(s); conduct original research and develop new scholarship that contributes to the advancement of the body of knowledge within their field(s) and within the museum profession as a whole.
- Make recommendations for acquiring and deaccessioning objects in the museum collection.
- Assume responsibility for the overall care and development of the collection, which may include artifacts, fine art, specimens, historic structures, and intellectual property.
- Advocate for and participate in the formulation of institutional policies and procedures for the care of the collection that are based on accepted professional standards and best practices as defined by AAM, CurCom, and other relevant professional organizations.
- Perform research to identify materials in the collection and to document their history.
- Interpret the objects belonging or loaned to the museum.
- Develop and organize exhibitions.
- Contribute to programs and educational materials.
- Advocate and provide for public use of the collection.

- Develop or contribute to monographs, essays, research papers, and other products of original thought.
- Represent their institution in the media, at public gatherings, and at professional conferences and seminars.
- Remain current on all state, national, and international laws as they pertain to objects in the museum collection.

In addition, AAMCC showed curatorial responsibilities as follows;

A. Research, Scholarship, and Integrity
B. Interpretation
C. Acquisition, Care, and Disposal
D. Collection Access and Use
E. Replication of Objects in the Collection

Thus curators have responsibilities for various aspects of exhibition activities. However, the most important activity will be a plan of exhibition. For that the above activities such as research, interpretation and acquisition are necessary. They should properly exhibit a truth which is result of their researches and interpretations.

3.2.2 e-Science Data Curation

The above curation is for actual museums. That is, curation is conducted mainly for actual works. However, curation in this section is for digital data. There are several differences between digital curation and analogue curation.

JISC pointed out an importance of curation as "promoting good curation and an information infrastructure to capitalise upon and preserve expensively gathered data means bringing together varied technical and managerial resources, and managing these over time. This activity needs to be supported by clear strategies for resourcing and support [20]."

They compare curation with archiving and preservation.

- Curation: The activity of managing and promoting the use of data from its point of creation, to ensure it is fit for contemporary purpose, and available for discovery and re-use. For dynamic datasets this may mean continuous enrichment or updating to keep it fit for purpose.
- Archiving: A curation activity which ensures that data is properly selected, stored, can be accessed and that its logical and physical integrity is maintained over time, including security and authenticity.

- Preservation: An archiving activity in which specific items of data are maintained over time so that they can still be accessed and understood through successive change and obsolescence of technologies.

That is, they pointed out that curation is more creative task. Then they showed aspects of curation as follows:

- Trust: Trust can be enhanced by the existence of qualified domain specialists who curate the data.
- Utility: Certain information about the data —where it came from, how it was generated, for example— is necessary to enable future users to gauge the utility and reliability of the data, and indeed any annotation of the data. Data utility also depends on the ability of users to manage and analyse it; data mining tools and algorithms, visualisation tools, user interfaces and portals will play a crucial role in accelerating research.
- Discoverability: How will future users find data, in particular data they do not know exists, in other domains, or archived according to terminology which has fallen out of use? Data access is often organised through portals; how will those portals be organised? What tools will users need to read or use the data, and who will provide these tools?
- Access management: A significant proportion of data involves confidentiality issues. Ownership and rights management also need to be taken into account.
- Heterogeneity: Not only is this data revolution creating a deluge of data, the data itself comes in very many different and often specialist formats, some created by the researchers themselves.
- Complexity: The data can be composite in nature, with links to external objects and external dependencies (such as calibration information), and be highly complex in structure. This complexity represents a significant challenge for the preservation of data.

They use "data curation" because they think data have value. Not only for keeping data but also usability of data for the public, they use the word "curation." Actually, most of data are neither art works nor archaeological artifacts. However, is is important to view data from the aspect of what should be preserved. The main difference between data and art works or archaeological artifacts is that data do not have a shape and cannot exist alone. It is necessary to prepare a container such as a cdrom and a hard disc drive system. Therefore for data curation, "Discoverability" plays a significant role.

3.2.3 Exhibition "Bacon and Caravaggio"

An exhibition "Bacon and Caravaggio" was held in Museo e Galleria Borghese, Roma, Italy during October 2 2009 and January 24 2010.

The display policy of this exhibition is rather different from the general special exhibition. First, the special exhibition was not separated from the space for permanent collections. Of course several Caravaggio's works were exhibited in their original places. The other Caravaggio's works and Bacon's works were exhibited between the permanent collections. For this type exhibition, usually exhibition is educational and two painters are compared in various point, for instance days and society painters lived. Before arriving at the exhibition, my expectation was that it would be an exhibition to address the contrast between the drawing policies of Bacon and Caravaggio. Caravaggio usually painted a perfect body of human beings. On the other hand, Bacon usually painted a flesh of human beings most of parts are removed to express the essence of human existence. However, in the catalogue of the exhibition "Bacon and Caravaggio," at first, Coliva wrote "This exhibition proposes a juxtaposition of Bacon and Caravaggio. It intends to offer visitors an opportunity for an aesthetic experience rather than an educational one... [14]." Then Coliva continued

> "An exhibition of generally conceived and prepared with a historicist mentality, but when it materializes, the simultaneous presence of the works — in the sense precisely of their hanging — opens up parallels and poses very complex and spontaneous questions, which may even be unexpected and not all stem exactly from questions initially posed by art-historical motives and theses. There are parallels that appear by themselves to the visitor's sensibility and are not imposed by a theory of the curator. This is certainly one aspect of the vitality of exhibitions, which make the works live and in this are necessary for the works. The display itself, in the sense of the presentation of the works that appear in an exhibition —the spectacle of their being on display — creates trains of thought that are independent of the interpretations provided by art-historical scholarship. And since for a profound experience of understanding a work these ramifications sometimes are more surprising and significant than the achievements of a specialized scholarship in its own field of action, an art raised to the status of an enigma like Bacon's seems to require the gamble of provoking these parallels. And since at the time, and again because of its qualitative greatness, Caravaggio's art deserves a similar provocation, the juxtaposition thus satisfies a legitimate aesthetic desire. On the other hand, the juxtaposition is a modest

and prudent solution, not so much for demonstrating, but for offering the attribute of "genius" — which the expressive common language attributes to the great artist of the past — opportunities to manifest itself. And the juxtaposition is induced by the Galleria Borghese itself, one of the most sensitive spaces with the simultaneous presence of genius."

Besides the importance in aesthetics and philosophy, I think the most important point is that "There are parallels that appear by themselves to the visitor's sensibility and are not imposed by a theory of the curator." That is, though actually a curator has a certain philosophy, he/she does not insist his/her philosophy but audiences will be able to discover additional meanings as well as the curator's intended philosophy.

3.2.4 Joseph Cornell / Jiří Kolář

From April 19 to May 26, 2007, Pavel Zoubok Gallery, NYC, USA organized an exhibition which combined Joseph Cornell and Jiří Kolář. Where I have also experienced the same situation as shown above.

In fact, both artists are collagists, but before visiting the gallery, I had not expect such combination. However the combination of well-known artists was very new and impressive and gave me additional perspective to the art.

Pavel Zoubok Gallery presented the combination based on the following concept. "In bringing these artists together we are confronted with two distinct traditions, one rooted in the fantastical visions of American Surrealism during the 1930s and 1940s and the other in the more politically charged spheres of the Central European avant-garde of the 1950s and 1960s, marked by social and cultural repression. Cornell's world, both inside and outside of the box, is one drawn primarily from the imagination. The worldliness and wonder of his art concealed a reality that was often fraught with sadness and an inability to connect directly with the world beyond Utopia Parkway and nearby Manhattan. By contrast, Jiří Kolàr's life and work reflected economic and political struggle and years spent in exile from his native Prague. The myriad collage techniques that he pioneered over fifty years formed an alternative language at a time when the artist/poet saw those in power employing words as an instrument of oppression and misinformation. This progressively led him to a purely visual means of expression.

In addition, Mullarkey pointed out "While Cornell remains an icon of 20th century art, Kolàr's profile as a collagist has diminished. This show suggests reason for the loss of momentum. Viewed side by side, Cornell stands as the more enduring of the two. Kolàr's work is smart, modish and fastidiously crafted. But Cornell's totems of enchantment achieve a disquieting beauty that transcends their moment."

Thus both artists have quite different social background and culture. It might be better to prepare such knowledge before enjoying the exhibition. However for some persons, it was not necessary to prepare such previous knowledge. Sometimes the previous knowledge is harmful because audiences might stop their thinking. The gallery did not explicitly prepare such information in the gallery. Without any previous information, audiences could enjoy the differences and similarities of those two types works from their own viewpoints.

Thus for me it was a simple juxtaposition of two different artistic things. Even such a situation, audiences would have discovered new or unintentional meanings from the juxtaposition. This type of juxtaposition can offer chances. In the above case, a "juxtaposition" itself can be a curation as a chance discovery application.

3.2.5 Exhibition in the Museum of University of Tokyo

In the museum of University Tokyo, they tried a unusual and tricky exhibition style. Where no panel for explanation is displayed. A director (curator) Endo pointed out that "We have intendedly organized a space without introduction, information, and educational objective. When audiences watch dead bodies which used to have lives and activities, they will conceive an importance of lives [26]."

In the above exhibition, small number of explanatory panels have been provided. However, in this exhibition, no explanatory panels is prepared. Actually, for (archaeological) museums specialized information will necessary. However sometimes reading such panels requires much times to audiences and removes opportunities of deep understandings of exhibitions from audiences. Thus properly few information gives audiences a chance of thinking.

Thus as shown above, originally a curation was an activity for offering an explicit education to audiences. However a contemporary curation offers audiences a certain freedom or opportunity such as deep thinking and new discovery, which can be regarded as a chance. Sometimes a situation without or with few information offers us chances which will become important factors for our future. Curation should be conducted with considering such implicit and potential possibilities. However, such possibilities should be rather easily discovered and arranged according to the user's interests and situations.

3.2.6 Curation in business

A "curation" for business in the internet age seems an interaction between customer (user) and products. There will not be a system to insist trends from big companies, but trends will be constructed or selected according to customers' interaction on (inter)networks. In addition, a (small) company or community can use this system to give rare products a certain trend. Thus the strategy of information delivery in business has changed in recent year and they call this type of information delivery as "curation." Curation in business means not only an information display system but also an information delivery strategy.

Rosenbaum pointed out as follows [38]:
Curation comes in many shapes and sizes. It is critically important to understand two things. First, curation is about adding value from humans who add their qualitative judgment to whatever is being gathered and organized. And second, there is both amateur and professional curation, and the emergence of amateur or pro-sumer curation isn't in any way a threat to professionals. He continued that "Curation is very much the core shift in commerce, editorial, and communities that require highly qualified humans." Accordingly he mainly discuss curation in the field of magazine and networks. He characterizes curation as the future of consumer conversation. He mentions that "as curated customer conversation take hold, there will not be a brand, a service, or a company that will emerge to give feedback and filter customer reaction to goods and services. [...] Indeed, reasonable and balanced communities curated to be about honest feedback and customer solutions will emerge as a new and powerful force in consumer-and-brand interaction." In addition, he seems to extend curation tasks to quite different type of jobs, for instance DJ. His definition of curation seems to cover quite a large field.

At the end of [38], he states "We are all curators. We all will be sharing into the ecosystem of our friend and families. For some, it will become part of who we are. And for a few of us, curation will become our livelihood. It's exciting for me to see that we're turning a corner. The network is built. The data center are in place. The next step will involve the human piece of the equation—humans are more-valuable machines."

This thought is very interesting to introduce in several applications.

4 Chance

4.1 What are a chance and chance discovery

Chance Discovery is a discovery of chance, rather than discovery by chance. Ohsawa defined chance (risk) as "*a novel or rare event/situation that can be conceived as either an opportunity or a risk in the future* [30]". It is naturally understood that a chance, which is either known or unknown, includes possibilities to cause unfamiliar observations. It can also be said that a chance is an alarm like an inflation of money supply or a big difference between future (estimated, reserved) and current stock prices that will change the middle or long term economic situation (Japan, in 1990). We sometimes ignore such critical factors, because we cannot understand that they are important factors. This is because the results or the factors are exceptions, and rare or novel events.

Chance discovery is also characterized as an explanatory reasoning, however since "chance" is defined as unknown hypotheses, some techniques to deal with an empty or an imperfect hypotheses base are required. If so, such an inference mechanism as usual abduction (hypothetical reasoning etc.) is not sufficient to achieve chance discovery. Chance discovery needs an explanatory reasoning that can deal with an empty or imperfect hypotheses base.

4.1.1 A chance in the financial crises in Japan (1990)

In the case of Japanese financial crises, it is usually said that it was quite unique event and it is not easy to determine the symptom for bubble breaking. Actually it might be difficult to determine beforehand, but a certain company could discover a chance (symptom) to utilize it.

Actually, in the December of 1989, there can be observed unusual big difference between the Nikkei average price of futures and spots. The difference was more than JPY1000. The S investment bank seemed to discover this situation as a chance to perform several buy and sell including put to control the Nikkei average price. Then the S investment bank could won a big money. However, since then Japanese economy has become worse and could not recovers even now.

Thus the symptom as a big difference between the Nikkei average price of futures and spots can be treated for both good and bad economical situation. That is, the S investment bank would regard the situation as their chance to make a big money. In addition, Japaneses analysts would regard the situation as a symptom to the continuous increasing stock prices.

Actually both are chances, but depending on how to deal with the chance, a result will become different directions. Thus a chance can be discovered and intentionally controlled according to the user's objectives. Therefore we should deeply think and consider how to deal with a chance. Because a symptom can become both a good chance or a bad chance.

As frequently pointed out, in chance discovery, an interaction between a chance and the user is very important.

4.2 Black Swan

In 2007 Taleb published "Black Swan" [40]. In the book, Taleb introduced a concept "Black Swan[1]" as an event with the following three attributes.

1. It is outlier, as it lies outside of the realm of regular expectations, because nothing in the past can convincingly point to its possibility.

2. It carries an extreme impact.

3. In spite of its outlier status, human nature makes us concoct explanations for its occurrence after the fact, making it explainable and predictable.

Thus Taleb discussed the similar event as a chance as black swan.

4.2.1 A Black Swan in the financial crises in USA (2007)

It is said that the USA's case was caused by the subprime crisis. Posner points out that in 2006 the first sign of the long-awaited showdown in U.S. housing market were just starting to materialize, although few expected the crisis...[35]. He then points out that by early 2007, newly originated mortgages to go bad quickly. [...] If 2% was really a measure of a one-standard-deviation variance in losses, then from 2006 to 2008 expectations had changed by improbably large 12 standard deviations. This surprise was the Black Swan at the heart of the economic crash.

Rajan also points out that subprime ZIP codes experienced an increase in default rates after 2006 that was three times that of prime ZIP codes, and much larger than the default rates these areas had experienced in the past [36]. In addition he uses the metaphor of "fault lines[2]" to explain the recent financial crises. For instance, he points out the rising income inequality in the U.S. was the first kind of fault line. Of course this is one of reasons that politicians have banks to expand housing credit (became subprime loans). He did not use the word Black Swan but fault lines are similar to Black Swan. Actually fault lines are rather difficult to discover, because they are hidden in the Earth.

Thus since 2006 certain symptoms which have not been experienced in the past have been observed in the U.S. economy. Since they were not visible or easily aware of, many persons ignored them. However some person including Rajan and Posner could realized the symptom to predict the future financial crises.

For the successful decision making, Posner shows interesting criteria including decision makers who rely on data mining to economize on human analysis may

---

[1] Black swans are native to Australia, but had never been seen in Europe.
[2] Fault lines are breaks in the Earth's surface where tectonic plates come in contact or collide.

overlook Black Swans whose force did not influence past data. Posner also points out the importance of mapping out problems to minimize complexity, understanding causal variables, and using judgment to weigh model output against other factors.

Similar to chance discovery, in order to be aware of Black Swan, it is necessary to perform a careful understanding of Black Swan and the future results.

For the better chance discovery, Posner's treatments of Black Swan seems promising. We would better to introduce some of his strategy to chance discovery.

4.3 Chance discovery

Though in various articles, the definition of a "chance" is described which was introduced by Ohsawa [30], I wish to introduce it here again. In fact, it rather differs from the original definition in [30] to reflect the recent research interests.

> A chance is rare, hidden, potential or novel event(s) / situation(s) that can be conceived either as a future opportunity or risk.

Then "chance discovery" research is a type of research to establish methods, strategies, theories, and even activities to discover a chance. In addition, it aims at discovering human factors for chance discoveries. Therefore not only researchers in computer science and engineering but also researchers with different expertise such as psychologists, philosophers, economists and sociologists take part in chance discovery research.

Thus it is very important to offer opportunities where receivers can feel and obtain chances in various situations. Many applications on chance discover have been proposed in these 10 years [6, 31]. For instance, visualisation systems for making users aware of unconscious preferences [13, 21], an analogy game which varies a construction of concepts according to perceptions, categorizations, and areas of focus derived from the expertise of the observer [28], a deposit overflow determination system to prevent various financial crises [44], ISOR-2, a combination of case-based reasoning and statistical modeling system which can deal with medical exceptions [42], and a web-based interactive interface which can check hidden or rare but very important relationships in medical diagnostic data sets [7] have been proposed in [6]. Those applications are real world applications where a discovery of chances plays an important role. However, strategies how to display chances have not been discussed in many applications. Strategy for discovering chances is of course important. In addition, strategy for an easy discovery interface of chances is more important. The above interface based application can be classified to curation type applications.

At the end of this section, I point out again that in chance discovery, an interaction between a chance and the user is very important. Accordingly the strategy for supporting interaction between human and computers in order to discover rare or novel events and think the meaning of the existing of them is very important. Curation is the one of solutions.

5 Abductive cognition

5.1 Affordance, abduction and chance discovery

It is important to deal with rare or novel phenomena which might lead us to risk or opportunity. We call this type of activity as chance discovery and discuss theories and methods to discover such chances. A chance is defined as *"a novel or rare event/situation that can be conceived either as an opportunity or a risk in the future"* [30]. Thus it is rather difficult to discover a chance by usual statistical strategies. We adopt abduction and analogy (Abductive Analogical Reasoning [2] which can also be regarded as an extension of CMS [37]) to perform chance discovery [3, 4]. Where chance discovery is regarded as an explanatory reasoning for the unknown or unfamiliar observations, and a chance is therefore defined as followings:

1. **Chance** is a set of unknown hypotheses. Therefore, explanation of an observation is not influenced by it. Accordingly, a possible observation that should be explained cannot be explained. In this case, a hypotheses base or a knowledge base lacks necessary hypotheses. Therefore, it is necessary to generate missing hypotheses. Missing hypotheses are characterized as chance.

2. **Chance** itself is a set of known facts, but it is unknown how to use them to explain an observation. That is, a certain set of rules is missing. Accordingly, an observation cannot be explained by the facts. Since rules are usually generated by inductive ways, rules that are different from the trend cannot be generated. In this case, rules are generated by abductive methods, so trends are not considered. Abductively generated rules are characterized as chance.

Magnani also discussed application of abduction to chance discovery. Especially, he pointed out "manipulative abduction happens when we are thinking through doing and not only, in a pragmatic sense, about doing. So the idea of manipulative abduction goes beyond the well-known role of experiments as capable of forming new scientific laws by means of the results (the nature's answers to the investigator's question) they present, or of merely playing a predictive role (in confirmation and in falsification). Manipulative abduction refers to an extra-theoretical behavior that aims at creating communicable accounts of new experiences to integrate them into previously existing systems of experimental and linguistic (theoretical) practices.

The existence of this kind of extra-theoretical cognitive behavior is also testified by the many everyday situations in which humans are perfectly able to perform very efficacious (and habitual) tasks without the immediate possibility of realizing their conceptual explanation" [22]. Then he pointed out that "in dealing with the exploitation of cognitive resources and chances embedded in the environment, the notion of affordance, originally proposed by Gibson to illustrate the hybrid character of visual perception, can be extremely relevant. [...] In order to solve various controversies on the concept of affordance, we will take advantage of some useful insights that come from the study on abduction. Abduction may fruitfully describe all those human and animal hypothetical inferences that are operated through actions which consist in smart manipulations to both detect new affordances and to create manufactured external objects that offer new affordances" [23]. Thus he suggests the application of abduction to detect affordances which can be regarded as chances embedded in the environment.

5.2 Curation and chance discovery

5.2.1 Is curation chance discovery?

My experience in a market store can be regarded as a type of chance discovery application, because the strategy generated a hidden or potential purchase chance to customers. Customers who were inspired by the combination of chicken and asparagus would have bought either or both of them for dinner. Actually, this is not a task in museums, but it can be also regarded as a curator's work (curation). Because the strategy includes philosophy in a combination of items, and based on the philosophy it will offer certain effects to audiences. Visualization strategies such that referred to above function as curation. Because they display candidate chances in a manner where important or necessary items or events can be easily or interactively discovered by the user.

Thus a new definition of curation in chance discovery is:

- Curation is a task to offer users opportunities to discover chances.

- Curation should be conducted with considering implicit and potential possibilities.

- Chances should not be explicitly displayed to users.

- However, such chances should be rather easily discovered and arranged according to the user's interests and situations.

- There should be a certain freedom for user to arrange chances.

For curatorial task, a serious problem is pointed out. Magnani and Bardone introduced an idea of chance-faking as a possible outcome of the activity of chance-seeking [24]. They discussed the problem by illustrating the idea of bullshit introduced by Frankfurt [15] as an activity promoting fake chances. Compared with lie, they illustrate the problem in bullshit. Since a lie is not informatively empty, people have various mechanisms for detecting lies. On the other hand, a bullshit (fake) can be a semantic attack which is concerning with the manipulation of the meaning a person assign to something that he is going to use in his/her decision-making process. That is, a bullshit has no intention to cheat and logically true. Accordingly, it is rather difficult to determine it as a fake. It will be necessary to be careful of such chance-faking situation or provide a mechanism to detect such chance-faking situation in curation.

5.3 Abductive cognition

In this section, I will introduce abduction based cognition.

5.3.1 Information offering strategies for dementia persons

In [5], based on abduction, I formalized a concept of affordance based support system for dementia persons. For a proper and an extended usage of a thing, it is necessary to present proper information of it. However, for dementia persons, ordinal information offering strategies cannot function. Therefore, it is necessary to prepare proper information offering strategies especially for dementia persons. However, for a progressive and promising system, it is not realistic to prepare all the necessary information to things. Sometimes such information is not correct and will change in the future. For instance, it is ridiculous to attach a sign such as "You can sit here." to tree stumps. It is rather realistic to suggest information about its hidden functions. Such hidden information can be presented as certain stimuli in such situations. Because, even for dementia person,if he/she receives certain stimuli, he/she sometimes achieve better performance. The problem is that what type of stimulus will be better to present and how to make it recognize. Actually such stimulus should be "afforded (selected from an environment)" by the user. That is, it can be regarded as an "affordance" in an environment. Accordingly we introduce concept of affordance to a dementia care system. Proper affordance might give a certain support to dementia persons understanding (meanings of) objects. Thus affordance is a fruitful concept for recognizing objects and using them as tools. According to Gibson's definition, affordance is hidden in the nature and it should be accepted by us naturally. For instance, if an object's upper side is flat and it has a certain height, the observer will be able to afford it as something to sit down, rest or sleep. Of course, the level of affordance will be change according to observer's acceptance ability. For a certain person a tree stump will function as a chair, but for the other person it will not. If they are able to regard a tree stump as a chair, it will be necessary to provide a proper guidance to discover an affordance as a something to sit down.

For that I introduced the concept of affordance which was ecologically introduced for perceptional phenomena by Gibison [16, 17]. Gibson defined the affordance of something as "a specific combination of the properties of its substance and its surfaces taken with reference to an animal."

Actually such stimulus should be "afforded" by the user. That is, it can be regarded as an "affordance" in an environment. Accordingly we introduce concept of affordance to a dementia care system. Proper affordance might give a certain support to dementia persons understanding (meanings of) objects. Though meaning exists inside of the *Object*, in this framework meaning is explicitly described. That is, meaning should be observed and affordance functions as a type of link to *Objects*. When meaning is fixed, the affordance determination situation will be logically described as follows (in the form of Theorist):

$$Object \cup affordance \models meaning \tag{14}$$

$$Object \cup affordance \not\models \Box \tag{15}$$

That is, affordance can be regarded as a hypothesis. We can select consistent affordance (equation (15)) in the environment (hypothesis base) to explain meaning. In addition, for understanding subset of or similar afforded objects (*Object'*), the affordance determination situation will be logically described as follows:

That is, affordance can be regarded as a hypothesis. We can select consistent affordance (equation (15)) in the environment (hypothesis base) to explain meaning. In addition, for understanding subset of or similar afforded objects (*Object'*), the affordance determination situation will be logically described as follows:

$$Object \cup Object' \cup M \cup affordance \models meaning \tag{16}$$

M is a mapping function [18] from *Object* to *Object'*. That is, to understand the same meaning of the subset of or similar afforded objects, an additional mapping function M is required. Thus if M can be determined and the usage of *Object* is known, *Object'* can also be understood. In fact, for normal persons, M is easy to understand. However, for dementia persons, it is pointed out that it is rather difficult to understand and determine M.

The above logical descriptions can be illustrated in Fig. 1.

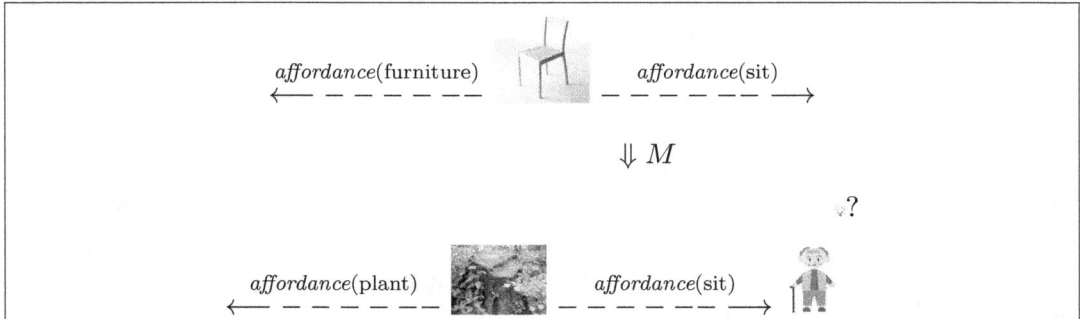

Figure 1: Affordance: communication between human and environment

The most important issue is how to suggest hidden information as affordance. My assumption was that complex situation can be transformed to a combination of simple situations. This type of combination can be achieved as curation. For such system, curation strategy should be introduced to offer understandable mapping

In [11], I discussed the importance of curation in the above system. Where I pointed put that by the introduction of the concept of curation, offering understandable mapping suggestion can be achieved. In addition, I introduced the notion of shikake. According to Matsumura's definition[25], a shikake is an embodied trigger for behavior change to solve social or personal issues. As a result of the action, all or part of problem will be solved. It sometimes may not be the person's will. Matsumura continues that the shikake should be properly designed. That is, the relationship between a problem to solve and a trigger to action should be properly designed. In addition, Matsumura uses a keyword "affordance" to explain such a trigger. His affordance is based on Norman's one [29], accordingly it may be different from Gibson's one. However, in the followings, discussion will be conducted based on Gibson's one.

A shikake can also be regarded as a strategy to lead us information. In the usual curation, information display is designed for audience to understand information easily. The curator's knowledge can be transferred to audience by his/her curation. By a shikake, the staff will not directly transfer his/her knowledge to the audience, but by a trigger of shikake, their knowledge will be transferred to the audience. Thus a shikake will be provided in some case in curation. In addition, a shikake can be explained by the concept of affordance. For instance, if people are aware of any affordance from a hidden Mickey which is a shikake, we can collect proper affordance which shows that a hidden Mickey is, for instance, very interesting and enjoyable. That is, they can select a proper affordance according to their better benefit.

A shikake is defined as an embodied trigger for behavior change to solve social or personal issues. In addition, as discussed in the previous section, a shikake should sometimes easy to discover, because it functions as a trigger, but the shikake itself may not easy to understand. In addition, by a shikake, the staff will not directly transfer his/her knowledge to the audience, but by a trigger of shikake, their knowledge will be transferred to the audience. Thus the feature of a shikake is suitable for suggestion of a mapping function M shown in the equation (16). Of course, the proper placement of a shikake is discussed in the concept of curation.

Currently we are dealing with a shikake in the situation of curation [39]. Tadaki pointed out that "[s]ince caption and artwork are usually displayed on the same wall so that visitors can easily see both caption and artwork in the same time. Although there are physical trigger, captions are not functioning as expected. In this paper, we try to add psychological triggers by adding some features to captions. The presence of change in how they see artwork was measured by time spent to see artworks, movement from caption to artwork, and participants' impressions to each artworks and each captions. The result of the experiment was suggesting nowadays captions are not the most effective tool to inform museum visitors. By our experiment, we can suggest the possibility of a Shikake displayed in text." Thus we consider a shikake in the situation of a museum exhibition. As curation we applied several types of shikake in order to make visitors to see artworks as well as reading captions.

6 Curation for abduction

As discussed in the previous section, I think that curation is very important in displaying information. Especially I focused on chance discovery based curation. Where information will not be shown explicitly. Accordingly curation is a task to offer users opportunities to discover hidden information (chances).

In fact, currently the computational igo, for instance Alphago Zero, has become extremely strong by applying deep learning and self learning. Even top professional player cannot win the game. Although the algorithm will be obtained by for instance deep learning, the strategy the computational igo uses is logical inference. The speed of calculation of such inference will be faster than humans. However, for the other side such as emotional side, computers cannot overcome humans. For instance, the professional player can point out the next suggestion place of stone perhaps without deep thinking. Such scheme will be intuitively achieved.

In the procedure of chance discovery, I can see the point where we should check.

However, for a novice will not perform this type of intuitive scheme. In order to help a novice, it will be necessary to prepare such interface as shown in Figure 2. The yellow circles can be regarded as affordance. We can select a proper circle. I regard this type of interface as curation. There will be several strategies in curation. This type of curation on one of solution. In addition this type of solution can be applied to the igo game.

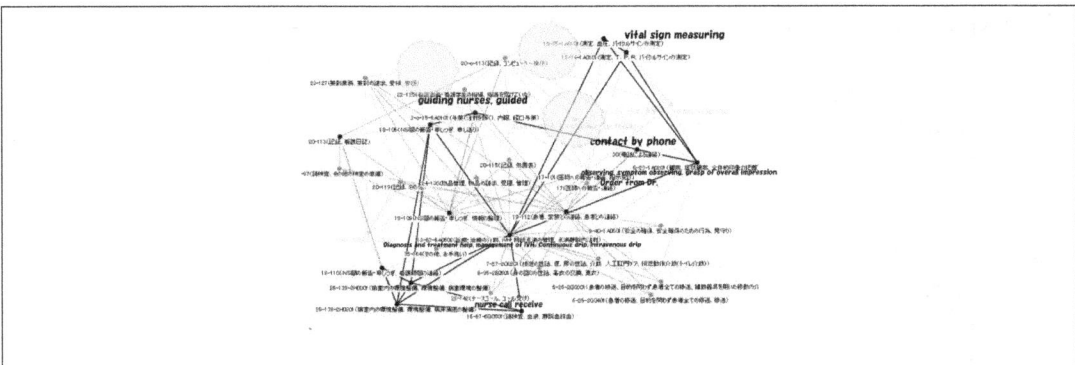

Figure 2: Affordance: communication between human and environment

In fact, the discovery can be conducted by abduction. However, it will be necessary to help the selection of extension by such a curation.

7 Conclusions

In this paper, first I illustrated abduction and computational abduction. Abduction is usually used for the discovery. In fact computational abduction is based on an explanatory reasoning. Accordingly the discovery is achieved as a selection of a possible hypothesis set. Usually a candidate hypothesis set is prepared as a hypothesis base as Theorist performs. In addition, a computational abduction such as CMS was proposed. Though the generated hypothesis set is not guaranteed, it can be used for the new discovery. One of the potential theme of this paper is chance discovery. Briefly speaking chance discovery is an intellectual activity to discover a novel or rare event/situation for the better future. This type of activity can be considered in the other activities such as a cognition of unknown matters. I discussed such activities by using the concept of affordance and curation. Curation is a very important strategy for the information display. In fact, after obtaining certain results (extensions) by, for instance abduction, it will be necessary to select the best or better extension. In such cases, the concept of curation will be necessary. In several papers (e.g. [8, 10]) I explained our previous applications from the viewpoint of chance discovery based curation. Where users can select the better result with deep consideration of the (candidate) results.

In this paper several concepts of the system were reviewed. In the future the proposal will be examined.

References

[1] Abe A.: On The Relation between Abductive Hypotheses and Inductive Hypotheses, in *Abduction and Induction (Flach P. A. and Kakas A. C. eds.), pp. 169–180 ,Kluwer* (2000)

[2] Abe A.: Abductive Analogical Reasoning, *Systems and Computers in Japan, Vol. 31, No. 1, pp. 11–19* (2000)

[3] Abe A.: The Role of Abduction in Chance Discovery, *New Generation Computing, Vol. 21, No. 1, pp. 61–71* (2003)

[4] Abe A.: Abduction and Analogy in Chance Discovery, *in [30], Chap. 16, pp. 231–248* (2003)

[5] Abe A.: Cognitive Chance Discovery, in *Universal Access in Human-Computer Interaction — Addressing Diversity (Stephanidis eds.), UAHCI 2009, Held as Part of HCI International 2009, Proceedings, Part I (LNCS5614), pp. 315–323*, Springer-Verlag (2009)

[6] Abe A. eds.: Special issue on Chance Discovery, *International Journal of Advanced Intelligence Paradigms, Vol. 2, No. 2/3* (2010)

[7] Abe A., Hagita N., Furutani M., Furutani Y., and Matsuoka R.: An Interface for Medical Diagnosis Support —from the viewpoint of Chance Discovery, in *[6], pp. 283–302* (2010)

[8] Abe A.: Curation in Chance Discovery, in *Ohsawa Y. and Abe A. eds.: Advances in Chance Discovery, SCI 423, pp. 1–18*, Springer Verlag (2012)

[9] Abe A.: Cognitive Chance Discovery: from abduction to affordance, in *Philosophy and Cognitive Science (Magnani L. and Li L. eds), SAPERE 2, pp. 155–172*, Springer Verlag (2012)

[10] Abe A.: Visualization as Curation with a holistic communication, *Proc. of TAAI2012, pp. 266–271* (2012)

[11] Abe A.: Cognitive Chance Discovery: from abduction to affordance and curation, *International Journal of Cognitive Informatics and Natural Intelligence (IJCINI), Volume 8, Issue 2., pp. 47–59* (2014)

[12] American Association of Museums Curators Committee: A code of ethics for curators, `http://www.curcom.org/_pdf/code_ethics2009.pdf` (2009)

[13] Amitani S. and Edmonds E.: A Method for Visualising Possible Contexts, in *[6], pp. 110–124* (2010)

[14] Colvia A.: Caravaggio Beckons to Bacon: The Beauty of Sorrow, in *Caravaggio Bacon (A. Coliva and M. Peppiatt eds.)*, pp. 17–22 (2009)

[15] Frankfurt H.: *On Bullshit*, Princeton University Press (2005)

[16] Gibson J.J.: The Theory of Affordances, *Perceiving, Acting, and Knowing (Shaw R. and Bransford J. eds.)* (1977)

[17] Gibson J.J.: *The Ecological Approach to Visual Perception*, Houghton Mifflin (1979)

[18] Goebel R.: A sketch of analogy as reasoning with equality hypotheses, *Proc. of Int'l Workshop Analogical and Inductive Inference (LNAI-397)*, pp. 243–253 (1989)

[19] Jones K.S.: What Is an Affordance?, *ECOLOGICAL PSYCHOLOGY*, 15(2), pp. 107–114 (2003)

[20] Lord P. and Macdonald A.: *e-Science Data Curation Report* https://www.cs.york.ac.uk/ftpdir/pub/leo/york-msc-2007/information/vsr-curation/science-dc-report.pdf (2003)

[21] Maeno Y. and Ohsawa Y.: Reflective visualization and verbalization of unconscious preference, in *[6]*, pp. 125–139 (2010)

[22] Magnani L.: Epistemic Mediators and Chance Morphodynamics, *Readings in Chance Discovery (Abe A. and Ohsawa Y. eds), International Series on Natural and Artificial Intelligence*, Vol. 3, Advanced Knowledge Intelligence, Chap. 13, pp. 140–155 (2005)

[23] Magnani L.: Chances, Affordances, and Cognitive Niche Construction: The Plasticity of Environmental Situatedness, *International Journal on Advanced Intelligence Paradigms* (2009)

[24] Magnani L. and Bardone E.: Faking Chance: Cognitive Niche Impoverishment, Proc. KES2010(LNAI6278), Part III,Springer Verlag (2010)

[25] Matsumura N.: A Shikake as an Embodied Trigger for Behavior Change, *Proc. of AAAI2013 Spring Symposium on Shikakelology*, pp. 62–67 (2013)

[26] Miyadai E.: Possibility of exhibition style, Eevening Asahi Shimbun, 11 March (2010) in Japanese

[27] Muggleton S. and Buntine W.: Machine invention of first-order predicates by inverting resolution, *Proc. of the 5th. International Workshop on Machine Learning*, pp.339–352 (1988)

[28] Nakamura J., Ohsawa Y., Nishio H.: An analogy game: toward cognitive upheaval through reflection-in-action, in *[6]*, pp. 220–234 (2010)

[29] Norman D.: *The Design of Everyday Things*, Addison Wesley (1988)

[30] Ohsawa Y. and McBurney P. eds.: *Chance Discovery*, Springer Verlag (2003)

[31] Ohsawa Y. and Tsumoto S. eds.: *Chance Discoveries in Real World Decision Making*, Data-based Interaction of Human Intelligence and Artificial Intelligence Series: Studies in Computational Intelligence, Springer Verlag (2006)

[32] Charles Sanders Peirce: Abduction and Induction, in *Philosophical Writings of Peirce*, **chap. 11**, pp. 150–156, Dover (1955)

[33] Poole D., Goebel R. and Aleliunas R.: Theorist: A Logical Reasoning System for De-

faults and Diagnosis, *The Knowledge Frontier: Essays in the Representation of Knowledge (Cercone N.J., McCalla G. Eds.), pp. 331–352*, Springer-Verlag (1987)

[34] Pople Jr. H. E.: On The Mechanization of Abductive Logic, *Proc. of IJCAI73, pp.147–152* (1973)

[35] Posner K.A.: *Stalking the Black Swan*, Columbia University Press (2010)

[36] Rajan R.G.: *Fault Lines*, Princeton University Press (2010)

[37] Reiter R. and de Kleer J.: Foundation of assumption-based truth maintenance systems: preliminary report, *Proc. of AAAI87, pp.183–188* (1987)

[38] Rosenbaum S.: *Curation Nation*, McGrau Hill (2011)

[39] Kotone Tadaki and Akinori Abe: Museum Visitors' Behavioral Change Caused by Captions, *Proc. of the 2nd. Int'l Workshop on Language Sense on Computer in IJCAI2017, pp. 53–58* (2017)

[40] Nassim Nicholas Taleb: *The Black Swan*, Allen Lane (2007)

[41] Thagard P.: Explanatory coherence, *Behavioral and Brain Sciences, 12, pp. 435–502* (1989)

[42] Vorobieva O. and Schmidt R.: Case-Based Reasoning to Explain Medical Model Exceptions, in *[6], pp. 271–282* (2010)

[43] Warren W.H.: Perceiving affordances: Visual guidance of stair-climbing, *Journal of Experimental Psychology: Human Perception and Performance, Vol. 10, pp. 683–703* (1984)

[44] Yada K., Washio T., and Ukai Y.: Modeling Deposit Outflow in Financial Crises: Application to Branch Management and Customer Relationship Management, in *[6], pp. 254–270* (2010)

[45] Zhang J. and Patel V.L.: Distributed cognition, representation, and affordance, *Cognition & Pragmatics* (2006)

Conjectures and Abductive Reasoning in Games

Ahti-Veikko Pietarinen
Tallinn University of Technology, TallinnEstonia
Nazarbayev University, Astana, Kazakhstan
National Research University Higher School of Economics
Moscow, Russia
ahti-veikko.pietarinen@ttu.ee

Abstract

There is abduction in games. Players deliberating about possible future histories take those positions, which according to the standard common knowledge and belief of rationality will never actually be reached, as the surprising facts that need accommodation. The need for such accommodation sets their minds in motion and trigger reasoning from effect to causes. Players are prompted to reason to an antedating action under which such positions would be rendered comprehensible, less surprising, or facile and natural. In games, reasoning abductively means to imaginatively look for where perturbations, such as trembles or quantal responses, could take place. Its conclusion is a conjecture about such perturbations.

1 Setting the Strategy

A game refers to an interactive situation involving two or more players engages in strategic decision-making. Its theory can be seen either as a branch of applied mathematics concerning optimal behaviour or social science concerning purposeful collective behaviour in situations involving certain ideas about decisions as rational. The history of game theory goes back at least to *New Essays on Human Understanding* (c.1702/1981), in which Gottfried Leibniz urged his colleagues to develop "a new kind of logic, concerned with degrees of probability, ... to pursue the investigation

---

Work supported by the Estonian Research Council (PUT1305: *Abduction in the Age of Fundamental Uncertainty*) and Russian Academic Excellence Project "5-100". My thanks to Woosuk Park for the organization of the 2016 KAIST Conference on the Logical Structure of Strategic Reasoning, and to all its speakers and participants.

of games of chance". Leibniz took the art of discovery to be thus improved, since the human mind "is more thoroughly displayed in games than in the most serious pursuits" (p. 467).

Modern game theory deals with concepts of rationality, probability and preferences, as well as player's beliefs about other players' actions, knowledge and intentions to act. It develops methodologies that apply in principle to all interactive situations, and is aimed at investigating such interactive processes that are governed by a coherent set of rules that would spell out the actions for a participant, given the actions of other participants. At the heart of this idea is the notion of a strategy, a non-constitutive rule intended to implement a mechanism or to provide a habit of behaviour that chooses from the set of possible actions the actual ones that the players should come up with in the course of the game, given the principles of rationality and related epistemic postulates.

In game theory one typically defines strategies for all histories of the game, even those that lie at the off-the-equilibrium path and will never lead a player to a win. A good strategy must thus prescribe reasonable moves also for histories that are reached, and are commonly known and believed to be reached by the players, no matter if with very low or zero probabilities.

A standard approach takes counterfactual reasoning in extensive-form games to be part of the causal-deductive and epistemic (involving the modalities of knowledge and belief) theories of the modellers and players. In this paper, I argue that strategic reasoning about future scenarios should essentially be seen as an *abductive* rather than a deductive task. The latter concerns the security of conclusions, no matter whether probabilistic or non-probabilistic. In abduction, in contrast, players deliberating about possible future histories take those positions, which according to the standard common knowledge and belief of rationality will never actually be reached, as the surprising facts that need accommodation. The need for such accommodation sets their minds in motion and trigger *reasoning from effect to causes*. Players are thus prompted to reason, inversely by a Modus Tollens, to an antedating action under which such positions *would be* rendered comprehensible, less surprising, or even facile and natural. In games, reasoning abductively means to imaginatively look for where perturbations, such as trembles or quantal responses, could take place. Its conclusion is a conjecture about such perturbations.

2 Science and Abduction

The motivation for the present inquiry is itself grounded in performing a piece of abductive reasoning. It is of the following kind. Begin first with the premise,

"Games occur everywhere in sciences."

I take this to be incontestable both theoretically and empirically ([19]). It is the activity that takes place when an experimenter interrogates nature, or when the logic of questions and answers is being modelled, or when the entire edifice of scientific inquiry was being erected contemplated and contemplated upon. From Plato to Kant, from Aristotle to *Abbreviatio Montana*, from *Dialectica Ludicra* to Leibniz, from Bacon to Newton and Popper to Hintikka, this way of theorizing about human inquiry has worked marvellous results.

Since Charles Peirce, we have also came to appreciate the fact that the mode of reasoning that sets the scientific mind in motion is abduction. Peirce gave the name for this prevalent mode that his studies in the history of science had brought to view; the mode that had led Kepler, Newton, and their legacies to some majestic insights and discoveries. Peirce did not claim that when studying the logic of abduction he would have discovered anything substantially new; he merely was aiming at identifying the salient features in the pattern of the inquiring mind could then be submitted to logical analysis to be teased out in detail. Thus:

"Abduction occurs everywhere in the sciences."

Thus from the previous two premisses, namely that "Games occur everywhere in the sciences" and "Abduction occurs everywhere in the sciences", we conclude that

"Abduction occurs in games."

This is of course itself an example of abduction, where the conclusion must be read in its tentative and interrogative moods and is not as such any known proposition. It is not even a believed one. The conclusion merely triggers further inquiry, delivering a recommendation 'This conclusion would be worth investigating further'—namely whether abduction occurs in games!

Luckily an answer to this co-hortative interrogation does not necessitate launching deep-space telescopes or building artificial brains. It is well worth funding research which is cheap. Thus our theoretical pathway takes us from games as omnipresent, complex strategic interactions to such issues as the game of Baduk, topology, computation, evolution and logic. Dismissing epistemic game theory as anything like an endpoint of this journey, we are led to discover what we started off with: that abduction solves inverse problems by reasoning from unexpected effects that occur in games to their causes. This is what is only to be expected, given that the justification of abduction is itself both an abductive, inductive and deductive problem ([22]): it is abductive as its leading principle is that *Nature is explainable* (CP 7.220). It is inductive, because its conclusions build up classes of hypotheses from which history of science takes its samples for the inductive test of time. The justification is also deductive, since as soon as the conclusion has been reached as a conjecture, its premisses would be inferrable by deduction.

Along the way, I will acclaim six heresies that these considerations give rise to.

3 Science and Games

Why would one resort to games as explanatory of human intellectual activities? Because they are no surface matter. "Games people play" is misleading in the colloquial senses of this phrase. We can observe complex strategic interactions as fundamental to the very existence of various disciplines and their phenomena and entities, as well as to the methods of inquiry they use, across as diverse and wide-ranging array of fields as the following list has us believe:

Computation The very idea of computation emerges from logical games. Traditionally, propositions are problems and their logical constituents operations on such computational problems. What is "algorithmically solvable" is the truth. This can be generalized: Computational problems are interactions where machines play games off against and with their environments. Solving the problem algorithmically means the existence of a winning strategy again any possible behaviour of the environment. The truth of the interaction comes out when the algorithm wins, and wins invariably ([6]).

Number Objects of mathematics emerge from games. In combinatorial domains, Conway's surreal numbers are an example in which numbers are themselves conceived as games. (In fact the game of Baduk provided a key insight to the discovery.) Model-building by games is another one.

Science Intellectual inquiry is an outgrowth of our neural tactics and strategic encounters. Take physics, where the process of extracting physical laws by means of acts of extracting information-measures has been proposed to be a process that can be modelled as a game between the Observer and Nature ([2]). Or, as another example: It was a question for von Neumann how games of imperfect information and logical foundations of quantum theory meet, something that remained as one of his unfulfilled goals.[1] A proposal for a solution has been

---

[1] "The first to explicitly raise the possibility of some fundamental connections between extensive form games of imperfect information, quantum logic and quantum mechanics ... [Pietarinen] [18] is the first to have studied the similarities between the logic of games of imperfect information and quantum theory ... There does not seem to be any evidence that von Neumann noticed a formal connection between these two kinds of 'indeterminism'" ([16]).

given in [18]. These theoretical strains are in accordance with the venerable tradition of inquiry being 'putting questions to Nature'. The general logical model of Interrogating Nature may well turn out to be a future development of this Peirce–Hintikka model of the logic of inquiry, and it continues to feeds into the discovery of hitherto unidentified hidden connections.

Universe The universe, or in the evolutionary cosmology the family members universes of the large multiverse class, what exists emerges from the copulation of the ideas of Khôra and Kosmos. Plato's description of thought as dialogical was his "greatest contribution to thought" (Peirce's marginal comment to *Plato's Logic*, Houghton Library). In the world of objective idealism, "[E]very reader will be able by experiments and observations of his own [...] reasonably to satisfy himself that all meditation, consideration, and thought consists or is embodied in talk with oneself" (MS 678, c. 1910).

Further, think of meaning and truth. The game-theoretic notion of truth is an immanent feature of games with their contexts and environments, devoid of allusions to transcendence. What does it mean that a simple probability of an event e is 1? It could mean several things:

PROB$(e) = 1$:

– Existence of a winning strategy for the players P.
– Proposition φ is true.
– Reality yields, and the event e is realized.
– P wins the game of Baduk.
– P makes an infinite amount of money.
– A winning trading strategy obtains for P.

All these are in certain senses equi-significant propositions. The existence of a winning strategy, for example, is an objective (non-transcendental) fact of the model in question.

The *game of meaning* is played on a model M that consists of a non-empty domain of the universe: The Opponent (O) aims at showing that the proposition is false in M, while the Proponent (P) aims at showing that it is true in M. Various constants in the proposition prompt moves by one of these players. Negations may prompt the change in the roles of the players, and the winning conventions (payoffs) would consequently also change. Each move reduces the complexity of a proposition and an atomic formula is finally reached. (Atomic games may still go on.) The truth-value of an atomic formula determines who wins. A strategy for any player is a mapping assigning to each significant part of the proposition a player. A winning strategy is a strategy by which a player can make operational choices such that every play results a win for him or her, no matter how the opponent chooses. Finally, the assertion is true in M if and only if there exists a winning strategy for the player who started the game off as P, and it is false in M if and only if there exists a winning strategy for the player who started the game off as O.

Is there anything more objective than a procedure such as this? At least Plato, Leibniz, Peirce, Wittgenstein and Hintikka thought not.[2] We are thus led to acclaim the following

HERESY 1 Divest science/logic/finance/reality from epistemology.

Then consider "Winning the game of Baduk". What does this mean? We now know exactly that the possibilities of finding an equilibrium play among the legal (19×19) positions L19 come from the large pool of options:

208 168 199 381 979 984 699 478 633 344 862 770 286 522 453 884 530 548 425 639 456 820 927 419 612 738 015 378 525 648 451 698 519 643 907 259 916 015 628 128 546 089 888 314 427 129 715 319 317 557 736 620 397 247 064 840 935 ([25])

For L2 this is 57. From there the growth of the function is fast. Truth is out there, and there is a lot of it.

On the other side of the fence, what does it mean that the probability of a simple event is zero?

Prob$(e) = 0$:

- There is a winning strategy for the player who started playing the game off as your opponent O.
- O wins, P loses.
- φ is false.
- Reality holds her ground; event e remains unrealized, for now.
- We make no money (or lose everything).

---

[2] "'Surely if he knows anything he must know that he sees!'—It is true that the game of 'showing or telling what one sees' is one of the most fundamental language games, which means that what we in ordinary life call using language mostly presupposes this game" ([27, item 149, p. 1]) See [19] on Wittgenstein's "one of the most fundamental language games", which shows that Wittgenstein knew about the developments on the theory of games.

These are the extremal cases. We can ask Nature a question. After each negative answer, which amounts to rejecting the initial working hypothesis and moving on to the next one, we invest increasingly more time, energy and money to the testing of those abductively concluded investigands. But as soon as Nature answers 'Yes', we can continue upholding it as far as practicable, and begin building our confidence (our world-view) upon the positive result. The expectation is that the losses of the past are recovered from the gains we get from the present hit (as the stakes are already higher and the question is getting big), plus the original investment on the research project in the first place. The probability itself emerges as the values of this betting strategy on our abductive interrogations in relation to the next answer that Nature provides.

We thus get our

HERESY 2 Between $1 - 0$, probabilities are payoffs of the game against Nature.

As a corollary, we have:

HERESY 3 Be prepared for $\text{Prob}(e) = 0$ events. Our opponent (Nature) can be clever, or even extremely clever.

But what does it mean to "be prepared"?

4 Deform and Perturbate

Take an object—any object—and study it under small perturbations. Divest the sheet of commutativity, and its geometry and logic will change. Deform the original object, and it becomes increasingly more specific under the light of a host of other but related questions, which previously did not apply to it. Deform and perturbate—this is the key to discovery in areas such as

Topology Meeting non-invariances, deform the problem to a generic case in the ambient space of $\dim(n)$.

Computation Risking the Church-Turing limit, ask an oracle. Performing Ω-computation is taking consultation or counselling. Games of inquiry involve abductive steps where you are compelled to ask a question before you move on and deduce something further.

Economics Unable to solve a decision-problem, look for non-linguistic advice. There may be nudges, norms of behaviour and Schelling-points at your disposal.

Game theory If your plan of action runs dry, look for ϵ-perturbations. (This is discussed in a separate section below.)

Evolution Go co-. Choose anti-fragile institutions, nurture genetically, expand your horizons. The process of evolution concerns both cosmic, biological and cultural forces. Looking back and trying to infer where and when the nearest common mitochondrial mother had lived is to infer from effects to causes. Backtracking the tree of life, think of what DNA and fossil records can reveal on the properties of historical species. Then take the whole phylogenetic tree of life as an extensive-form game, with the root where the game starts as the last universal common ancestor to bacteria, archaea and eukarya. Inferring the 355 genes that the LUCA presents indeed was an abductive undertaking ([26]).[3] The players of that game are teams consisting of drivers that change the gene frequencies over time: descent, variation, migration, drift, co-evolutionary forces (mutual interaction, genetic nurturing, niche construction etc.), and last but not least, natural selection.[4]

John Maynard Smith famously extended game theory to biological phenomena and population genetics, by applying the concept of evolutionarily stable strategies to animal and population behaviour. Small perturbations do not change the epigenetic landscape of the population structure, as the neighbourhood exhibits sufficient similarity or isomorphism to it. A population consisting of actors occupying a type of position they themselves might not be

---

[3]When Darwin in the *On the Origin of Species* wrote "Therefore I should infer from analogy that probably all the organic beings which have ever lived on this earth have descended from some one primordial form, into which life was first breathed" ([1, p. 484]), the only word not fully to the point was "analogy", which he would now replace with "abduction".

[4]Horizontal gene transfers between the histories of the tree can further be modelled as non-trivial information sets. They add imperfect information to the game. Later stages of selection and mutation who may not know some of the earlier decisions mean that horizontal gene transfer has occurred some point earlier in the game. Universality of the genetic code and its preservation is thus guaranteed when information is hidden from subsequent choices in this way. When actions are hidden, at the later decision points choices are made not knowing precisely what had happened. In other words, the players have to respond while not knowing exactly where they are in the structure of the game: development of antibiotic resistance would be an example of such information hiding. Thus the structure of life's evolution would be a more organized structure than an unconstrained directed graph or a mosaic; imperfect-information (and perhaps imperfect-recall) extensive-form trees permit the inverse reasoning by backtracking.

To trace the evolutionary histories in the agent-normal extensive-form modelling of the phylogenetic tree is to reason abductively to the causes (moves) of these teams. Since there are obvious mistakes and dead ends in the speciation, the game of life is also a perturbed game, where players can at any time play moves that lead organism away from the phylogenetically optimal equilibrium path.

aware of have an access to such strategies that become more prevalent as the game goes on. There is no way of distinguishing the players from each other insofar as they do not play any roles. The winner is the one who goes on longer whatever the task is. It is a common feature that games are played over and over again by agents drawn from large populations, guided by an evolutionary selection process affecting their behaviour. It may happen that neither of the populations, nor their individual members, can defeat each other, and both will retreat.

So if the law of the excluded middle (LEM) does not apply to life, how does that character of a logical law emerge? This leads us to

Logic Are you stuck with a paradox? Then look for some generic, stochastic, multi-valued, or noise-tolerant deformations of the original situation. Tarski's undefinability of truth, for one, is an artifact of the infinite precision demanded by reasoning about complete certainty. The standard laws are assumed to hold—for the reasoning would break down otherwise. But what if the LEM is not a logical truth? What if no reasoning is absolutely certain?

These considerations compel us to identify another heresy:

HERESY 4 Idealizations are specifications; deformations generalizations.

This is a heresy only in the light of contemporary philosophy of science and logic, where it is commonplace to think that in scientific modelling, idealizations are losing something of the essence of the original domain and thus run the risk of becoming just a convenient fiction. But models are hypotheses, and no good hypothesis that is a conclusion of abduction and thus a candidate for testing is in any realistic sense non-ideal.

4.1 Strategic Reasoning and Epistemic Game Theory

We see epistemic game theory as much like the Holy Roman Empire: at bottom, it is neither about knowledge, strategic interactions, nor a particularly successful attempt to articulate a general perspective or method to the field it wants to investigate. Typical presumptions include the following: rational, Bayesian decision-making, problem spaces as context-free, and payoffs independent of 'irrelevant' alternatives. The notion of knowledge involved is interpreted with strong introspection axioms, typically yielding equivalence structures (S5) and as such never the preferred playground of *Knowledge & Belief* ([4]). Players do not play weakly dominated strategies, and degrees of beliefs, given the priors, are adjusted according to the update protocols.

All this makes epistemic game theory more of a deductive exercise on risk assessment than an insight to advance inquiry. When there is uncertainty, it is about the 'known unknowns'. Decision makers are non-procedurally rational, the framework is probabilistic, and the decision-making looks away from genuine uncertainty (the 'unknown unknowns'). Further, the structure of the game and the available actions are given *ex ante*, and so there is unlimited foresight enabling one to perform the Sherlock–Moriarty pattern. This suggests that

- the idea of the maximization of expected utility is an estimate on the probability of opponent's strategy profiles.

What would be the alternative? Could we rather device anti-fragile decision-making protocols under fundamental uncertainty that embraces for the impacts of unknown unknowns? Could we, instead of Bayesian probabilities, speak qualitatively of potential surprises, shocks, plausibilities and contingencies of actions? Could we have epistemics as a representation of sub-belief modalities (conjectures, surmises, guesses, presumptions, hopes, wishes) rather than equivalence structures in the partitional and possible-worlds fashion?

Here the involvement of ill-posed problems and under-structured problem spaces turn out to illuminate the meaning of these questions.

5 Actions as Interrogations

5.1 Interrogative Abduction in Ill-Posed Inverse Problems

According to Peirce, abduction is a Modus Tollens from the premises "If A is true, C is not true" and "But C is not true", where the conclusion is an interrogative, "Is A not true?". He described the process as "Reasoning from Surprise to Inquiry", where the mood of the conclusion is a mixture of interrogative and imperative moods: "It is to be inquired whether A is not true" (MS L 463, 1905). Peirce termed this the "investigand" mood ([12]).

It has been observed that the branch of applied mathematics that studies inverse problems deals successfully with abductive types of inference ([14]). This view takes abduction to be the mode of reasoning when we move from effects to causes. It is a limiting view of abduction as it reduces certain classes of inverse problems, especially the well-posed, continuous and parametric-model ones, to matters of deductive inferences.

Peirce's interrogative construal suggests a broad view of abduction fitted for situations in which strict cause-effect relationships may be unobtainable. This may not be known at the moment when the problem is being identified, as those situations concern severely under-structured problem contexts. In the area of inverse problems, such contexts give rise to ill-posed problems: the converse of a continuous mapping is discontinuous so that analog samples do not work, models are non-parametric, etc. Inference in such contexts calls for abduction in its interrogative or investigand mood, to 'guess at the unknown unknowns'. For example when forming confidence regions or choosing parameters tend to be under-smoothing, a guess is the best bet.

If the inverse problems are well-posed, that is, if the relevant parameters or properties of models are known so that the solution depends continuously on the available data, the predominant mode of reasoning is indeed the most secure, deductive one. It may thus be that the predominant mode of inference in those inverse problems that are well-posed but ill-conditioned is in fact inductive, and deductive and abductive stay in the background.

5.2 Epistemic Game Theory

Back to the Holy Roman Empire, we also know that Bayesian rationality is not sufficient to ensure that an iteratively non-dominated strategy profile is going to be played in a generic case. And we also know that having the first-order level of beliefs concerning rationality is not sufficient for that to happen either. In fact, how many levels of beliefs are needed to guarantee an existence of the rational play remains an open question.

Some proposals to approach the issue introduce methods of the lexicographic probability systems, where one admits that irrational choices occur as subordinate Martian options, and departures from the common assumptions of Nash equilibrium, common belief of rationality, backward-induction algorithms, or maximisation of expected utilities altogether. In these alternative worlds, one encounters terms such as rationalizability, unawareness, and iterated regret minimization.

But the received remedies leave something to be desired. When Prob(e)=0, reality holds her ground. How are we to take zero-probable events? Is it that agents are *unaware* of such actually live possibilities? Or has the game just not been long enough? Either attempts seem inadequate. Our awarenesses and unawarenesses are psychological mechanisms impotent when the task is to invite reality to yield something useful. Continuing to play finite games would not seem to help either: In the game of Baduk, the longest game on 19×19 is 10^{48}, with the upper limit of games bordering on nothing less than $10^{10^{171}}$.

5.3 The Limits of Counterfactual Reasoning

Further, what does it then mean that in counterfactual reasoning, "deliberation about what to do in any context requires reasoning about what will or would happen in various alternative situations, including situations that the agent knows will never in fact be realized" ([23])? Despite the fact that counterfactual reasoning is thoroughly studied, situations that in fact are not realized can be controlled well in the method of perturbations. But the middle-part of the sentence, "reasoning about what will or would happen" in the future scenarios, introduces a novelty, albeit unbeknownst to its author. The fundamental mode of reasoning in games is *abductive*, and not a deductive, causal-epistemic and omniscient decision problem that would aim at bringing closure to the inferences under the derivation relation.

Generically, counterfactual scenarios belong to the reality, not to the existence. The *would-be*s and *could-be*s exert non-causal influence upon our decisions. Our flirt with them happens through strategies that are *habitual*. That is, behavioural strategies do not compute the given probabilities, they create them by the very fact that they are part of our interrogations of Nature.

In game theory, plans of action make a subclass of all strategy profiles. (The distinction becomes material in games with *imperfect recall*, for example, in which not all strategies could be taken to be plans according to which the players would in fact play the game.) The generic class of strategies that would adequately accommodate the *would-be* character of real possibilities and counterfactual reasoning suggest another generalization, namely the class of habits, of which all strategy profiles form a subclass.

It is this latter expansion that has not yet been fulfilled in a theoretically satisfying matter. Why not? The reason is that there is a blank interval—an indeterminacy—between contingent payoffs (such as the histories of the play, values, money, happiness) and contingent claims and propositions (making an assertion, writing a derivative, landing a contract, or engaging in self-controlled activities). No psychological mechanism explains how this gap is to be filled. Expectations, intentions to act, or various singular desires and wishes are not generalized forms of behaviour that could bring goal-directed payoffs to bear on propositions that profess to reach them.

By making plausible guesses we contribute to the making of probabilities. These guesses may concern truth-assignments, asset-trading behaviour, or success in ascertaining or failing to ascertain human well-being. *What-could-have-been*s are hindsight narratives; the habits of action fill the gap, the blank interval exhibited by the off-equilibrium behaviour, between the payoffs and what the antecedents of our counterfactuals are interpreted to say.

Hence ϵ-deviations occur outside standard probability distributions. This already rules out expectations and the usual refinements of Nash equilibria as plausible mediators between the contingencies of the payoffs and the contingencies of the cheap talk.

6 Abduction in Perturbed Games

6.1 The Abductive Schema in Games

The schema of abduction that we find in games is this:

1. A surprising event is imagined (say, the player P_2 would play the off-equilibrium history h_2)

2. If ϵ were to be the case, then h_2 would be a matter of course.

3. Therefore, there is a reason to suspect that ϵ is immanent in the game.

Infinitesimal perturbations make a difference. What will be argued below is epitomized in our fifth heresy:

HERESY 5 The very presence of perturbations is the effect of our reciprocal abductions.

A deviation from the cozy equilibrium path is that surprising event: our ignorance is exposed to cognitive irritation and doubt. We can identify two kinds (or levels of intensity) of irritations:

Weak irritation This happens when the surprising event ϵ could be explained by the internal epistemics of the game situation, that is, having the relevant beliefs, expectations, or probabilities at our disposal.

- This is what Magnani [11] terms *selective abduction*. Here the epistemic analysis of the informational structures, as the standard methods have suggested, may suffice. But there is also

Strong irritation This happens when facing ϵ we draw blank from our epistemic repositories.

- This is the *creative* side of abduction. Abduction is creative in the sense that it serves as the progenitress of probability distributions from totally mixed strategies. In Peirce's terms, we get from tychism, which is the state of absolute randomness devoid of laws, to habits of action as those laws. The state of randomness exhibits only habit-taking tendencies, and they are not laws, law-like relations or habits.

6.2 The Problem of Maximal Information Structures

Does a maximal information structure exist for epistemic games? In so far as we trust the Brandenburger–Keisler Paradox, there really is not:

- Ann believes that Bob assumes that Ann believes that Bob's assumption is wrong.

Does Ann believe Bob's assumption is wrong? YES iff NO. Hence not every constellation of beliefs is representable, ever.

So adding epistemic structures to the games has its natural limitations. There is a paradox, or else its revenge. It is not surprising that formalizations run amok this way. The efficiency and meaning of deductions, just as computations and algorithms, have to be assessed in terms of the resources that they expend and the economy that is wasted. Mechanism design for algorithms that would contain no mechanism to prevent or control a flash crash would clearly be a highly uneconomical undertaking, yet it is an area that currently claims billions of unregulated research funds. A Vickrey auction can mitigate the harmful effects of the Winners Curse, while similar biases are allowed to replicate in modern algorithmic industry.

6.3 Surprising Facts and Factoids

The nature of strategic reasoning can be seen under an essentially different manner, or so I will argue. Strategic reasoning about future scenarios is primarily an abductive, and not a deductive, task. This starts off with the observation that players deliberating about possible future histories take those positions, which according to the standard common knowledge and belief of rationality will never actually be reached, as the *surprising facts* they encounter in the plays of the game.

Players are thus prompted to reason, inversely by a Modus Tollens, to an antedating action under which such positions would be rendered comprehensible, less surprising, or even facile and natural. In games, abductive reasoning refers to the tasks of the players to imaginatively look for where perturbations, such as trembles or quantal responses, could take place. The conclusion of those abductions is then a conjecture about such perturbations.

The really surprising fact is that the abductive logic inherent in strategic reasoning in games has not been pointed out in the literature. The reasons for the absence may be manifold. Three points stand out.

1. The deductive paradigm has been that the only kind of security of reasoning we can be assured of is deductive, and that deductive reasoning equals necessary reasoning. The security of the conclusion can be no more than the security of the premises. This deductive paradigm obviously inundates also the structures of epistemic game theory.

2. The second reason is the assumption that the security of reasoning would match that of the rationality of players, and that there would thus be no reason to question either. Common belief in rationality or common alignment of beliefs are staples of this conception of rationality. There nevertheless is no intrinsic reason why reasoning and rationality should become correlated only in the deductive realm, as the whole edifice of strategic reasoning is much more about uncertainties and the elimination of them than hoarding surface information on game-theoretic and interactive structures with fixed action sets and unlimited foresight.

3. The third and the main reason seems to be, unsurprisingly, that abduction just is not a well-known type of reasoning compared to deduction and induction—and even when it is claimed to be so, it is easily conflated with inference to the best explanation or various forms of induction, or misunderstood in some other ways.

Trying the rectify the prevalent misunderstandings about abductive reasoning are not our concern. Briefly, the obvious one has been to take abduction as tantamount to IBE. This has been conclusively refuted in the literature and there is no need to review the issue here. Second, even if one would take Peirce's sketches of the logical schemas of abduction at their face value, his late refinements have been mostly overlooked.

6.4 Conjecture-making and Perturbations

In any event, in game theory, abductive reasoning refers to reasoning in the sense of the players making informed guesses about perturbations, such as trembles or other deviations, forming hypotheses concerning the nature, location and weight of those perturbations, and the assessments of their implications to their overall interactive strategic decision-making settings. The conclusion of those abductions is thus a conjecture about such perturbations. This leads to the view that

- Maximization of expected utilities is, among others, a conjecture on plausible hypotheses on opponent's strategy profiles.

The major types of perturbations in the games include trembles and their refinements, as well as equilibria extensions such as correlations. These are taken up next.

6.4.1 Trembling Hands

The motivation for the *trembling-hand equilibria* is that we need not rule out weakly dominated strategies in order to remain rational, that is, in order to remain on the trajectory of a Nash equilibrium play. Such plays and histories can well include weakly dominated strategies. The question is whether players would ever be expected to play them? Should players be expected to be absolutely sure about the choices that other players would make, under the common knowledge of rationality? If there is a risk, even a most minuscule one, that other players make mistakes, how should a player be prepared for that scenario? How should the theory of games model such situations?

The famous answer to these worries has been the *trembling-hand perfection*, namely the modelling of a perturbed game in which every strategy is played with a minimal probability. The strategy profile would thus include strategies that lead to off-equilibrium plays and would never be played by a rational player, should absolute certainty obtain that a super-intelligent machine would not make any mistakes. The minimal probability is that mistakes would always be unavoidable although they would be highly unlikely. This refinement of the Nash equilibrium concept then states that a sequence of games that is so perturbed exists, and that it converges to the game for which there exists a Nash equilibrium.

This is a nice re-statement of Peirce's fallibilism, namely the senses in which our ordinary notions of knowledge (which we need not separately formalize as a superstructure) involved in inquiry and information-seeking fail to reach certainty. Anti-radical-scepticism is not an option in 'epistemology without knowledge', to borrow Hintikka's turn of phrase ([5]).

The trembling-hand refinement defines a probability distribution to all strategies. Does this make this type of perturbation itself an abductive problem? No, since probabilistic reasoning would still be a form of deductive reasoning, in which conclusions would follow deductively and apodictically with some probability measure. One could try to model trembles not with probabilities but with plausibility, for example, in which case the players would need to take into consideration whether certain trembles occurring in certain situations would be more plausible than in some others. Since plausibility measures are not additive, the convergence of perturbed sequences with Nash sequences would not follow, however.

But even when modelling perturbed games with probabilistic trembles, the abductive problem that does remain in the strategies of the players is to guess whether the trembles—knowing that they could occur at any time—would be the kinds of mistakes that lead the opponent off the 'rational' path to the Nash equilibria that would not otherwise be played. If so, this needs an explanation in the sense of a novel hypothesis or formulation of an investigative question about them. Should the player expend energy (cognitive, logical and economical) to formulate a co-hortative interrogation that 'it would be wise to investigate further, whether this perturbation is more consequential than others'? No, because all perturbations are equally likely, with a small positive probability occurring to realize them. No weakly dominated pure strategy would be played with any positive probability.

This result does not rule out that trembles could be dominated by zero-probable strategies. Such strategies are possible, although they would not occur in actual situations in the play. But games are about the reality, not actuality or existence. As a game of inquiry, where the experimenter (obviously a team of scientists) is interrogating Nature, tremble-style perturbation is a suitable idealization as all Nash equilibria with weakly dominated strategies in two-player games (assuming Nature as an active decision maker) is trembling-hand perfect. This does not hold for $n > 2$ players.

6.4.2 Proper Equilibrium

Another perturbation to refine the solutions of games is to add a hierarchy of trembles. This is an obvious recommendation from the limitations of the above considerations, as they showed that trembles alone would not irritate doubt or instigate genuine abductive rumination. The idea of *proper equilibrium* is to incorporate the values of 'more likely trembles' into the strategies. The measure of trembles is here how likely, in the sense of how often, they would occur. The recurrence makes them more costly, and others can exploit that feature.

6.4.3 Correlated Equilibrium

Third, an extension and not a refinement that perturbs the situation into another direction, away from minimizing the amount of Nash equilibria, is to *correlate* the scenarios with an extraneous signal. This could be a norm of the society, or your aunt who enters the room of quarreling children. The third party flips a coin and announces the result. The players play their next two moves along that announcement. The signal is public. In this symmetric play, the plays and best responses to them are common knowledge. Generalizing the situation, one gets a correlated equilibrium that dominates the mixed solution. The crux of the generalization is that while in mixed-strategy solutions the way players randomize over their choices (say, the probability distribution over trembles) are independent from each other, in correlated case they may be dependent such as conditioned on a public signal. Either you choose your actions according to the odd or even number of sneezes you had yesterday or take note of what your dean had to say on Tuesday morning. On matters of vital importance, it may be your doctor who seeks to run her new placebo-controlled clinical trial that is your best bet.

6.4.4 Quantal Response

Yet another option is to assign small chances for entire strategies to be wrong, not just individual actions. This has some good and some bad news. The good news is that the future is not discounted: Players treat their future selves as independent decision makers. Derek Parfait would have been delighted to see the personal identity go away in this way in which it is at once being replaced by a theoretically pleasing method that supports both the ethics and the rationality of that move.

The bad news is that, lest there are some further restrictions, perturbations over entire strategy profiles can come as too strong: the error structures cannot be all eliminated and nothing at the end would in fact be so very wrong, yielding infallibilism and even relativism. All forms of behaviour would be rationalizable in some remote and fanciful ways. Which constraints of these strategy-laden perturbations are reasonable and would yield interesting predictions is largely to be seen ([3]).

6.4.5 Self-Confirming Equilibria

Last, take *self-confirming equilibria*: No one can by telling you that your beliefs are wrong to hope to have made you change your mind at once. But if you could induce the players to themselves observe that their mental models to be skewed (say, by having them to look at their beliefs and contrast them from the point of view of the equilibrium play), they would revise them the next time around. Assertions produce entrenchment and complacency, nudges perturbancy. Sufficiently strong nudges, both external and internal, might trigger experimentation and manipulation of the relationships that the agents have formed in their thoughts about the other players' types and their strategies, with a hope to have them observe that some of those thoughts are off-path.

This presupposes that something like a genuinely exploratory thought is a reality of agents' mentality. In order to make this transformative shift one cannot seek to satisfy propositions according to assessment of whether they conform to beliefs that one has, or whether they conform to beliefs that one expects others to have, and so on, in order to cheaply increase one's own credibility.

6.4.6 Implications

The very idea of perturbations of the strategic decision problem, such as deformations that cause generalizations, stems from the modes of reasoning that we need to ascribe to strategically interacting agents. That reasoning is by no means exhausted by deductive or inductive inferential schemas and rules. The creative element involved in realistic scenarios is to take into account surprising events. And what else could be more surprising that the players' observations that there are deviations from the equilibrium play under the common constraints of rationality, and the realization that they have to be prepared for such unexpected events? As the methods to accommodate the element of surprise and sub-normal facts involve seeking for presumptions, explanations and rational hopes that would or could explain the observed events to indeed be the matters of course, what is involved in any realistic game-playing scenario is the presence of abductive reasoning. Thus abduction shows itself as strategic reasoning from effects to causes. Tell the tale as inner locutions such as:

> "As my opponent deviated from the rational path, what accommodations would I have to make to my modalities so that the matters of course would make such deviations rational after all...".

There is another and no less important lesson concerning the implications of such perturbations: as soon as they are introduced on systems that are already unstable, even minimal additions can exhibit a tipping-point behaviour and transform decision situation into a qualitatively different one. Naturally most of the real situations are like this, including scientific experiments, replicator dynamics, evolutionary arms-races, and in general all the dynamics that we exhibit in the physically and topologically modelled vector spaces.

Thus it is also not to be expected that the normal-form representations of the game would always serve our needs. Under perturbations, extensive forms of the game may be irreducible to normal, strategic forms. This is to be kept in mind when faced with complex situations. Thus the counterfactual formalization of abductive reasoning, among its premises the conditionals as "If A were to be the general class of actions, then C would be the matter of rational play" is not something preserved well in normal forms. Extensive forms fare better on this front. The reason is that there are unreachable information sets (unreachable with any positive probability) under an equilibrium play. Introducing behavioural strategies with perturbations results in various forms of probabilistic plays, and those uncertainties make the behaviours structurally unstable.

6.4.7 Rationalizability

By 'rational' we mean aligned with the beliefs that we used to hold. The very idea of rationality may be in flux given what could happen in perturbed games. A doubt may arise and a new belief arises to fill the vacuum. Nevertheless, one of the most expansive way to stretch the concept of the equilibrium and yet remaining in the borderlands of rationality is indeed *rationalizability*: our best responses are responses to *beliefs* that we hold about our actions being good. The measure of good here is roughly the evidence that can be gained from the opponent's actions.

All of these proposals above, including rationalizability, insist players' beliefs to be the central modality. We indeed saw how reasoning loses its certainty and the rational process is flipped from deduction to effects-to-causes modes. Since abduction is insecure, mere belief systems come out as too strong, however. A conjecture is nothing but putting a hypothesis on probation. What the epistemic game theory would look like under something informationally much weaker, something that has to cope with sub-beliefs such as guesses and surmises, remains to be seen. Players' conjectures, which are weaker than beliefs, mean that players need not act upon those conjectures whenever the cognitive expenditure grows inconveniently high. But these conjectures are stronger than blind guesses or *may-be*s, so that they can yield evidential support and informational value about other players actions, about other players conjectures, and about the support gained for those actions and conjectures. These sub-belief modalities have not been taken into account or formally modelled in game theory and its interactive epistemology. Yet the matter of guesses is eminently congenial in situations calling up cost-benefit analyses. Abduction is a complexity of reasoning as it defects from the pattern of security and wishes us to err on the side of *uberty*.

7 Conclusions

Here, in summary, are the heresies that have been advocated in this paper:

Heresy 1 Save science/logic/finance/reality from epistemology; let them thrive in uncertainty.

> *Motivation*: Unknown unknowns are known to populate under-structured search spaces, with limited foresight and genuine problems genuinely ill-posed.

Some reverse engineering is needed: get approximate shapes from proximate shades. Infallibility remains an illusion.

Heresy 2 Probabilities are payoffs of the interrogative game against Nature.

Motivation: Values that govern choices are more valuable than choices that govern values. Placing a well-thought-out question before reality is expected to be reciprocated by the universe. What the value of the reciprocation is, on the other hand, not something that we subjectively would expect. Nature has a habit to surprise.

Heresy 3 Be prepared for $\text{Prob}(e)=0$ events.

Motivation: These are the unlucky ones, as they are the events never to be born. Theory does not recommend them. On all the others, we can believe whatever we want, because we do not want players to outsmart the theory.

Heresy 4 Idealizations are specifications; perturbations specify by generalizing first.

Motivation: Idealizations strive to model reality by precissive abstraction. Models that are idealized are the very antagonists of fiction, despite voices to the contrary that think models resonant with imitation and sham. The idea of idealization is to form a hypothesis simple enough to be a proposition that corresponds exactly to the facts. That we can also actually show it to stand in such a relation is a bonus. Deformations and perturbations strive to grasp reality as well, and they do so by dropping the rope just to get some more light later. Deformed objects and structures can regain the lost precision later on, and with a vengeance.

Heresy 5 The very presence of perturbations is the sign of players' reciprocal abductions on off-equilibrium plays having been set into motion.

Motivation: Players want more light, and the way to throw it is to deform the problem so as to be accessible to a range of new questions.

On the question of what the game is to be for, one is reminded of Wittgenstein: the right way to study games is to vary and play them. Answers trickle in from various perspectives, and can be collated to yield approximate, idealized and abstracted models of meaningful interaction, as well as an abductive one.

References

[1] Darwin, C. (1859). *On the Origin of Species*. London: John Murray.

[2] Frieden, B. R. (1998). *Physics from Fisher Information: A Unification.* Cambridge, Mass.: Cambridge University Press.

[3] Haile, P. A., Hortaşu, A. and Kosenok, G. (2008). On the Empirical Content of Quantal Response Equilibrium. *The American Economic Review* 98(1), 180–200.

[4] Hintikka, J. 1962. *Knowledge and Belief: An Introduction into the Logic of the Two Notions.* Ithaca, NY: Cornell University Press.

[5] Hintikka, J. (2007). *Socratic Epistemology: Explorations of Knowledge-Seeking by Questioning,* Cambridge, Mass.: Cambridge University Press.

[6] Japaridge, G. (2018). Elementary-base Cirquent Calculus I: Parallel and Choice Connectives. *Journal of Applied Logics – IfCoLoG Journal of Logics and their Applications* 5(1), 367–388.

[7] Leibniz, G. W. (1981). *New Essays on Human Understanding.* Translated and edidetd by P. Remnant and J. Bennett. Cambridge: Cambridge University Press.

[8] Luce, R. D. and Raiffa, H. (1957). *Games and Decisions.* New York.

[9] Lupia, A., Levine, A. S. and Zharinova, N. (2010). When Should Political Scientists Use the Self-Confirming Equilibrium Concept? Benefits, Costs, and an Application to Jury Theorems. *Political Analysis* 18, 103–123. doi:10.1093/pan/mpp026

[10] Mabsout, R. (2015). Abduction and Economics: The Contributions of Charles Peirce and Herbert Simon. *Journal of Economic Methodology* DOI: 10.1080/1350178X.2015.1024876

[11] Magnani, L. (2009). *Abductive Cognition: The Epistemological and Eco-cognitive Dimensions of Hypothetical Reasoning.* Berlin: Springer.

[12] Ma, M. and Pietarinen, A.-V. (2018). Let Us Investigate! Dynamic Conjecture-Making as the Formal Logic of Abduction. *Journal of Philosophical Logic.* https://doi.org/10.1007/s10992-017-9454-x

[13] Maynard Smith, J. (1979). Hypercycles and the Origin of Life. *Nature* 280, 445–446.

[14] Niiniluoto, I. (2011). Abduction, Tomography, and Other Inverse Problems. *Studies In History and Philosophy of Science Part A* 42(1), 135–139.

[15] Peirce, C. S. 1967. *Manuscripts in the Houghton Library of Harvard University.* Identified by Richard Robin, *Annotated Catalogue of the Papers of Charles S. Peirce,* Amherst: University of Massachusetts Press, 1967, and The Peirce Papers: A supplementary catalogue, *Transactions of the C. S. Peirce Society* 7 (1971): 37—57. Cited as MS followed by manuscript number.

[16] Pelosse, Y. (2016). The Intrinsic Quantum Nature of Nash Equilibrium Mixtures. *Journal of Philosophical Logic.*

[17] Pelosse, Y. (2018). The Intrinsic Quantum Nature of Classical Game Theory. In Haven, E. and Khrennikov, A. (eds.). *The Palgrave Handbook of Quantum Models in Social Science: Applications and Grand Challenges.* London: Palgrave, 59–74.

[18] Pietarinen, A.-V. (2002). Quantum Logic and Quantum Theory in a Game-Theoretic Perspective. *Open Systems & Information Dynamics* 9(3), 273–290.

[19] Pietarinen, A.-V. (2003a). Games as Formal Tools vs. Games as Explanations in Logic

and Science. *Foundations of Science* 8, 317–364.

[20] Pietarinen, A.-V. (2003b). What Do Epistemic Logic and Cognitive Science Have to Do with Each Other? *Cognitive systems research* 4(3), 169–190.

[21] Pietarinen, A.-V. (2005). Cultivating Habits of Reason: Peirce and the *Logica Utens* versus *Logica Docens* Distinction. *History of Philosophy Quarterly* 22(4), 357–372.

[22] Pietarinen, A.-V. and Bellucci, F. (2014). New Light on Peirce's Conceptions of Retroduction, Deduction, and Scientific Reasoning. *International Studies in the Philosophy of Science* 28(4), 353–373. DOI 10.1080/02698595.2014.979667

[23] Stalnaker, R. (1996). Knowledge, Belief and Counterfactual Reasoning in Games. *Economics and Philosophy* 12, 133–163.

[24] Tohmé, F. and Crespo, R. (2013). Abduction in Economics: A Conceptual Framework and Its Model. *Synthese* DOI 10.1007/s11229-013-0268-2

[25] Tromp, J. (2016). Number of Legal Go Positions. https://tromp.github.io/go/legal.html

[26] Weiss, Madeline C., Sousa, Filipa L., Mrnjavac, Natalia, Neukirchen, Sinje, Roettger, Mayo, Nelson-Sathi, Shijulal and Martin, William F. (2016). The Physiology and Habitat of the Last Universal Common Ancestor, *Nature Microbiology* 1: 16116. doi:10.1038/nmicrobiol.2016.116

[27] Wittgenstein, L. (2000). *Wittgenstein's Nachlass, The Bergen Electronic Edition*. The Wittgenstein Trustees, The University of Bergen, Oxford University Press.

Enthymematic Interaction in Baduk

Woosuk Park
Korea Advanced Institute of Science and Technology, Korea
e_wspark@kaist.ac.kr

Abstract

In this paper, I propose to view each move in a game of Baduk (Go, Weiqi) as presenting an enthymematic argument. It is largely inspired by Paglieri and Woods, who suggested parsimony rather than charity as the driving force of enthymematic argumentation. Since their theory is not the final word in the history of enthymeme, my interpretation of Baduk as enthymematic interaction in terms of their fine distinctions may shed light not only on strategic reasoning in Baduk but also on the study of enthymeme itself.

1 Introduction

In this paper, I propose to view each move in a game of Baduk (Go, Weiqi) as presenting an enthymematic argument. It is largely inspired by Paglieri and Woods 2011, in which they suggested parsimony rather than charity as the driving force of enthymematic argumentation. Since their theory is not the final word in the history of enthymeme, my interpretation of Baduk as enthymematic interaction in terms of their fine distinctions may shed light not only on strategic reasoning in Baduk but also on the study of enthymeme itself.

1.1 What Is an Enthymeme?[1]

What is an Enthymeme? And how could I think that enthymemes are everywhere in Baduk? Why is it so obvious to me that Baduk players are constantly presenting

---

[1] Inspired by Paglieri and Woods [13], Park examined the possibility of interpreting as hic hypothesis generation as enthymeme resolution. For basic exposition of their theory of enthymeme, I heavily draw from it. Please notice that both the problem of ad hoc hypothesis and the problem of enthymemes may be understood as sub-species of the problem of anomaly resolution (see [6, 7, 8, 5, 1, 23, 22]).

enthymemes in playing games? In order for my interpretation of Baduk as essentially consisting of enthymematic interaction to get off the ground, some preliminary discussion is needed.

Textbooks of elementary logic usually provide us a short section on enthymeme. For example, Copi deals with it in a chapter entitled "Arguments in ordinary Language", and in a section preceding another section for sorites. According to him, an enthymeme is

> an argument that is stated incompletely, part being "understood" or only "in the mind" (Copi [2, p. 253]).

After pointing out that enthymemes are everywhere in everyday discourse as well as in scientific practice, he turns to the problem of how to test their validity:

> In testing an enthymeme for validity, two steps are involved. The first is to supply the missing parts of the argument; the second is to test the resulting syllogism (ibid., p. 255).

Given this basic understanding of enthymemes, we may understand what a sorites is. According to Copi,

> when "an argument is expressed enthymematically, with only the premises and the final conclusion stated, it is called a sorites". (Ibid., p. 259)

When we are facing an argument, which has more than two premises, we cannot test its validity by usual method for categorical syllogisms. But we may reconstruct it as a chain of categorical syllogisms, some of which are enthymemes. If so, it seems natural to accept the following:

> An argument of this type is valid if, and only if, all its constituent syllogisms are valid. (Ibid, p. p. 258)

1.2 Baduk as Enthymematic Interaction: Basic Idea

The frequent use of enthymemes in our ordinary life seems no news to anyone. Let us think about driving a car on a highway over 100 miles per hour on a rainy night. I firmly believe and openly claim that such an activity is simply crazy. It would be, however, an arduous task for me to prove that such a belief is a justified true belief, for I am ignorant of college physics, not to mention mechanical engineering or hydrodynamics. Likewise, in playing a game of Baduk, we need to present and examine a huge number of enthymemes. Yes, while there are only a few moves played on the board, we have to deal with so many different possible variations. Just as in our dialogues or conversations in ordinary life, the interaction between the two players of a Baduk game must include manipulation, threatening, persuasion, seduction, cheating, information sharing, mocking, compromising, and communication. Further, just as in our ordinary life, it is possible for us to do all these interactions in Baduk simply because we share extremely many things in common, which include common sense, knowledge and experience. Since we do not interact in playing Baduk by using any natural language, as we do in our ordinary life, it seems rather remarkable that we can have very effective enthymematic interaction in Baduk.

Let me use the famous challenge match between Lee Sedol and AlphaGo as an example to elaborate my enthymematic interaction interpretation of Baduk.

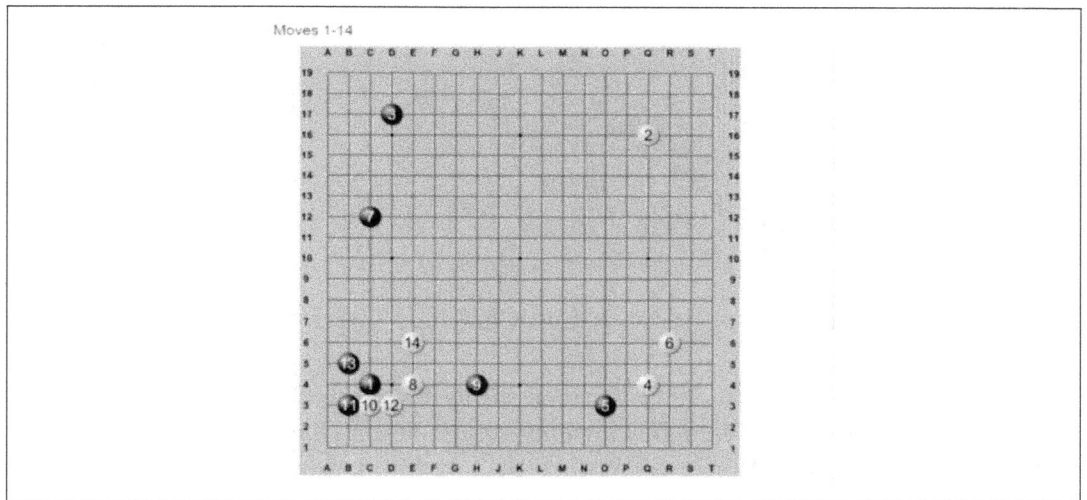

Figure 1: Google DeepMind Challenge Match Game 1

According to Fan Hui's commentary, at the moment of White 14, "AlphaGo already felt that the game had begun to favor White". The feeling of the players in the commentary room, who were skeptical about AlphaGo's moves at the left bottom corner on the ground that "that this way of playing is very heavy for White, not only strengthening Black's corner, but turning 9 into the perfect attacking move", also suddenly changed. As Fan Hui aptly points out, "[t]he black stones on the left were too low, and Black's formation was nothing to be proud of after all!"

Now, in terms of my enthymematic interpretation, we may understand the situation in the following way. At the time of playing W10, AlphaGo was presenting an enthymeme consisting of all previous moves, and the entire boad situation as the conclusion. And, we may interpret this enthymeme as virtually claiming that "No matter how you respond to W10, the game is already favorable to White". In other words, no matter how Black responds to W 10, the future sequences of moves will turn my enthymeme into a better and better argument rather than refuting it. Possible future moves are not yet played, and in that sense missing right now. The next sequence of B11, W12, B13, and W14 seem to confirm AhphaGo's prediction as a correct one.

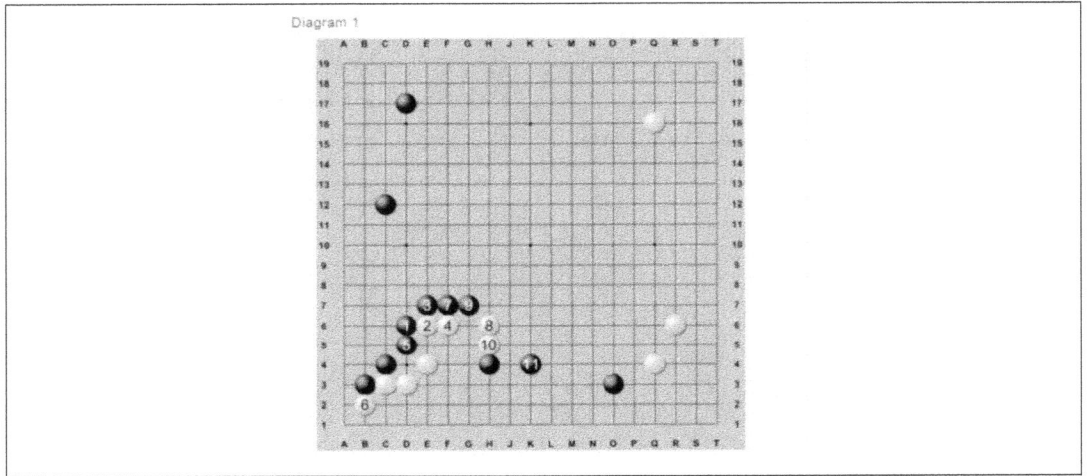

Figure 2: Google DeepMind Challenge Match Game 1

According to AlphaGo's analysis, reported by Fan Hui, "the problem lay in the tiger's mouth at Black 13, and Black should have chosen Diagram 1: If Black plays the knight's move at 1, White will attach at 2 to avoid being enclosed, and Black can continue to press with 3, 7, and 9. Through 11, Black's formation on the left is clearly better than the game". In other words, according to AlphGo's opinion, Lee Sedol's enthymeme at the move 13 was poor, let alone valid.

Now, let us return to logic in order to sharpen our conceptions of enthymeme, thereby fastening our analogy between logic and Baduk later.

1.3 Definition of Enthymeme

Copi's brief characterization of enthymeme seems to represent what Paglieri and Woods call "the modern conception of enthymeme", which is "something of a hybrid of elements drawn from" Epictetus' and Aristotle's definitions (Paglieri and Woods [13, 463, 465–67]). One essential feature of the modern conception is reported by Paglieri and Woods as follows:

According to the modern idea, there is some property Q that it is desirable for arguments to have, such that an enthymeme lacks it yet its completion has it. Consider validity as an example. Validity is what the enthymeme lacks and its completion has. An enthymeme of this sort is a valid argument-in-waiting (ibid., 463).

Paglieri and Woods count *incompleteness* "as a key feature of enthymeme worth preserving in the modern conception":

We agree with the dominant view that something is missing and yet understood in enthymemes; moreover, that something is essential to their interpretation (ibid., 468).

However, there are many features of the modern conception they found somewhat troublesome. For example, they want to avoid a possible confusion between "interpreting an enthymeme" and "assessing its value". Also, they believe that "something is essential" in interpreting enthymemes does not mean "that, by understanding the missing element, the hearer will be inclined to accept the argument as good" but merely "that the missing element is crucial to make an informed assessment of the enthymeme, be it positive or negative" (ibid.). There are, of course, many other problems and issues in defining an enthymeme, as there are many different varieties of enthymemes. Paglieri and Woods enumerate at least the following different varieties:

(a) Valid and sound elliptic argument: "Socrates is a man, therefore he is mortal".

(b) Purely formal elliptic argument: "All P are Q, so some R are Q".

(c) Unsound elliptic argument: "The mackerel is a fish, so it is colour-blind".

(d) Crazy elliptic argument: "Today I am happy, therefore Mars is not a planet".

(e) Invalid elliptic argument: "Every Catholic priest is male, so John is a Catholic priest".

(f) Defeasible elliptic argument: "Ozzie is an ocelot, therefore Ozzie is four-legged".

(g) Materially valid argument: "The shirt is red, therefore the shirt is coloured".

(h) Complete argument, either valid or invalid: "Socrates is a man and all men are mortal, therefore Socrates is mortal", and "Lassie is mortal and every man is mortal, therefore Lassie is a man".

(i) Isolated statement: "Socrates is mortal" (Paglieri and Woods [13, p. 469–72]).

Though it is apparently trivial, one interesting idea is that any invalid argument might become a valid one by adding some appropriate premises. We may consider any arbitrarily selected invalid argument, e.g., P/∴S. As Paglieri and Woods point out, by adding to the premise-set a proposition, i.e., the conditional proposition with P as antecedent and S as consequent, we can produce a valid argument. No one would deny that an enthymeme cannot be "just any invalid argument validated by addition of its corresponding conditional as premiss". But how are we to distinguish between valid and invalid enthymemes? What Paglieri and Woods call the *demarcation problem*, which is a central task for a theory of enthymemes, is nothing but "to preserve the distinction, and to bring it to a decent level of theoretical articulation" (ibid., 467).

Paglieri and Woods suggest the following as the definition of enthymeme:

> ENTHYMEME: A is an enthymeme if and only if A contains at least one explicit premiss explicitly linked to an explicit conclusion, and yet A can be assessed according to some standard of argument evaluation if and only if A is first supplemented with some additional premiss P that preserves the relevance of all A's premisses to A + P's conclusion and is selected by applying a general reconstructive principle to A (ibid., 468).

According to them, they included five main conditions in this definition:

(i) A is an argument in a minimal sense, i.e. it contains at least one (explicit stated) premiss that carries the presumption of supporting at least one (explicitly stated) conclusion.

(ii) A is not assessable on some standard of argument evaluation.

(iii) A + P is assessable on some standard of argument evaluation.

(iv) Adding P to A does not make irrelevant any of A's premisses to A + P's conclusion.

(v) The transition from A to A + P is not arbitrary, but rather governed by some general principle (yet to be determined; see Sects. 2 and 3) (ibid.).

We can appreciate how judiciously Paglieri and Woods selected the conditions for an enthymeme to satisfy in this definition. As intended by them, we may focus on the problem of "determining the appropriate criterion that should guide the reconstruction of enthymemes".

1.4 Enthymemes in Baduk

In order to make my analogy between logic and Baduk at this stage, two projects seem most pertinent and promising. First of all, we may indicate that all those different varieties of enthymemes enumerated by Paglieri and Woods can be found by citing enough number of examples from Baduk. Secondly, we may demonstrate that what I claim to be enthymematic interactions in Baduk indeed fit the definition of enthymeme suggested by Paglieri and Woods. Even though the first project does have its intrinsic value, I will not discuss it in this paper. For, it seems that the second one is much more urgent for my present purpose.

Let us take a look at an elementary life and death problem in Baduk.

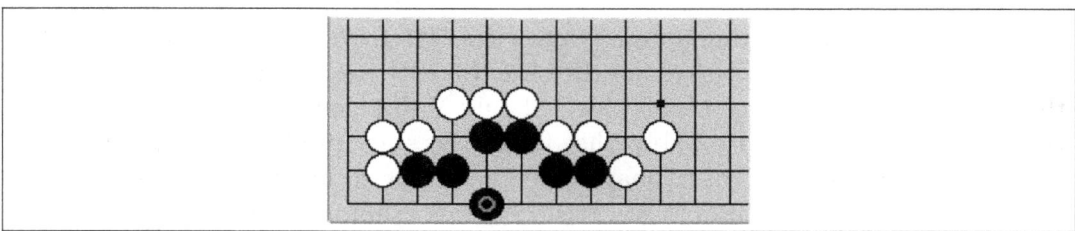

Figure 3: LD1 (http://senseis.xmp.net/?BeginnerExercise50)

In playing the move marked by red circle (bottom line), what enthymeme is being presented? The explicit premises are all there: i.e., they would amount to A. Though there is no explicit conclusion, the context indicates that that would amount to "No matter what the next move of White is, this group of Black is safe". But this enthymeme is not valid in the sense that no matter what P is added White has a way to kill this group of Black. Once White 1 in LD2 is played, there is no possible way of saving the Black group. In other words, the enthymeme presented by White at W1, i.e., "No matter what P is added, there is no way for Black to save this group" is a valid one.

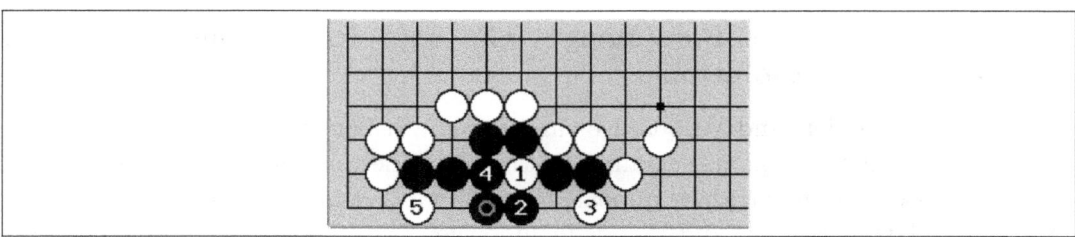

Figure 4: LD2

Except for condition (1), which requires at least one (explicitly stated) conclusion, I think, all conditions for Paglieri and Woods' definition of enthymeme can be satisfied rather easily. Now, the most important issue at this stage must be "how to evaluate the enthymemes in LD1 and LD2?": By what principles, criteria, or algorithms, can we evaluate them? In considering this problem, one focal issue must involve the question as to whether we should fathom our opponent's intentions. To make sense of this issue, let me use a variation of LD1. At the moment of presenting the enthymeme by LD1, Black might have the possible sequence of White 1 and Black 2 in LD3 as P. In other words, Black was anticipating W1 in LD3 as White's response to its move marked red in LD1 and planned to meet it by B2 in LDS. And, indeed A+P is valid, if P is <W1, B2> in LD3. But the problem is that White's choice at LD1 may not be W1 in LD3 but W1 in LD2. That means, Black was in serious trouble in presenting an enthymeme in LD1 by failing to consider another enthymeme, i.e., White's enthymeme in LD2. LD3 betrays typical wishful thinking, thereby presents us a very useful clue. Should White consider LD3 in evaluating the enthymeme in LD1? Probably, that is not necessary insofar as White is able to present LD2. In other words, it may not be necessary to understand the intentions of our opponent all the times. I will return to this example later.

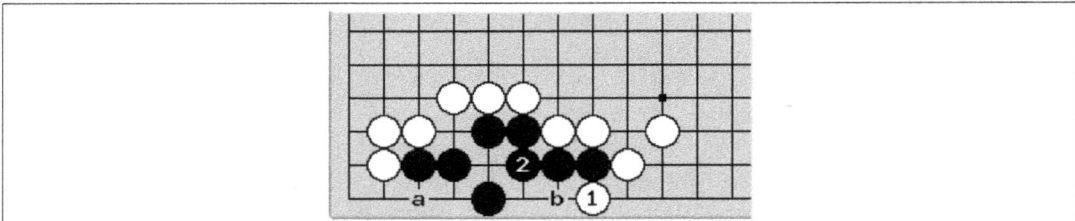

Figure 5: LD3

2 The Problem of Enthymeme Resolution

2.1 Charity Versus Parsimony: The Two Rival Theories of Enthymeme Resolution

According to Paglieri and Woods, we can distinguish between approaches that are committed to the completion-as-amelioration doctrine and those that aren't. Approaches based on unrestrained appeals to charity are clear examples of the former. Among the latter, Paglieri and Woods include their own theory based on parsimony and other analyses of enthymemes that are not committed to the completion-as-amelioration thesis, and yet differ also from their own proposal in terms of parsimony, such as *contextualism* and *anti-reconstructionism*. However, it is obviously beyond the scope of this paper to do justice to each and every one of the theories identified by Paglieri and Woods. For my present purpose, it would be enough to focus on approaches based on unrestrained appeals to charity as the foremost example of completion-as-amelioration doctrine and Paglieri and Woods' theory based on parsimony. Further, insofar as the former is the mainstream dominant theory of enthymeme (see ibid., 496) and the idea of counting enthymeme as a sort of valid argument-in-waiting is behind the notion of charity (see ibid., 463), what is needed is understanding why and on what ground Paglieri and Woods criticize the theory based on charity and present their own theory based on parsimony as an alternative.

2.1.1 Against Charity

Paglieri and Woods devote an entire section for criticizing the theory based on charity (ibid., Sect. 2 "Against Charity", 473-476). However, the more serious and important criticisms were made even before turning to the section. For they argue that the following widely accepted idea cannot be true:

> GOOD COMPLETIONS AS VALIDATING: A good enthymeme-completion selection will be a premise that validates it (ibid., 472).

They believe that the existence of cases (e) above is enough to conclude that way: (e) Invalid elliptic argument: "Every Catholic priest is male, so John is a Catholic priest". Good completion of this elliptic argument would need to supplementing it with "John is male". But that does not make the argument any more valid. The only possibility left for the defenders of modern conception of enthymemes that Paglieri and Woods can figure out is as follows:

> In contrast, the modern conception of enthymemes would force us to either exclude these cases from the definition of enthymemes (because their completion is invalid, when reconstructed in the most plausible way), or reconstruct them in ways that ensure validity by adding bizarre and false premises (in this example, the conditional "If every Catholic priest is male, then John is a Catholic priest") (ibid., 471).

As a consequence, Paglieri and Woods claim that the doctrine of completion-as-validation conflates two quite different tasks:

> One is the task of spotting a premiss that completes an incomplete argument. The other is the task of finding a premiss that will validate it (ibid., 472).

Encouraged by the possible liberating effect of separating the two tasks, Paglieri and Woods even try to generalize:

> GOOD COMPLETIONS/BAD ARGUMENTS: It cannot be typical of enthymemes that they be properly complete only by premises that make them good arguments, in whatever sense of good fits the particular case. In brief, completion is not amelioration (ibid., 473).

As Paglieri and Woods point out, there is still "the problem of elucidating what guides selection of the appropriate premiss to supplement the enthymeme with". Again, as they do not forget to mention, that is nothing but fleshing out the condition (v) of their definition of enthymemes.

2.1.2 For Parsimony

Paglieri and Woods counts "the pragmatic question of why enthymemes are so frequent in human communication" as another fundamental question that cannot be answered by invoking charity:

> Even assuming charity in interpreting each other's arguments, why do we so often indulge in enthymematic argumentation, and why are we so favourably disposed towards incomplete arguments? If charity were the rule, why would it be so? And why do we systematically fail to speak or write our arguments in a more complete and explicit fashion, relying instead on the charity of the audience? (ibid., 477)

In order to answer these questions, Paglieri and Woods start with Herbert Simon's notion of bounded rationality (see Simon [15, 16]). They explain that, from the vast literature on bounded rationality, thy draw upon on a single aspect: "how the fact that agents are cognitively resource-bounded affects their dialectical capacities", and "how this same fact should inform a good theory of argumentative rationality" (ibid., 477). As they count resource-boundedness as a simple fact of life, they find the need to be parsimonious in using the finite and scant resources as inherent to the rationality of all our actions (ibid.).

So, Paglieri and Woods' answer to the questions raised above is utterly simple:

> In a nutshell, our hypothesis is that both the frequent use of enthymemes and the principles governing their reconstruction are ultimately motivated by an attempt, by the arguer as well as by the interpreter, to save valuable cognitive resources, without injury to the performance of their respective tasks. Accordingly PARSIMONY: It is parsimony, rather than charity, that inspires our enthymematic inclinations (ibid., 477-478).

2.2 Parallel Problems in Baduk

One important reason why I try to interpret Baduk in terms of enthymeme is that Baduk may provide us with some crucial tests for the rival theories of enthymeme. There must be grain of truth in each of these rival theories. If so, we may also highlight what exactly are the strengths and weaknesses of these theories by using Baduk. Of course, it would need a huge data in order to arrive at a convincing and reasonable conclusion. My aim here is merely indicate that in our necessary comparative task Baduk may play an important role by providing us somewhat visible illustrations.

In fact, it is not crystal clear in what ways the two theories of enthymeme, i.e., charity theory and parsimony theory, are entirely different, and for what reason such difference, if any, are of so much interest and significance. For that reason, again, rehearsing the issue in the context of Baduk could be meaningfully useful. So, what would it mean to take charity theory or parsimony theory in playing Baduk as a game of enthymematic interaction?

Let us start with a situation in Baduk. There may be some differences in degree depending on the different phases and different developments of the game. Nevertheless, in considering the best possible move, for any particular move we need to fathom our opponent's intention, strategy, and assessment of the current overall situation. In other words, what is at stake is nothing but whether we should take charity or parsimony as the guiding principle in determining the validity, inductive strength, or fruitfulness of opponent's enthymematic reasoning. And, as every Baduk player knows, it is not so easy to settle this issue. To what extent, for example, should we reconstruct our opponent's enthymeme sympathetically? If we find a way of resolving the enthymeme with relative ease, according to which it turns out to be valid, we should avoid the sequence of moves our opponent expects. However, it is not always the case that our opponent's enthymeme is valid. Nor is it always inductively strong or abductively promising. If so, why should we be sympathetic to our opponent in such a way that we would be willing to spend time and energy to figure out a resolution of the enthymeme? Furthermore, it is one thing that the enthymeme is valid, quite another that such a way of resolving the enthymeme was intended by our opponent.

I do not believe that it is easy to decide between charity theory and parsimony theory. It is especially so in dealing with enthymemes in Baduk. There are indubitably attractive features in both. But, as it must have been apparent to the perceptive readers, I definitely incline to parsimony theory. So, my problem is how to argue for parsimony theory without being blind to the merits of its rival theory. The most salient strength of charity theory seems to lie in the fact that, by definition, it tries to pay due attention to the intention of the arguer. My strategy to solve my problem is simple. I would suggest that, since it is all important to understand the intention of our opponent (in argumentation or in playing Baduk), a fortiori we should opt for parsimony theory. But, before fleshing out such an argument, let us appreciate a bit further how important it is to understand our opponent's intention.

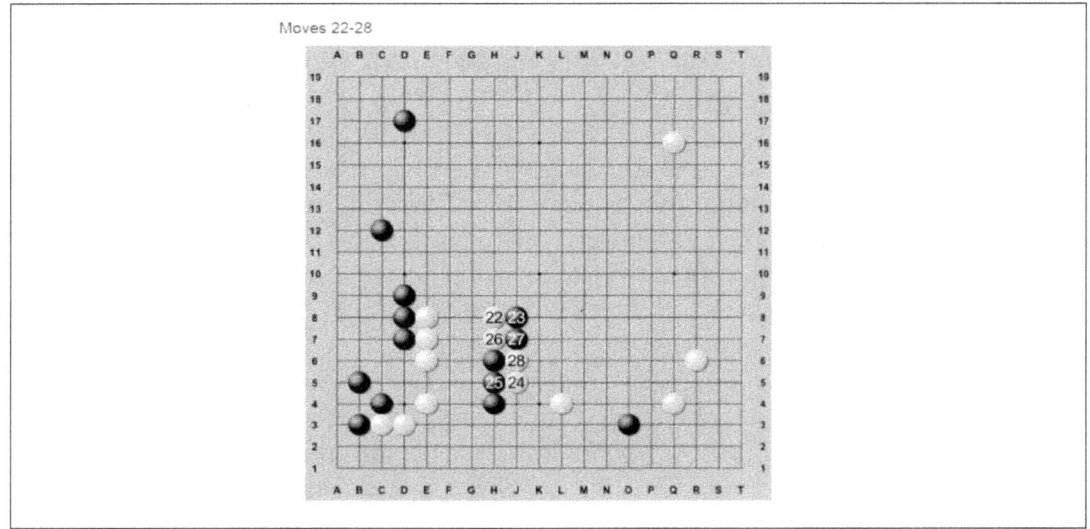

Figure 6: Google DeepMind Challenge Match Game 1

Let us use again the first game between AlphaGo and Lee Sedol. In his commentary on the moves 22-28 and moves 28-38, Fan Hui is particularly sensitive to the intentions of the players. For example, as for the moves 22-28, he writes:

> For most players, the decision is a combination of personal style and a feeling for shape. At the same time, it is important to consider the opponent's intentions. For instance, did White pause before playing 26? Did White look pleased or dissatisfied with the previous position? Normally, reading the opponent is one of Lee's strong points. Thanks to his exquisite perception, he can pick whichever way of playing gives him the greatest advantage. This time, however, he hesitated. Since the start of the game, AlphaGo had played at a constant rhythm, with no sign of doubt or confusion. It felt as if every move were inevitable. This was the first time in the match that Lee found himself at a crossroads, and he had no intention of backing down before a machine. So, after three minutes of thought, Lee resolutely blocked at 27, and White cut at 28. At this point, AlphaGo's win rate rose to 56%. (Emphasis is mine) [3, Game 1].

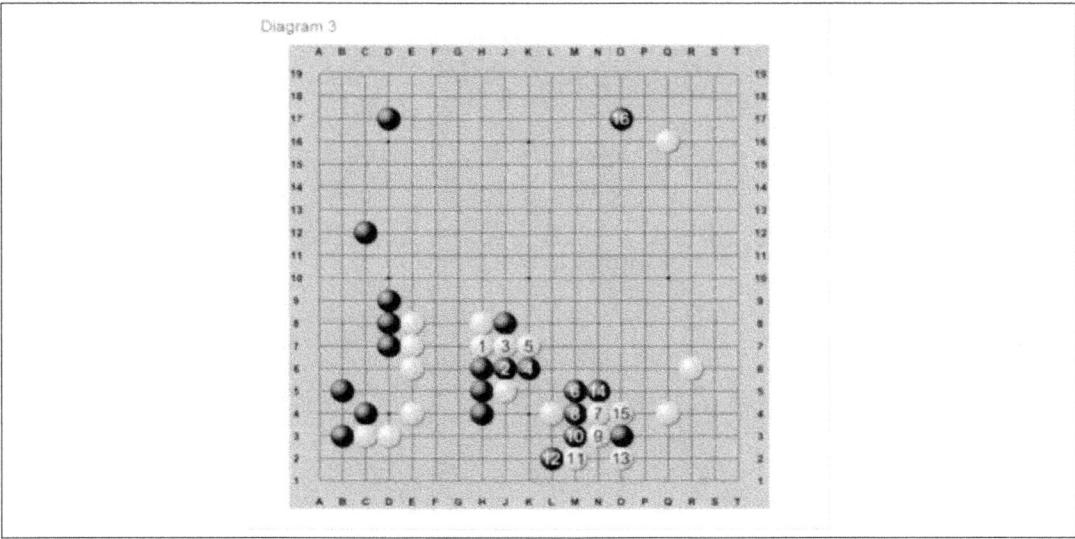

Figure 7: Google DeepMind Challenge Match Game 1

In AlphaGo's judgment, reported by Fan Hui, "Black should have followed diagram 3 and turned at 28 after all". With this post mortem assessment in mind, let us try to fathom what message Faun Hui wanted to convey to the readers in the paragraph quoted at length. Apparently, he wanted to point out that See Sedol had extreme difficulties in identifying AlphaGo's intentions. Also, he wanted to remind us of typical tendencies or habits of strong players of Baduk: (1) trying to resist against opponent's intentions, and (2) trying to be persistent in one's own intentions. Indeed, these typical tendencies or habits of stronger players of Baduk can be adopted by all Baduk players as excellent strategies or tactics. If so, what was wrong in Lee Sedol's part? The key for understanding this might be found in Fan Hui's commentary on the subsequent development of the game.

Again, Fan Hui refers to the intention of the players in commenting on moves 28-38:

> Black peeped at 29 and White connected at 30. At this point, Lee Sedol had 1 hour and 43 minutes remaining, AlphaGo 1 hour and 44 minutes. Black 31 continued to probe White's intentions. Lee was asking AlphaGo, "Dare you block?" Without hesitation, AlphaGo did. The curtain rose on the first major battle, and the moves to 38 were a one way street. However, AlphaGo only grew more and more confident, its win rate reaching 58%.

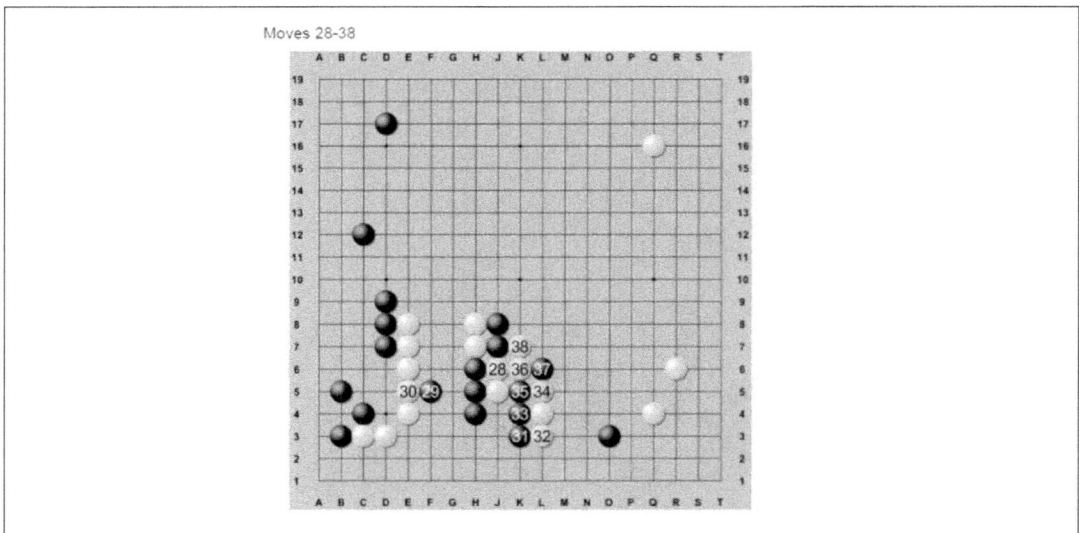

Figure 8: Google DeepMind Challenge Match Game 1

What is intriguing here is that Fan Hui (or AlphaGo) is psychoanalyzing Lee Sedol in terms of folk psychology of Baduk players. According to Fan Hui's report, AlphaGo suggests diagram 4 as a better alternative variation for Lee Sedol.

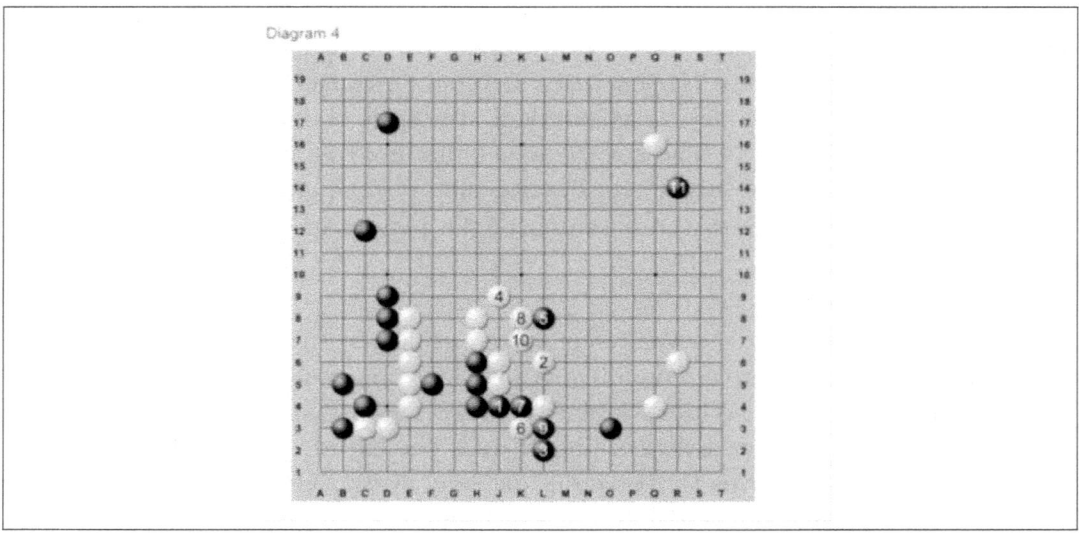

Figure 9: Google DeepMind Challenge Match Game 1

Interestingly, when Fan Hui showed diagram 4, they exclaimed "If you show this to Lee Sedol, he'll think you're joking!" What is so shocking about this variation in such a way that Lee Sedol could not have ever thought about it? Fan Hui continued to pin down the psychological blind spot in the idea of "sacrificing two critical stones in the center". It is by now clear enough that Lee Sedol and AlphaGo had entirely different opinions about the value of the two critical stones: Compared to AlphaGo, Lee Sedol overestimated its value. As we noted above, in evaluating enthymemes in Baduk, being clear about intentions of the players are all important as a useful heuristic device. But not the other way around: Evaluating an enthymeme is not for understanding the opponent's intention. In passing, we may also note that by avoiding the variation in diagram 4 Lee Sedol was revealing a secret to AlphaGo, i.e., that he overestimates the value of the two crucial stones at issue. This is important, because in the later development of the game AlphaGo seemed to exploit this secret information in somewhat fancy ways.

Have we made any progress in solving the problem of "how to argue for parsimony theory without being blind to the merits of its rival theory"? Without ignoring the most salient strength of charity theory, i.e., its emphasis on the intention of the arguer, how are we to show the superiority of parsimony theory over charity theory? Can I execute my strategy of arguing that "since it is all important to understand the intention of our opponent (in argumentation or in playing Baduk), a fortiori we should opt for parsimony theory"? The answer seems to be "Yes", for the intentions of Lee Sedol and AlphaGo exposed by Fan Hui's commentary may be explained in terms of the lessons from Paglieri and Woods' comparison of charity theory and parsimony theory. As was mentioned above, for example, Paglieri and Woods criticizes the doctrine of completion-as-validation, presupposed by charity theory, as conflating the task of spotting a premiss that completes an incomplete argument with the task of finding a premiss that will validate it [13, p. 472]. Thanks to this distinction, we may contrast AlphaGo's success and Lee Sedol's failure in assessing the opponent's intention and the resulting enthymemes they present: Unlike AlphaGo, who never conflates the two different task, Lee Sedol seems to fluctuate between the two task. In other words, AlphaGo seems to reject the doctrine of completion-as-validation once and for all, thereby subscribing to parsimony theory rather than charity theory. On the other hand, despite his instinctive preference for parsimony theory, Lee Sedol has not completely freed himself from the doctrine of completion-as-validation.

In order to hint at what I have in mind in my purely speculative comparison of Lee Sedol and AlphaGo in terms of their strategies in handling opponent's enthymemes, let me use a scene in the second game between them. Black 37 was one of the most remarkable moves AlphaGo played in this game.

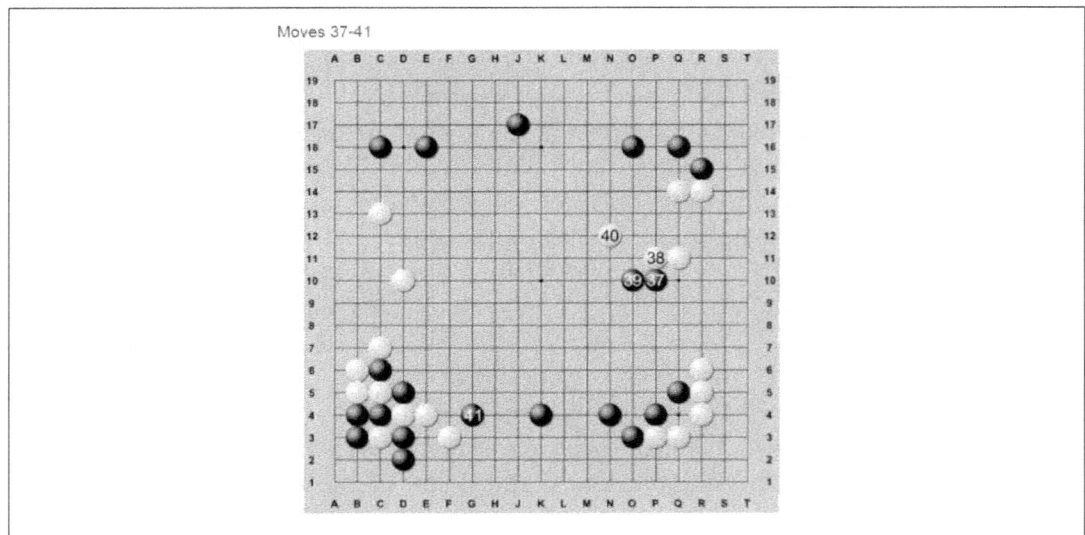

Figure 10: Google DeepMind Challenge Match Game 2

Fan Hui's commentary on this scene is revealing in many respects:

> After twelve minutes of thought, Lee finally pushed up at 38. Perhaps he was feeling the pressure, as the direction of this move is clearly problematic: White is helping Black build up the bottom. See diagram 7. Even after 39, AlphaGo thought the knight's move at 40 was inappropriate, and recommended diagram 8 instead. It seems that 37 not only helped Black on the board, but also threw Lee off balance psychologically. At this point, AlphaGo's win rate reached 55%. Black 41 further restricted White's potential on the left, while enlarging Black's in the center.

It is interesting that Fan Hui appeals to psychological explanation such as "Perhaps he was feeling the pressure, as the direction of this move is clearly problematic" or "37 not only helped Black on the board, but also threw Lee off balance psychologically". When Black played 37, White has just two options: W38 or W2 in diagram 7. Since diagram 7 seems to be better than Moves 37-41, as testified most professional Baduk players, the only way for Fan Hui to explain Lee Sedol's unfortunate choice is to blame Lee's psychology. We might rephrase this psychological explanation in terms of intention: In refusing to take diagram 7, Lee Sedol was too much obsessed by the dictum that we should resist our opponent's intention. Except for the fact that by playing W38 Lee Sedol was refusing to acquiesce to AlphaGo's request, i.e., diagram 7, however, Move 37-41is not particularly superior to diagram in nullifying

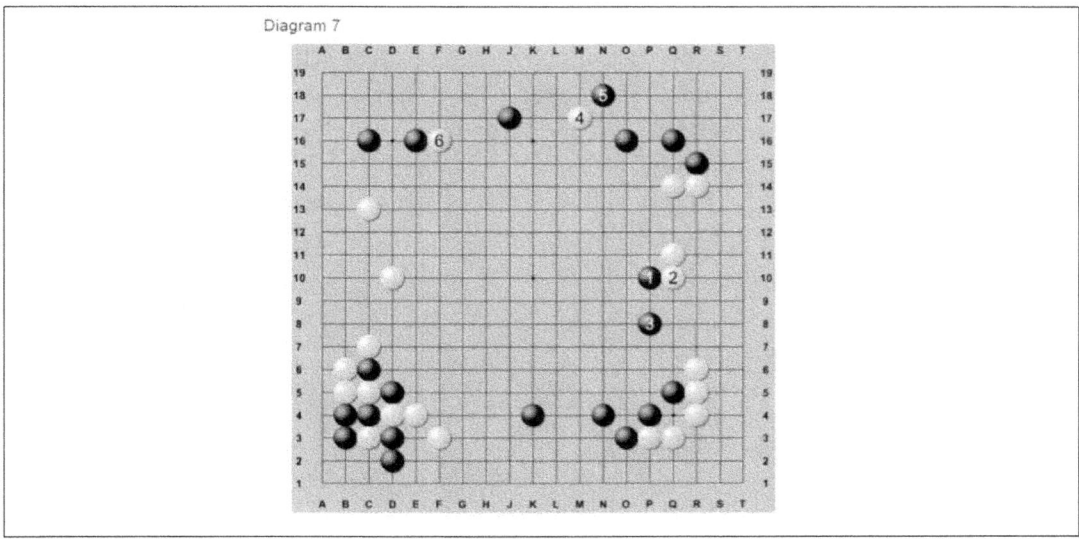

Figure 11: Google DeepMind Challenge Match Game 2

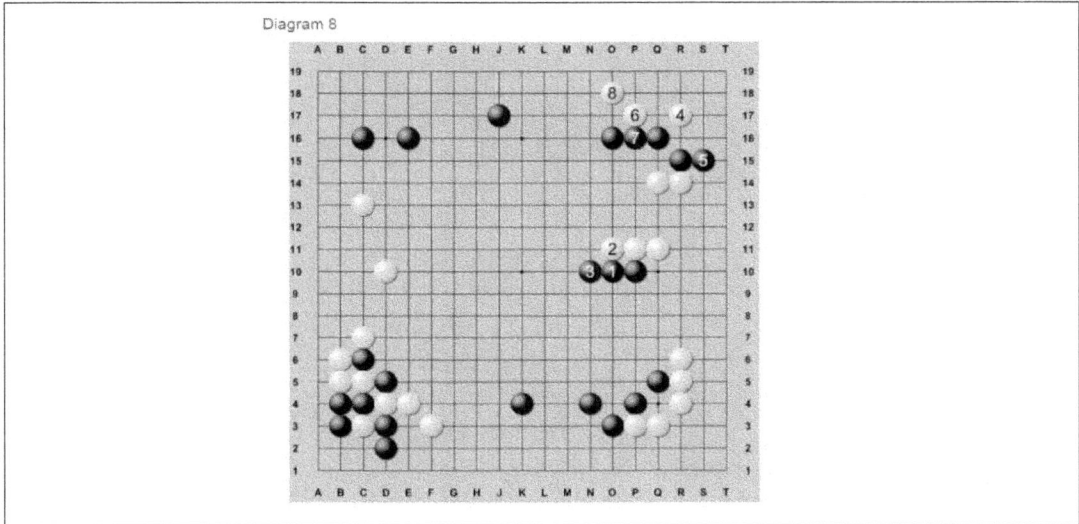

Figure 12: Google DeepMind Challenge Match Game 2

Black's intention, i.e., strengthening Black's influence in the center. The lesson here seems to be this. When our opponent's intention is good, and the resulting enthymeme presented is good enough, we should be extremely prudent. Paglieri and Wood's principle of balance seems pertinent at this stage:

BALANCE: In communication in general, and enthymematic argumentation in particular, it is necessary to strike a delicate balance between the cognitive resources that we use to interpret each other's messages, and the informational resources that we extract from them. (483)

Adopting the sequence of moves 37-41 rather than diagram 7, Lee Sedol seems to have wasted cognitive resources too much.

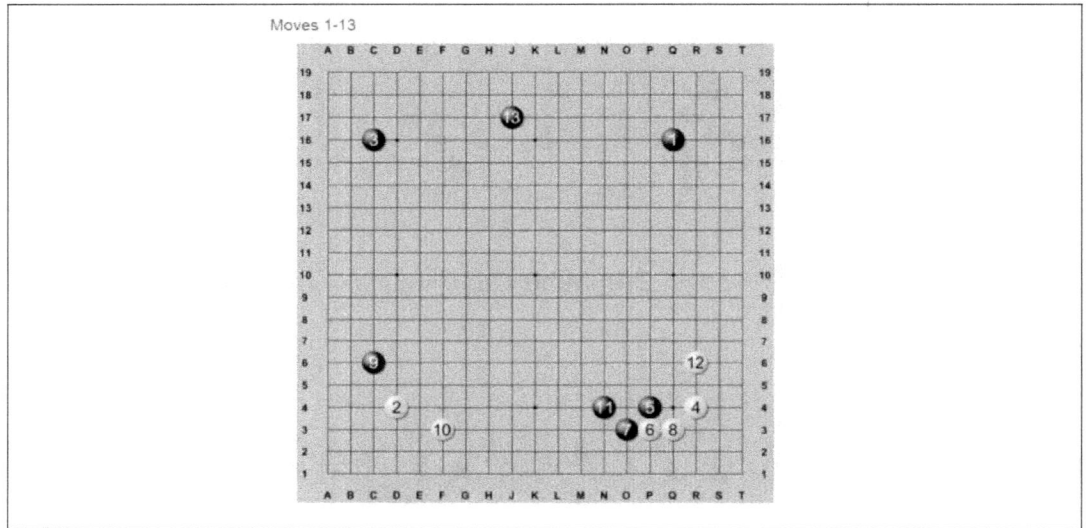

Figure 13: Google DeepMind Challenge Match Game 2

On the other hand, it is possible to interpret AlphaGo's moves in terms of parsimony theory at its best. Black 13 in Game 2 must have been shocking to Lee Sedol, for it ignores the usual pattern (joseki) on the right bottom corner. AlphaGo's opinion, later reported by Fan Hui, seems even more shocking, for even White 12 is severely criticized. Diagram 1 is what AlphaGo suggests as a better alternative.

That means, at the moment of playing Black 11, thereby presenting an enthymeme, AlphaGo already evaluated the possible consequences of both Moves 1-13 and Diagram 1. Whether it is correct or not, it is simply amazing that AlphaGo can not only figure out some such possible sequences of moves and but also evaluate their relative appropriateness. It is a well-known fact that computer Baduk programs are extremely powerful in dealing with life and death problems and ending games. Here, however, we are facing problems in the opening. How could AlphaGo be so sure about the relative values of all the different scenarios, each of which consists of a huge number of enthymemes?

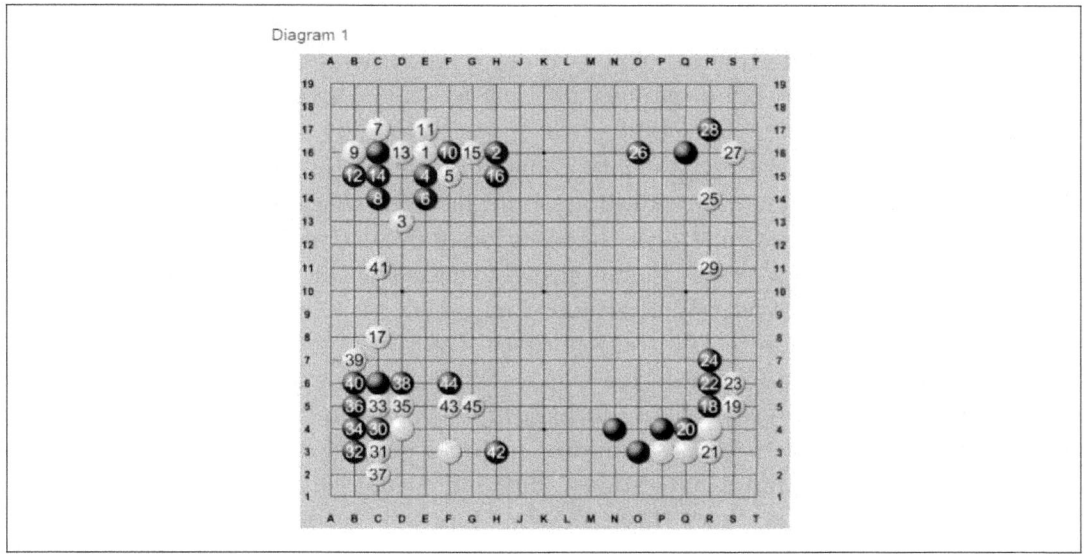

Figure 14: Google DeepMind Challenge Match Game 2

3 Glimpse Beyond

One of the important factors that might uncover the secret of success of AlphaGo can be found in its capability to present and evaluate enthymemes. That is my belief, and I hope that I have been successful by this paper to convince the readers to share it. I also hope that I have made the possibility of enhancing our understanding of the problem of enthymeme itself by interpreting each and every move in a game of Baduk as enthymeme more realistic. In lieu of conclusion, now I would like to discuss another possible major factor that enabled the success of AlphaGo. For, that is nothing but the most difficult problem in the resolution of enthymemes.

I think the doctrine of completion-as-validation, which is a fundamental assumption of charity theory, is a stumbling block for understanding the problem of enthymemes. Above all, it represents too narrow concept of validity, i.e., deductive validity. One terrible consequence of this is evident in dealing with enthymemes in Baduk is evident. In assessing the value of our opponent's enthymemes, we tend to underestimate their strength, for, as they stand, they are by definition deductively invalid. The problem is that still they could be very fine arguments. Further, in trying reconstructing enthymemes according to charity theory, we would unnecessarily try to make them deductively valid. But in most cases such an attempt would be futile but also failing to capture the intention of our opponent. Once we take parsimony theory instead, a door is open to save all those good arguments that are deductively invalid. If so, in reconstructing enthymemes, rather than trying to make them deductively valid, we can just try to make them inductively strong or abductively promising.

In the case where we are appreciating a very long chain of moves, such a broadening conception of enthymeme resolution would have far-reaching implications. Such a long chain of reasoning must be a combination of deductive, inductive, and abductive arguments. As Charles S. Peirce already realized clearly, these three different logical inferences are intricately intertwined in human reasoning. Even though there seems to be great advance quite recently (see, e.g., [11, 12]), our understanding of abduction-deduction-induction circle is not yet far superior to that of Peirce. Now, if AlphaGo has such a great ability in handling enthymemes as we saw above, it is highly likely that it has been programmed to perform deductive, inductive, and abductive inferences at appropriate stages. Further, it may have the ability to exploit the combined force of all these three different types of reasoning.

Let us just take a look at the opening of Game 3.

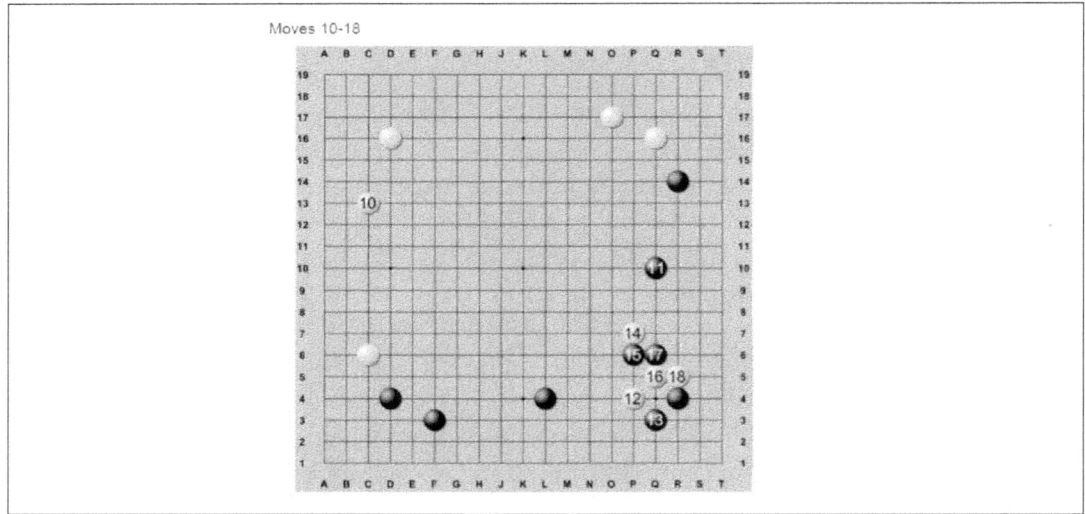

Figure 15: Google DeepMind Challenge Match Game 3

According to Fan Hui, not only from the perspective of Go fundamentals but also in AlphaGo's thought, Black 15 was overplay.

When White jumped at 28, according to Fan Hui's report, AlphaGo's win rate stood at 59%.

According to Fan Hui, Black 31 was a definite overplay. And, as Fan Hui claims,

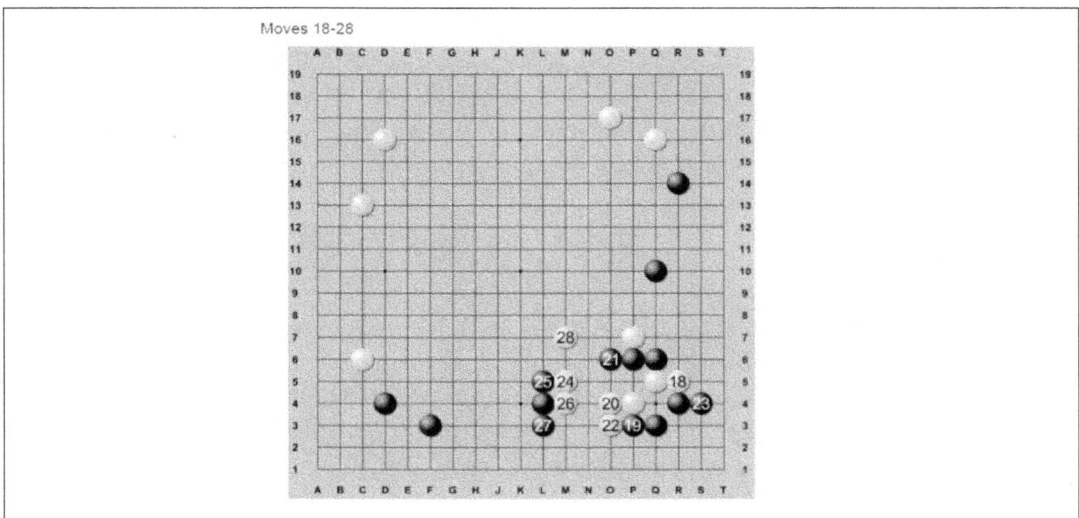

Figure 16: Google DeepMind Challenge Match Game 3

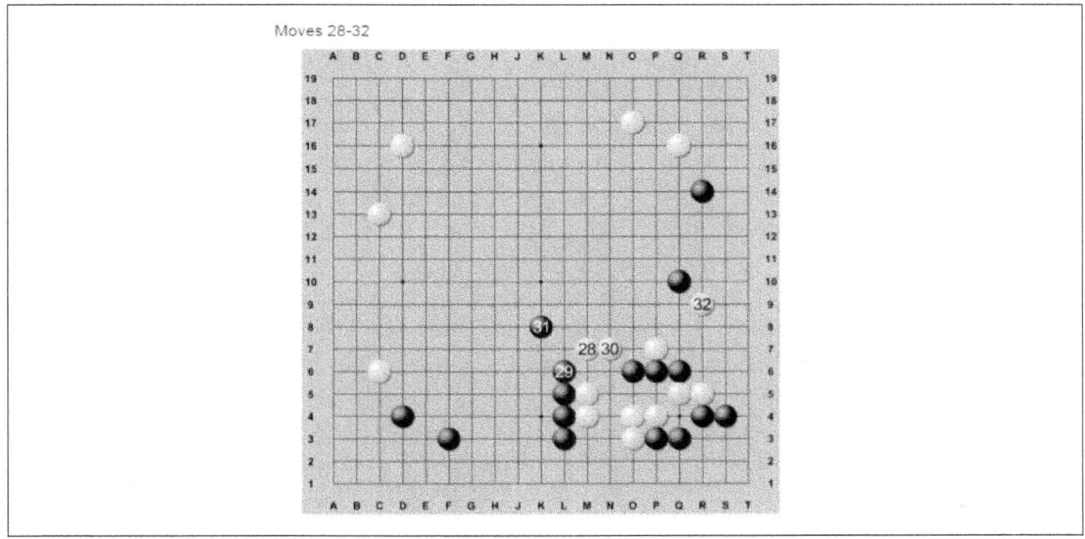

Figure 17: Google DeepMind Challenge Match Game 3

"When move 32 appeared onscreen, everyone agreed it was a beautiful, perfect move!" Also, he is right in saying that "Although White's local shape was strange, the situation already looked out of Black's control". At this point, he reports that AlphaGo's win rate had reached 62%.

The long chain of moves from White 12 to White 32 culminates at such a beautiful move as White 32. One natural and obvious question must be exactly when it was conceived. If it was conceived at the time of White 12, then there is no doubt that AlphaGo knows how to handle the problem of enthymeme as well as the problem of abduction-deduction-induction circle.

References

[1] G. Brun and H. Rott. Interpreting enthymematic arguments using belief revision. *Synthese*, 190, 4041–4063, 2013.

[2] I. Copi. *Introduction to logic (6th ed.)*. New York: Macmillan, 1982.

[3] Fan Hui. Commentary on DeepMind Challenge Match between Lee Sedil and AlphaGo https://deepmind.com/research/alphago/, 2016

[4] D. M. Gabbay and J. Woods. *The Reach of Abduction: Insight and Trial*. Amsterdam: North-Holland, 2005.

[5] W. C. Humphreys. *Anomalies and Scientific Theories*. San Francisco: Freeman, Cooper & Company, 1968.

[6] L. Magnani. Inconsistencies and creative abduction in science. In *AI and Scientific Creativity. Proceedings of the AISB99 Symposium on Scientific Creativity*, pp. 1–8. Edinburgh: Society for the Study of Artificial Intelligence and Simulation of Behaviour, Edinburgh College of Art and Division of Informatics, University of Edinburgh, 1999.

[7] L. Magnani. *Abduction, Reason, and Science: Processes of Discovery and Explanation*. New York: Kluwer, 2001.

[8] L. Magnani. *Abductive Cognition. The Epistemological and Eco-cognitive Dimensions of Hypothetical Reasoning*. Berlin: Springer, 2009.

[9] L. Magnani. The eco-cognitive model of abduction 1 $\alpha\pi\alpha\gamma\omega\gamma\acute{\eta}$ now: Naturalizing the logic of abduction. *Journal of Applied Logic*, 13, 285–315, 2009.

[10] G. Minnameier. Peirce-suit of Truth-Why Inference to the Best Explanation and Abduction Ought Not to Be Confused, *Erkenntnis* 60, 75-105, 2004.

[11] G. Minnameier. Abduction, Selection, and Selective Abduction. In L. Magnani and C. Casadio (eds.), *Model-Based Reasoning in Science and Technology. Models and Inferences: Logical, Epistemological, and Cognitive Issues*, Sapere, Springer, Heidelberg/Berlin, pp. 309-318, 2016a.

[12] G. Minnameier. Forms of abduction and an inferential taxonomy. In L. Magnani & T. Bertolotti (Eds.), *Springer Handbook of Model-based Reasoning*, Berlin: Springer, 2016.

[13] F. Paglieri and J. Woods. Enthymematic parsimony. *Synthese*, 178, 461–501, 2011a.

[14] F. Paglieri and J. Woods. Enthymemes: From reconstruction to understanding. *Argumentation*, 25, 127–139, 2011b.

[15] H. Simon. A behavioral model of rational choice. *Quarterly Journal of Economics*, 69, 99–118, 1955.

[16] H. Simon. Rational choice and the structure of the environment. *Psychological Review*, 63, 129–138, 1956.

[17] W. Park *Philosophy of Baduk*, Seoul: Dongyun. (in Korean), 2002.

[18] W. Park. Counterfactual Reasoning in Baduk: A Preliminary Survey, *Journal of Baduk Studies* 1.1, 125-143, 2004.

[19] W. Park. Belief revision in Baduk: A Preliminary Discussion *Journal of Baduk Studies* 2.2, 1-11, 2005.

[20] W. Park. Abduction and Thought Experiment in Baduk, Unpublished paper read at International Conference on Applying Peirce (Helsinki), 2007.

[21] W. Park *Ad hoc* Hypothesis Generation as Enthymeme Resolution, In L. Magnani and C. Casadio (eds.), *Model-Based Reasoning in Science and Technology Models and Inferences: Logical, Epistemological, and Cognitive Issues*, Sapere, Springer, Heidelberg/Berlin, pp. 507-529, 2016.

[22] G. Schurz. Abductive belief revision in science. In E. J. Olsson & S. Enqvist (Eds.), *Belief Revision Meets Philosophy of Science* (pp. 77–104). Dordrecht: Springer, 2013.

[23] F. Zenker, Lakatos's challenge? Auxiliary hypotheses and non-monotonous Inference. *Journal for General Philosophy of Science*, 37, 405–415, 2006.

Received 1 November 2017

WHEN IS A STRATEGY IN GAMES?

WOOSUK PARK
Humanities and Social Sciences, KAIST
Woosukpark@kaist.ac.kr

Abstract

Strategic reasoning is everywhere, as it has been a focal issue in many scientific disciplines. But what is strategy? What is logic of strategy? In recent years, the dominance of game theory can be witnessed in all this. However, there are many serious problems with the concept of strategy in game theory. Not to mention the classical game theory, which aimed at the highest mathematical abstraction, it is rare to find serious attempts to capture the essence of strategic reasoning even in more recent trends in game theory, such as evolutionary or epistemic game theory. It is good news that logicians and game theorists are becoming more enthusiastic about their collaborations. Starting with the active interaction between epistemic logic and game theory, new research fields such as game logic or strategy logic have appeared. I shall argue, however, there is an unbridgeable gap between the concept of strategy in game theory and that in real games. As an antidote, I propose to analyze the concept of strategy in Baduk (Weichi, Go). For, in this ancient Asian board game, which has become famous by the recent success of AlphaGo, we can get lessons for both theoretical and practical reasoning. Admittedly, the previous discussions of strategy in Baduk literature are not thorough enough to secure a rigorous definition of strategy. However, there is one important clue: What is salient in usual approaches to strategic reasoning in Baduk is that strategy is always discussed together with tactics. Ultimately, I aim at a concept of strategy, according to which (1) it is not necessarily the case that a strategy is found in any game, (2) there has to be an intriguing interaction between a strategy and tactics, (3) it is inconsistency-robust. I shall present an analysis of a historical game record as an example that satisfies all these desiderata. Insofar as this preliminary attempt deserves more careful examination, it would be interesting to raise questions such as "Does AlphaGo have Any Strategy?" or "Could There Be a Strategy in a Mirror Game?". By discussing these questions, I will be able to hint at some implications to some crucial concepts, such as backward induction or common knowledge, in game theory

Strategic reasoning is everywhere, as it has been a focal issue in many scientific disciplines. But what is strategy? What is logic of strategy? In recent years, the dominance of game theory can be witnessed in all this. However, there are many serious problems with the concept of strategy in game theory. Not to mention the classical game theory, which aimed at the highest mathematical abstraction, it is rare to find serious attempts to capture the essence of strategic reasoning even in more recent trends in game theory, such as evolutionary or epistemic game theory. It is good news that logicians and game theorists are becoming more enthusiastic about their collaborations. Starting with the active interaction between epistemic logic and game theory, new research fields such as game logic or strategy logic have appeared. I shall argue, however, there is an unbridgeable gap between the concept of strategy in game theory and that in real games. As an antidote, I propose to analyze the concept of strategy in Baduk (Weichi, Go). For, in this ancient Asian board game, which has become famous by the recent success of AlphaGo, we can get lessons for both theoretical and practical reasoning. Admittedly, the previous discussions of strategy in Baduk literature are not thorough enough to secure a rigorous definition of strategy. However, there is one important clue: What is salient in usual approaches to strategic reasoning in Baduk is that strategy is always discussed together with tactics. Ultimately, I aim at a concept of strategy, according to which (1) it is not necessarily the case that a strategy is found in any game, (2) there has to be an intriguing interaction between a strategy and tactics, (3) it is inconsistency-robust. I shall present an analysis of a historical game record as an example that satisfies all these desiderata. Insofar as this preliminary attempt deserves more careful examination, it would be interesting to raise questions such as "Does AlphaGo have Any Strategy?" or "Could There Be a Strategy in a Mirror Game?". By discussing these questions, I will be able to hint at some implications to some crucial concepts, such as backward induction or common knowledge, in game theory.

1 A Critique of the Concept of Strategy in Game Theory

1.1 Classical Game Theory

Cudd [27] claims that game theory (as a part of rational choice theory) should be distinguished from individual decision theory and social choice theory. According to her, game theory is inspired by the following three ideas:

1) the idea that rationality is utility maximization; (2) the idea that

rational beliefs and rational expectations (that is, of utility) can be formalized using probability theory; and (3) the idea that rational interaction, or interaction among rational agents, is strategic. [27, p. 102]

Unlike the first two of these three ideas, the third idea "distinguishes game theory from individual decision theory". She elaborates the idea as

That in order to act rationally in situations of interaction with other rational agents one must act strategically. [Ibid., p. 103]

If Cudd is right, then the importance of the concept of strategy in game theory cannot be too much emphasized. For, it is the differentia of game theory.

Curiously, however, the concept of strategy in game theory has never been seriously examined. In any standard textbook of game theory, of course, we can find virtually the same definition of strategy. For example, Perea [93] defines a strategy for a player i as "a complete plan of his choices throughout the game":

Definition 1 (Strategy). *A strategy for player i is a function s_i that assigns to each of his information sets $h \in H_i$ some available choice $s_i(h) \in C_i(h)$, unless h cannot be reached due to some choice $s_i(h')$ at an earlier information set $h' \in H_i$. In the latter case, no choice needs to be specified at h.*

[p 358]

What should be noted is that some such definition of strategy in game theory might have been originated from von Neumann's paper "On the Notion of Games of Strategy (1928). As Cudd reports, the first formal treatment of strategic games was presented by von Neumann there:

Von Neumann formalized the notion of strategy by first reducing games of chance, that is, games in which there is a risky event, to games of pure strategy by calculating the expected outcome for each player and for each possible outcome of the risky event. Then a strategy for each player consists in a set of decisions that he makes, one action for each possible decision point contingent upon the information that he has at that point. [27, p. 121]

Virtually the same definition of strategy is found in the monumental book co-authored by Von Neumann and Morgenstern published in 1944. In a sub-section entitled "11.1. The Concept of a Strategy and Its Formalization", we read:

> Imagine now that each player k = 1, ... , n, instead of making each decision as the necessity for it arises, makes up his mind in advance for all possible contingencies; i.e. that the player k begins to play with a complete plan: a plan which specifies what choices he will make in every possible situation, for every possible actual information which he may possess at that moment in conformity with the pattern of information which the rules of the game provide for him for that case. We call such a plan a strategy. [77, p. 79]

The formalization appears on the next page:

Theorem ((11:A)). *A strategy of the player k is a function $\Sigma_k(k; D_x)$ which is defined for every $k = 1, \cdots$, and every D_x of $\mathcal{D}_x(k)$, and whose value*

$$\Sigma_k(k; D_x) = C_x$$

Has always these properties: C_x belongs to $\mathcal{C}_x(k)$ and is a subset of Dx. [Ibid., p. 80]

Of course, what is most important in von Neumann and Morgenstern (1944) is the so-called maxmin approach. As Perea points out, the focus of this approach is whether the given strategy guarantees "a player a certain minimum outcome, irrespective of what the opponent does". [93, [1440001-5]] Some salient characteristics of this approach is duly noted by Perea as follows:

> Note that this approach is basically free of any reasoning about the opponent, because it is interested in outcomes that can be guaranteed by a player even if he has no clue about the opponent's choice. Indeed, the maxmin-criterion makes no distinction between more reasonable and less reasonable choices by the opponent, but simply looks at the "word" strategy that the opponent could choose for you, no matter whether this strategy is plausible or not. [Ibid.]

There is no doubt that the maxmin approach set the research agenda in game theory for subsequent years. Nor there is anyone who would deny that the so-called Nash equilibrium is the most significant innovation in game theory after von Neumann and Morgenstern [77]. Bradenburger even contrasts the maximin criterion with equilibrium criterion. [20, pp. 60-62] Nash triumphantly announces in his monumental 2 page paper:

> In the two-person zero-sum case the "main theorem" [77] and the existence of an equilibrium point are equivalent. In this case any two equilibrium points lead to the same expectations for the players, but this need not occur in general. [74, p. 49]

In a subsequent a bit more detailed exposition he notes:

> The notion of an equilibrium point is the basic ingredient in our theory. This notion yields a generalization of the concept of the solution of a two-person zero-sum game. It turns out that the set of equilibrium points of a two-person zero-sum game is simply the set of all pairs of opposing "good strategies". [75, p. 286]

According to Nash, an n-person game consists of three components: n players, finite set of pure strategies for each, and "payoff function, pi , which maps the set of all n-tuples of pure strategies into the real numbers". [Ibid., 286] Then, he defines mixed strategy as "a collection of non-negative numbers which have unit sum and are in one to one correspondence with his pure strategies". [Ibid., p. 287] And, he finally defines equilibrium point:

Definition 2 (Equilibrium Point). *An n-tuple $\mathcal{S}$ is an equilibrium point if and only if for every i*

$$(1) P_i(\mathcal{S}) = max[P_i(\mathcal{S}; r_i)] \quad for\ All\ \ r_i$$

[Ibid.]

He further introduces some useful concepts such as "dominated strategy" or "equilibrium strategy", thereby making obvious some truisms such as "no equilibrium point can involve a dominated strategy". [Ibid., p. 292]

Unlike Brandenburger, Perea claims that Nash's concept of equilibrium "can be seen as a product of the maxmin approach to games, as it yields precisely von Neumann's maxmin strategies when applied to two-person zero-sum games". [94, [144001-5]] It is simply beyond the scope of this paper to do justice to this small controversy. Let it suffice to note that Perea's observation seems to support rather strongly my standpoint in this paper:

> Its original definition – stating that a player's strategy must be optimal given the opponents' strategies – suggests that players are somehow able to correctly foresee the strategies by their opponents. This makes it hard to place the concept of equilibrium within a model of reasoning, because in such models it seems natural to allow players to have incorrect beliefs about the opponents' choices. [94, [144001-5–6]]

1.2 Epistemic Game Theory

There are at least a few historical overviews of origins of epistemic game theory. However, to the best of my knowledge, no one explicitly identifies a crucial event as the beginning of epistemic game theory. After all, the term "epistemic game theory" seems to be only a recent comer. And Perea [93] is announced to be the first textbook on epistemic game theory. Perea hints at what is going on as follows:

> As important characteristic of human beings is that they reason before making a decision. Indeed, before we make a choice we typically think about the possible consequences, and we look for the choice that yields – at least in our expectation – the most favorable outcome. This reasoning aspect is even more prominent in game theoretic situations, in which the consequence of a choice also depends on the choices made by others. In such situations it is natural to reason about the possible choices that our opponents may make. And in order to reason our way towards sensible predictions about the opponents' choices, it may be helpful to also reason about the possible desires and beliefs of our opponents. This naturally leads to the emergence of belief hierarchies which do not only describe what one believes about the others' choices and desires, but also what one believes about the beliefs that others have about their opponents' choices and desires, and so on. [94, 144001-1–2]] (Perea's emphasis)

Here, Perea seems to be simply puzzled by the historical fact that the classical game theory ignored all these epistemic issues that cannot be overlooked easily. So, he continues:

> However, it took game theory a very long time before it finally incorporated the aspect of reasoning into its analysis. The question that we wish to answer is why? [Ibid.; Perea's emphasis; See also Brandenburger [20, p. 60]

According to him, we need to pay close attention to the early history of game theory in order to answer this question. Roughly speaking, the earliest results in game theory, i.e., von Neumann [76] and von Neumann and Morgenstern [77] "shaped the classical approach to game theory – an approach that would set the research agenda for many decades to come". [Ibid.] He explicitly concedes that only quite lately "the important epistemic notions such as belief hierarchies, common belief, and common belief in rationality, arose" and "slowly but surely provided an alternative to the classical approach". [Ibid.] But, following Brandenburger's lead, he wants to highlight the pioneering insights of Morgenstern for epistemic game theory.

As Brandenburger and Perea claim, Morgenstern's discussion of a "battle of wits between Sherlock Holmes and Professor Moriarty" [71, 102] clearly demonstrates that he was fully conscious of the epistemic and reasoning aspects of game theory. So, Brandenburger suggests

> That the answer, at least in part, is that von Neumann, the intellectual giant with whom Morgenstern embarked on the systematic construction of game theory, put different considerations center-stage. [20, p. 60]

In order to understand why it took such a long time for epistemic game theory to emerge, another excellent topic would "common knowlege" or "common belief". Usually the first formulation of this concept is credited to David Lewis' book Convention (1969). However, Robert Aumann used the same word for this concept without knowing the existence of Lewis' precedence in 1976 [3] (see [6, pp. 24–25]). And it seems to be Aumann's legacy that has more influence among game theorists. For, as Cubitt and Sugden (2003) note, it is only after Aumann and others' formulation of this concept when common knowledge entered game theory. Cubitt and Sugden re-examine Lewis' theory by reconstructing more formally its key parts. For they believe that "the most distinctive and valuable features of Lewis' game theory have been overlooked" [26, p. 175]. Whether it be through Lewis or Aumann, the entrance of "common knowledge" or "common belief" in rationality and other cognate issues such as belief hierarchies clearly shows the indispensability of the epistemic concerns. (For these issues, see [4, 11, 13, 17, 32, 115].)

In explaining the transition from Bayesian equilibrium to epistemic game theory, Brandenburger [20] identifies "the idea of uncertainty about the strategies in a game" and "the introduction of irrationality" as the main factors. According to him, "a clear break" is evident in Bernheim [12] and Pearce [87], for

> Written during the height of the equilibrium refinement program, while many people were working on trying to narrow down the set of Nash equilibria in a game, these two papers challenged the view that Nash equilibrium was the inevitable starting point of analysis in the first place. [20, pp. 66–67]

Further, he elaborates how challenging their views by writing that

> rather than banish uncertainty about strategies (as Nash did), Bernheim and Pearce make this uncertainty central. (But, they did not treat irrationality.) [Ibid.]

Perea seconds Brandenburger's evaluation of the role of Bernheim and Pearce's in the emergence of epistemic game theory. He also appreciates their contributions as presenting "a more basic and natural alternative to Nash equilibrium. According to Perea, their notion of rationalizability, which is equivalent to the idea of common belief in rationality, is a constructive answer to their own critique of Nash equilibrium. [94, [14001-15–16]]

In order to sense how spirited Bernheim's and Pearce's criticism of Nash equilibrium tradition, as Perea does, it would be helpful to quote from their groundbreaking papers. Bernheim starts with the following observation:

> While analyses of Nash equilibria have unquestionably contributed to our understanding of economic behavior, it would be unreasonably optimistic to maintain that Nash "solved" the problem of on cooperative strategic choice [12, p. 1007]

Of course, but for what reason could he claim that "the notion of an equilibrium has little intrinsic appeal within a strategic context"? [Ibid.] His answer is as follows:

> When an agent reaches a decision in ignorance of the strategies adopted by other players, rationality consists of making a choice which is justifiable by an internally consistent system of beliefs, rather than one which is optimal, post hoc. [Ibid.]

I think, this remark is to the point, i.e., that there is no genuine deliberation in Nash approach, or in other words, the so-called equilibrium strategy is not a genuine strategy at all, though bluntly and abruptly presented. I will come back to this point soon. Even if it is indeed not so appealing, what exactly is wrong with Nash strategies? Bernheim replies that

> The economist's predilection for equilibria frequently arises from the belief that some underlying dynamic process (often suppressed in formal models) moves a system to a point from which it moves no further. However, where there are no equilibrating forces, equilibrium in this sense is not a relevant concept. [Ibid., p. 1008]

The diagnosis of the economist's problem seems quite understandable. However, what situation does he have in mind in referring to "where there is no equilibrating forces"? He immediate elaboration seems to answer this question directly:

> Since each strategic choice is resolved for all time at a specific point during the play of a game, the game itself provides no dynamic for equilibration. Further, there is no sensible way to introduce a dynamic while still preserving individual rationality. [Ibid.]

Pearce's criticism of Nash tradition is no less radical than that of Bernheim's. He writes:

> The position developed here, however, is that as a criterion for judging a profile of strategies to be "reasonable" choices for players in a game, the Nash equilibrium property is neither necessary nor sufficient. Some Nash equilibria are intuitively unreasonable, and not all reasonable strategy profiles are Nash equilibria. [87, p. 1029]

After reminding us of "the fact that a Nash equilibrium can be intuitively unattractive", for "the equilibrium may be "imperfect"', he draws our attention to the fact that "the idea of imperfect equilibria has prompted game theorists to search for a narrower definition of equilibrium". [Ibid.] His criticism of the standard justification of the Nash approach is also extremely harsh:

> The standard justifications for considering only Nash profiles are circular in nature, or make gratuitous assumptions about players' decision criteria or beliefs. [Ibid.]

According to him, such a justification asserts

> that a player's strategy must be a best response to those selected by other players, because he can deduce what those strategies are. Player i can figure out j's strategic choice by merely imagining himself in j's position. [Ibid., p. 1030]

However, as Pearce points out,

> this takes for granted that there is a unique rational choice for j to make; this uniqueness is not derived from fundamental rationality postulates, but is simply assumed. [Ibid.]

1.2.1 John van Benthem on Strategy

As we saw above, game theorists finally became interested in logical and epistemological issues after a long tortuous path. In view of the fact that both Hintikka's Knowledge and Belief [48] and Lewis' Convention [58] were already there, it is hard to understand why epistemic logicians failed to interact with game theorists much earlier. Be that as it may, epistemic logic began to have close connection with game theory, as the latter took the epistemic turn from the classical game theory to epistemic game theory, say around 1990. In fact, there are so many reasons logicians and game theorists should collaborate. Let me skip to discuss what motivates logicians to study game theory. Pacuit explains what motivates game theorists to study logic as follows:

> Game theory is full of deep puzzles, and there is often disagreement about proposed solutions to them. The puzzlement and disagreement are neither empirical nor mathematical but, rather, concern the meanings of fundamental concepts ('solution', 'rational', 'complete information') and the soundness of certain arguments (that solutions must be Nash equilibria, that rational players defect in Prisoner's Dilemmas, that players should consider what would happen in eventualities which they regard as impossible). Logic appears to be an appropriate tool for game theory both because these conceptual obscurities involve notions such as reasoning, knowledge and counterfactuality which are part of the stock-in-trade of logic, and because it is a prime function of logic to establish the validity or invalidity of Disputed arguments. [8, p. 317] (see [79, p. 742]).

Not to mention Galeazzi and Lorini [35] that intends to discuss the interaction between epistemic logic and epistemic game theory, there is already huge literature that is relevant to this issue. But I shall focus on John van Benthem's work on the borderline between epistemic logic and game theory.

Among epistemic logicians pursuing active interaction with game theorists, John van Benthem is the one who has emphasized the importance of strategy as a key concept in game theory. (See especially [108–113].) According to him,

> Much of game theory is about the question whether strategic equilibria exist. But there are hardly any explicit languages for defining, comparing, or combining strategies as such: the way we have languages for actions and plans, maybe the closest intuitive analogue to strategies.

> True, there are many current logics for describing game structure – but these tend to have existential quantifiers saying that "players have a strategy" for achieving some purpose, while descriptions of these strategies themselves are not part of the logical language (cf. [80, 115]). In contrast with this, I consider strategies 'the unsung heroes of game theory', and I want to show how the right kind of logic can bring them to the fore. One guide-line of adequacy for doing so, in the fast-growing rain forest of 'game logics', is the following. We would like to explicitly represent the elementary reasoning about strategies underlying basic game-theoretic results, starting from, say, Zermelo's Theorem or Backward Induction. Or in more general terms, we want to explicitly represent agents' reasoning about their plans. [111, p. 96] (Emphasis mine).

As far as what motivates van Benthem to highlight strategy in game theory, I am absolutely sympathetic with him. On the other hand, I am rather skeptical about his ways of treating the unsung hero of game theory. Whether it be called "game logic" or "strategic logic" (see, for example [23, 24, 42, 47, 50, 70]), it would be just wonderful to let logic to bring strategies to the fore. Instead of rectifying game theory by logic of strategy, he merely wants it to serve game theory. As is implicit in his mention of strategies underlying game-theoretic results, he doesn't seem to have any intention to criticize the concept of strategy in game theory. For this reason, I believe, this quotation already betrays the certain limitations involved in his project.

Of course, van Benthem is fully aware of the essential characteristics of his project. He must have made a judicious decision as to what kind of research he wants to do, as is clear from his following remark:

> If we take strategies seriously, what sort of logical analysis will make most sense? ... Should the subject of the logical analysis be strategies themselves, or the way we reason about strategies (surely, not the same thing), or even just modeling reasoning about strategies, as done by real agents in games, thereby placing two layers of intellectual distance: 'reasoning' and 'modeling', on top of the original phenomenon itself? [113, p. 321]

As some perceptive readers must anticipate, I believe, van Benthem should have opted for the last alternative, i.e., "modeling reasoning about strategies, as done by agents in games". Ironically, van Benthem [113] is the closing piece in van Benthem et al. [114], entitled Models of Strategic Reasoning. As we'll see below, the title nicely captures the most recent trend in game logic or strategy logic, which is led by his collaborators. Van Benthem's preference is rather apparent. He is primarily interested in the logical analysis be strategies themselves. All too probably, he seems to believe that strategies themselves have been studied by game theorists. If he is right in that belief, what is needed is just adding more advanced logic. He seems reluctant, but at least willing to swallow studying the way we reason about strategies. Probably, he has the impression that such a study has been welcomed by the younger colleagues in his research community recently. There is some evidence to surmise that he even anticipates what is near to come, for he writes:

> Let me add on a positive note all the same. Sometimes, when theoretical analysis seems to make things more, rather than less complex, there is a last resort: consulting the empirical facts. [113, p. 329]

I do not believe that it is an easy concession to make for a leader of game logic or strategy logic. So, the question is what enforces him to take such a concession. Partial answer may be extracted from his concluding remarks:

> I feel that strategies reflect an undeniable human practice: social interaction has been claimed to be the human evolutionary feature par excellence ([28, 30, 31]). It would be good then to also listen to cognitive studies of strategic behavior [118], since that is where our subject is anchored eventually. Now, making a significant connection with cognitive psychology may not be easy, since strategic structure with its delicate compositional, generic, and counter-factual aspects is not immediately visible and testable in actual psychological experiments. But that just means that, in addition to its logical, computational, and philosophical dimensions that have been mentioned here, the study of strategies also invites sophisticated empirical fact gathering [113, p. 329]

We should note two points here. For one thing, Van Benthem must have been so much impressed by the achievements of one of his collaborators, i.e., Verbrugge, on the border between logic and cognitive psychology. In fact, Ghosh, Meijering, and Verbrugge sharply contrast the game-theoretic concept of strategy with that of cognitive scientists as follows:

> From the game-theoretic viewpoint, a strategy of a player can be defined as a partial function from the set of histories (sequences of events) at each stage of the game to the set of actions of the player when it is supposed to make a move [78]...

> In cognitive science, the term 'strategy' is used much more broadly than in game theory. A well-known example is formed by George Polya's problem solving strategies (understanding the problem, developing a plan for a solution, carrying out the plan, and looking back to see what can be learned) [96]. [40, p. 2]

Another thing to keep in mind is that van Benthem is convinced that his life-long work on strategy logic still has something to contribute on the future study of strategies. What could be the possible contribution of logic in studying strategies themselves? As we pointed out in earlier sections, van Benthem also finds the term "strategy" itself troublesome:

> Even when we decide to give strategies themselves their due, another issue arises. The core meaning of the very term 'strategy' is contentious, reflecting a broader issue of where to locate the essence of this notion [113, p. 322]

Above all, it is good news that van Benthem, who has been prominent in the interaction of logic and game theory, finally finds fault with the fundamental concept of game theory, i.e., strategy. But, in order to capture the essence of the notion of strategy, what concepts of strategy does he consider important? There seem to be two layers in his approach: first, he notes some different concepts of strategy in various scientific disciplines, and second, in our ordinary life. For the former, he writes:

> Some people think of what is strategic in behavior as typically involving some structured plan for the longer term, in line with the crucial role of programs in computer science, or plans in AI [2] and philosophy [21]. But others, for instance, cognitive scientists and social scientists [22, 38], see the heart of strategic behavior in interest- or preference-driven action, often with ulterior goals beyond what is immediately observable. In the latter case, standard computational logics, no matter how sophisticated (cf. the survey of modern fixed-point logics in [117]) may not suffice as a paradigm for studying strategies, as agents' preferences between runs of the system now become of the essence, something that has not been integrated systematically into computation (but see [101, 116] for some attempts). [113, p. 322]

To be sure, such a perceptive observation of how the notion of strategy is understood in all these scientific disciplines provides us with a nice point of departure. However, if we include the meaning of strategy in our ordinary life the pursuit of the core meaning of strategy seems to a formidable, if not an impossible, task. He writes:

> It may be significant here that the linguistic terminology used around the notion of strategy shows a great variety, both in ordinary language and in academic research. People talk of strategies, tactics, plans, protocols, methods, agent types, and the like, which often amount to similar things. For instance, is a 'Liar' a type of person, a program producing a certain behavior, a method for dealing with other people, or a strategy? One can find instances of all these views in the literature, and in professional talk. Clearly, these terms are not all formally well-defined, though some cues for their use might be culled from natural language. [113, p. 323]
> (emphasis mine)

In some sense, van Benthem is confessing that he is simply ignorant of what strategy is. It is as if we now face more and tougher questions at the end of a Socratic dialogue among game theorists, which started with a simple question as to what strategy is. I do believe that any pair-wise comparison of cognate terms van Benthem enumerated would be a meaningful study in the logic of strategies. Van Benthem hints at the meaningfulness of doing so by using strategy/tactics pair:

> In daily discourse, tactics means strategy writ small (a 'strategette'), while strategy is tactics writ large – and one feels that they are similar notions of modus operandi, but operating at different levels of describing activities. It might be worth aiming for further conceptual clarification here, and reserve terms for various uses in a natural family of descriptions for interactive behavior. [113, p. 323]

But how are we to specify and sharpen the questions regarding the intriguing relationship between strategy and tactics in order to define them appropriately?

2 At the Crossroad

Before plunging into the problem of strategic reasoning in Baduk, we need to summarize the lessons from our discussion of previous game-theoretic strategic logic. For, only then, we may be clear about what exactly we need to find or establish in the notion of strategy in Baduk, as a prime example of strategy in general. In this regard, John Woods' Key-note Lecture at the KAIST/KSBS International Workshop "Logical Foundations of Strategic Reasoning" could be instrumental [121].

For Woods' lecture clearly indicates the naturalistic turn in the pursuit of the logical foundations of strategic reasoning:

> The proposed partnership of logic and epistemology, together with epistemology's partnership with the natural sciences of cognition, has a naturalizing effect upon logic, in which its sometimes rightful leanings toward the mathematical are balanced by the obligations of empirical sensitivity. The corresponding shift of logic's preoccupation with truth-preserving consequence relations back to the founding interest in how human beings manage to think straight in real time helps restore logic to its founding origins as a humanities discipline. All this helps set the stage for a principled discussion of strategic reasoning which, whatever its details, is something that humans do in real time under the press of life's shifting variabilities. Since those involved in it are information-processing beings with cognitive agendas, and the knowledge they achieve is an extraction from information under the right conditions, information is bound to play a foundational role here. (Ibid.; See also the more detailed discussion in [119])

Woods' call for the naturalized logic of strategic reason is much more radical than one might expect. In order to sense this feature, it would be sufficient to quote again from [122]:

> Like all grand alliances, this one between logic and empirical science must be more circumspect than heartfelt. A naturalized logic of human reasoning cannot flourish without a well-disciplined empirical sensitivity. But not anything we happen to like goes here. Some of the least attractive features of cognitive psychology have been borrowed from command and control epistemology, especially its embodiments in formal epistemology.* (*See, for example, Vincent F. Hendricks, Mainstream and Formal Epistemology, New York: Cambridge University Press, 2006. See also Paul Gochet and Pascal Gribomont, "Epistemic logic", in Dov M. Gabbay and John Woods, editors, Logic and the Modalities in the Twentieth Century, volume 7 of their Handbook of the History of Logic pages 99-195, Amsterdam: North-Holland, 2006. [122]

It is not entirely clear exactly what are involved in Woods' naturalistic logic of strategies. But it is evident that he is extremely critical of all previous influences of CC (command and control) epistemology, and thereby that of epistemic logic, to cognitive psychology. If Woods is right, that means, even though I welcomed van

Benthem's concession to the necessity of empirical studies of strategies, it would be too early to celebrate strategic logicians' turn to cognitive psychology.

What would be the obvious consequences of turning from strategic logic based on abstract and mathematical game theory to the new naturalistic strategic logic? One obvious benefit, I believe, would be that we can have a much more realistic and useful concept of strategy by deleting some ungrounded assumptions of mathematical game theory. For example, we don't have to assume that every game must have a strategy. Nor is it necessary for every move of a game to have a strategy. So far, game theorists have never made it clear what could be the bearer (or the owner) of a strategy. Sometimes, they talk as if it is the games themselves that have strategies. At other times, they invoke the strategies taken by each move of a game. No matter which they would prefer, the concept of strategy any future strategic logician should aim at doesn't have to assume the existence of strategy, either at the level of games or at the level of moves. For convenience's sake, let us call that strategy*.

Contrary to one might think, such an understanding of strategies (as strategies*) is rather consonant with what great authors of logic of war have claimed throughout the history. Woods calls what is common in Carl von Clausewitz, Edward N. Luttwak, and Henry Minzberg, as the CLM approach. And the basic spirit of CLM approach seems to be manifest in Clausewitz, who "was simply uninterested in defining things in generic [= universal] abstract terms; he regarded as such attempts as futile and pedantic'. ([59]; quoted from [122]). Woods continues to point out:

> What this suggests is a Clausewitzian resistance to formalized approaches to strategic practice, certainly to the idea that the best theoretical language for strategies is a formal one whose formulae carry no propositional content. Frege's way of foundationalizing arithmetic would be the wrong kind of way to approach the theory of strategic reasoning. This matters for game theoretic logics. Although not tethered to Frege's mission, they are tethered to the mathematical treatment of entities constructed from uninterpreted formal languages. For Clausewitz, this would be a step too far. [122]

Another nicety of naturalistic strategic logic would be the possibility of scrutinizing the subtle relationship between strategies and tactics. For example, we may study whether a strategy* must include at least one tactics, or not. If tactics turns out to be a necessary component of a strategy, more systematic and inductive study of tactics would be prerequisite of studying strategies. My discussion of some actual examples of strategies and tactics in Baduk will clarify what I have in mind here.

Furthermore, there is still another possible merit of adopting Woods' naturalistic logic of strategy. Unlike the previous strategy logics based on game theory, where consistency is one of the cardinal vitures, it is meant to be inconsistency-robust. According to Woods,

> An information-system is inconsistency-robust when it is big in ways that require multiple millions of lines of code to computerize, as with climate modelling and modelling of the human neural system. Its inconsistencies are perpetual, pervasive, expungeable in localized contexts but irremovable over-all. Although IR systems do indispensable practical work, they are imperfect and costly. Over-zealous efforts to spare them inconsistency's ignominy seriously damage their practical utility. [121]

What is so nice about inconsistency-robustness? Woods' favorite example of an inconsistency-robust information-system is SHAEF (The Supreme Headquarters Allied Expeditionary Force) established in 1943. Woods writes:

> The Supreme Headquarters Allied Expeditionary Force commanded the largest number of formations ever assigned to a given operation on the Western Front: including First Airborne Army, British 21st Army Group (First Canadian Army and Second British Army), American 12th Army Group, and American 6th Army Group (French First and American Seventh). It was purposed to discharge Operation Overlord against occupied France. SHAEF was a large and complex multi-agent: A multi-agent is a composite of sub-agents, often themselves multi-agents in their own right, working interactively according to some operational agenda or in fulfillment of some conventional arrangement. In some cases, multiagency is an additive combination of its separate parts. In others, it is an emergent fusion of subsets of its parts, a cohesion of "the mangle of practice." [Ibid.]

Now the implication of all this to sufficiently complicate games like Baduk is rather obvious. In such an adversary game, if we simply adopt game-theoretic conception of strategy, what would happen? In other words, if one player's strategy is a set of his strategies at each move, and if consistency of the strategy is all-important for it to be a good one, it would be simply impossible to have one. As a consequence, any good strategy in Baduk must be inconsistency-robust. In the next section, I shall give an example to demonstrate this point.

To sum up, we have found three desiderata for any strategy*, i.e., any good strategy, should satisfy.

(1) It is not necessarily the case that a strategy is found in any game.

(2) There must be an intriguing relationship between it and the tactics supporting it.

(3) It should be inconsistency-robust.

Now we turn to the problem of strategic reasoning in Baduk in order to check whether we can have strategy* in Baduk satisfying all these desiderata.

3 Strategy in Baduk (Weichi, Go)

If we turn to the players of real games in order to model strategic reasoning, what would be the purpose and benefit of doing so? As we saw above, the game-theoretic approaches simply failed to model the strategic reasoning of real agents in real life situations. They turn out to be too abstract and irrelevant for achieving that goal. So, one might turn to the opposite direction, i.e., logic of military strategies. To be sure, there is huge literature on strategies of war. As was demonstrated by Woods' discussion of CLM approach, there are indeed lots of lessons to learn from strategies in wars. What is the point of invoking the strategic reasoning in games rather than wars?

The rough answer might be something like this. The many real games humans have played in history are, unlike the abstract games in game theory, near enough to real life events such as wars. On the other hand, they are already abstractions and idealizations from reality. They are neither too abstract nor simply irrelevant. Be that as it may, I propose to examine what has been meant by players and scholars of Baduk (Weichi, Go) in what follows. After briefly scheming some previous work on strategies in Baduk, where the distinction between strategy and tactics looms large, I shall uncover some desiderata any good strategy in Baduk should satisfy as a preparatory step toward the definition of strategy in Baduk. Then, I will test these desiderata by analyzing a game record from history of Baduk. Finally, I will return to some of the main issues in game theory such as common knowledge or belief and backward induction strategy in order to hint at wherein lies the possibility of more fruitful study of strategic reasoning in the near future.

There is an unmistakable analogy between Baduk and war. Furthermore, interaction between the strategies in Baduk and the strategies in war were so extensive, sometimes it is hard to distinguish between the two. Sun Tzu's The Art of War is the legendary classic of the study of strategies in the eastern world, about which Woods writes:

> If logic is an ancient subject, strategics is a century older, arising not in Greece but in China in the 5th century BC. Sun Tzu's The Art of War is widely considered its founding document.* The name if not its nominatum is Greek, deriving from stratēgia, meaning the arts of a troop leader, the office of a general or the exercise of that office. (* The Art of War, http://ctext.org/art-of-war, Chinese-English bilingual edition, Chinese Text Project.) [121]

Baduk players throughout the history of Baduk have tried to seek analogues not only of Sun Tzu's strategies but also of other famous strategy books such as 吴子 - Wu Zi [Warring States (475 BC - 221 BC)] Wu Qi, 六韜 - Liu Tao [Warring States (475 BC - 221 BC)], 司馬法 - Si Ma Fa [Spring and Autumn - Warring States (772 BC - 221 BC)], 尉繚子 - Wei Liao Zi [Warring States (475 BC - 221 BC)], 三略 - Three Strategies [Western Han - 100 BC-9] (See the texts at http://ctext.org/school-of-the-military) Also, as the historical fact that masters of Baduk were advisors of the lords and generals indicates, the use of Baduk as a simulation of real wars was widely accepted. The so-called "ten golden-rules of Baduk: 圍棋十訣" are in fact abstract enough to be applicable to any field of human life:

- Tan Bu De Sheng （贪不得胜）- The greedy do not get success
- Ru Jie Yi Huan （入界宜缓）- Be unhurried to enter opponent´s territory
- Gong Bi Gu Wo （攻彼顾我）- Take care of oneself when attacking the other
- Qi Zi Zheng Xian （弃子争先）- Discard a stone to gain sente
- She Xiao Jiu Da （舍小就大）- Abandon small to save big
- Feng Wei Xu Qi （逢危须弃）- When in danger, sacrifice
- Shen Wu Qing Su （慎勿轻速）-Make thick shape, avoid hasty moves
- Dong Xu Xiang Ying （动须相应）- A move must respond to the opponent's
- Bi Qiang Zi Bao （彼强自保）- Against strong positions, play safely
- Shi Gu Qu He （势孤取和）- Look for peace, avoid fighting in an isolated or weak situation [http://senseis.xmp.net/?TheTenGoldenRulesList]

Compared to the extensive study of strategics in the western world, as we can witness from CLM approach, one might feel that its counterpart in eastern world seems rather minimal. Embedded in widely used proverbs, such as "Make a feint to the east while attacking in the west (聲東擊西), which is one of the "thirty six strategies (三十 計)", the ancient wisdom related to strategic thinking is found everywhere in human life. However, possibly for that very reason, systematic scientific study of strategy might have been hindered. If we confine our interest to the study of strategy in Baduk, it is disappointingly rare to find serious discussions. On the other hand, the opposite point of view is already at hand. Peter Shotwell [103] reports that in 1941

> A tattered reprint of a pamphlet first published around 1700 A,D. was bought from a street-side book vendor in Sichuan. Falling into the hands of the Chinese army, Thirty-Six Strategies: The Secret Art of War, was considered so potentially disruptive that it was not released to the general public until after the Cultural Revolution calmed down more than thirty years later. [103, p. 166]

According to Shotwell, "since the paphlet's release, more than five hundred books have been published" in Asian countries "to apply the teachings of the the strategies to business practices alone". [Ibid.] Of course, Shotwell does not forget to mention that there are many other books applying the teachings to other fields including Baduk. (See [64].) One possible way of reconciling these apparent conflicting views might be to claim that theoretical and scientific study of strategies themselves is still in its infantile stage, even though applications of the legendary strategies are rather popular, in the eastern world.

3.1 Between Strategies and Tactics

Still, there is one aspect in which one might learn something significant from those rare discussions of strategies in Baduk. The relatively careful and ample discussion of the intriguing relationships between strategies and tactics is what I have in mind. Broadly speaking, strategies in Baduk have been thought to be a higher and more profound concept than tactics. The point is that strategies are not merely the tactics writ large. It is interesting to note that, while there seems to be only a few book-length studies devoted to strategies in Baduk, there is a huge literature on tactics in Baduk. Of course, it is simply impossible to understand strategies in Baduk without understanding its tactics.

Yoshiaki Nagahara and Richard Bozulich's monograph Strategic Concepts of Go [72] seems to be one of the few books devoted to strategies in Baduk. It consists of eight chapters dealing with the following eight concepts: Miai, Aji, Kikashi, Thickness, Korigatachi, Sabaki, Furikawari, Yosu-miru. It is interesting to note that, except for "thickness", all these terms were still waiting to be translated into English in 1972. Of course, there have been attempts to find appropriate English translation of these terms into English. In Nam's Contemporary Go Terms: Definitions and Translations [73], most of these terms have English counterparts. For example, Miai (見合; 맞보기) is translated into English as "paired moves". Nevertheless, we still have to concede that all these terms related to strategies in Baduk tend to be harder than terms related to tactics to translate into English.

Much more troublesome, however, is that there is room for doubt whether Nagahara and Bozulich's selection of strategic concepts in Baduk is reliable enough to capture the essence of strategy in Baduk. First of all, their list may not be exhaustive. Sensei's library (http://senseis.xmp.net/?StrategicConcepts) enumerates the following concepts under the rubric of "strategic concepts"

> aji, amarigatachi = overdeveloped, attack, capture, capturing race , connection, construction, cut, defense, destruction, efficiency, eyespace, flexibility, gote, haengma, honte = proper move, furikawari = exchange, ko strategy, influence, initiative = a player's successive sentes, invasion, investment, karami = splitting attack, kikasare = having been pushed around, kikashi = forcing, korigatachi = overconcentrated, leaning attack, miai = two equal options, mobility, multi-purpose plays, option?, probe, reduction, sabaki, sacrifice, sente, stability, territory, thickness, tedomari = last play of significantly greater value, tenuki = playing elsewhere, timing, urgency.

Apparently, Sensei's library's list is superior to that of Nagahara and Bozulich. Unfortunately, this list still shares many serious problems with its ancestor. It does not distinguish between strategy and tactics. Some terms seem to be more appropriate to be categorized as a strategy, while others seem to belong to the class of tactics. It is still extremely difficult, if not impossible, to extract the what is common in all these strategic concepts. Even in the problem of exhaustiveness, it may not be complete. This suspicion stems from the fact that there is no principled attempt to classify the strategies enumerated.

Jeong and Trinks [53] is a remarkable paper, which presents an original model, i.e., ASPIRE model of strategic reasoning in Baduk. According to this model, problem-solving in Baduk consists of the following six steps:

- Analyze the situation
- Set up the strategic goal
- Presume possible means
- Identify a few plausible moves
- Reason the future events
- Evaluate the results. (Ibid. p. 33; See also [52])

Since it is very well-informed of the recent concepts and theories in Western cognitive psychology, it is destined to be influential in cross-cultural studies on strategies of Baduk in the future. Unfortunately, it does not pay particular attention to the distinction between strategy and tactics.

To the best of my knowledge, Lee [57] is truly exceptional in that it at least tries to uncover the subtle relationships between strategies and tactics. He starts with defining strategy and tactics tentatively as follows:

> Strategy: It refers to the aspects of attack and defense. We need to consider both its principles and management.
>
> Tactics: It refers to viewing the moves of Baduk in terms of functions. It is subdivided into location (vital parts), time (order), and the state of power (hangma: 행마) [57, p. 21]

In some sense, Lee's entire architectonic plan depends on the distinction between strategy and tactics. He assigns a separate chapter for each. Even though there are some terrific insights in his discussion of tactics, let me focus on the chapter on strategies.

The best part in the chapter on the strategies in Baduk is found at the beginning, where Lee tries to contrast strategies and tactics in several different ways. It is not entirely clear whether in all these attempts he is consistent by sustaining the tentative definitions of strategy and tactics quoted above. The remaining parts of the chapter are, of course, quite informative, and shed light on different aspects of the issues at hand. However, these parts are mostly about the strategies in wars. Even when the strategies in Baduk are the major topics, the discussion largely depends on the lessons from the strategies in wars, not the other way around.

Each of Lee's attempts to compare and contrast strategies and tactics deserve careful attention, for many of his claims appear to be truths in Baduk. Let me quote only some of these:

If we call the principles of power to acquire victory in the game of war tactics, we may view "꾀 (wit, trick)", i.e., the principles of ideas, as strategy.

Strategy is the art of scheming tactics,, while tactics is the art of execution.

Strategy is for the victory in a war, tactics is for the victory in combats.

Strategy acquires the ultimate victory by planning the entire states in unified fashion, by knowing both oneself and the opponent, and by uncovering the opponent's schemes, thereby pushing him to the corner.

Tactics alone, no matter how it is distinguished, cannot be identical with the victory in the entire war. If there is an error in the strategy that is the higher order concept over tactics, we cannot avoid the consequential errors in the tactics to realize the strategy.

On the other hand, even though strategy is obviously in higher order than the tactics, it can be actualized only through the tactics. Strategy coordinates variety based on consistency, while tactics promotes consistency based on variety. That is the gist of the management of strategies and tactics. [57, p. 281]

One possible complaint to Lee is this. As already touched upon, all this is applicable to strategies and tactics in every field. Though Lee has a particular sub-section on strategies in Baduk, again he discusses general principles rather than special lessons from the strategies in Baduk. Another possible criticism is that, as a consequence of his preoccupation of contrasting strategy and tactics vividly, he fails to scrutinize exactly how strategies and tactics interact and intertwine with each other.

3.2 An Insightful Example

Now I suggest to start with an insightful example of a strategy*, i.e., a good strategy, that satisfies the three desiderata introduced above: (1) It is not necessarily the case that a strategy is found in any game; (2) There must be an intriguing relationship between it and the tactics supporting it; (3) It should be inconsistency-robust. As one might guess rather easily, it may not be so easy to find out a particular strategy that evidently satisfies all these three desiderata. So, I focus on one game record that instantiates a sacrificial strategy (捨石作戰). Above all, this particular choice is mandated by the third desideratum. Sacrificial strategies would rather naturally show that incosistency-robustness is satisfied by it. However, as was pointed out several times above, there is no clear dividing line between sacrificial strategy and sacrificial tactics. As a consequence, I have the burden to show that a sacrificial strategy is not merely a sacrificial tactics writ large. The first desideratum is motivated by the strong resistance against the ungrounded assumption of game theory that allows any game or any move a strategy. This desideratum indicates that it is rather a sequence of moves that can have a strategy*. I hope that the following example would make clear how we can satisfy the first desideratum by satisfying both the second and the third desiderata.

The game shown in diagram 1 and 2 is a well-known legendary game between Hon'inbō Jōwa (本因坊丈和; 1787-1847) and Inoue Gennan Inseki (井上幻庵因碩; 1798-1859) on December 2, 14, and 24, 1815. Though the winner turned out to be Jōwa, I am so impressed by Inseki's large-scale sacrificial strategy in this game. As I see it, Inseki's sacrificial strategy in diagram 1 and 2 includes 2 examples of sacrificial tactics. At the right bottom corner of diagram 1, we can find one example of a sacrificial tactics. However, there is another example of a sacrificial tactics shown in the top and the center parts of diagrams 1 and 2. What I claim is that Inseki's large scale sacrificial strategy includes both of these examples of sacrificial tactics as its parts. Probably, the genius of Inseki in this sacrificial strategy lies in the creative way he coordinates and synthesize these two examples of sacrificial tactics into one large-scale sacrificial strategy. As the term "sacrificial tactics" indicates, by sacrificing a large group of his stones, Inseki had to put up with inconsistency locally, but in the game as a whole he secured a huge influence in the center and the left bottom corner.

I believe that this game records clearly shows that Inseki not only had a large-scale strategy but also actually executed it. Arguably, his strategy satisfies the three desiderata for a strategy*, i.e., a good strategy.

As for the third desideratum, I gently ask the readers to remember that I am simply assuming that sacrificial tactics must be inconsistency-robust. We would need to explain how paraconsistent logic can be applied to understanding sophisticated reasoning in Baduk in order to prove it. But, I believe, the idea behind the assumption is intuitively acceptable. For, the basic idea of sacrificial tactics presupposes that the opponent can attack and kill one's stones (or groups of stones) because there is inconsistency in the set of one's all previous moves. In other words, a sacrificial tactics can be considered because a player can anticipate that his opponent would detect the inconsistency and probably attack and kill a particular group of stones of his. Of course, the problem of anticipation would introduce a host of challenging issues to our discussion of strategic reasoning. Further, the problem of common knowledge or common belief will arise again rather naturally. For, almost at the same time when one player is scheming a particular tactics to execute, the opponent must detect that possibility. Let us assume that a particular tactics t

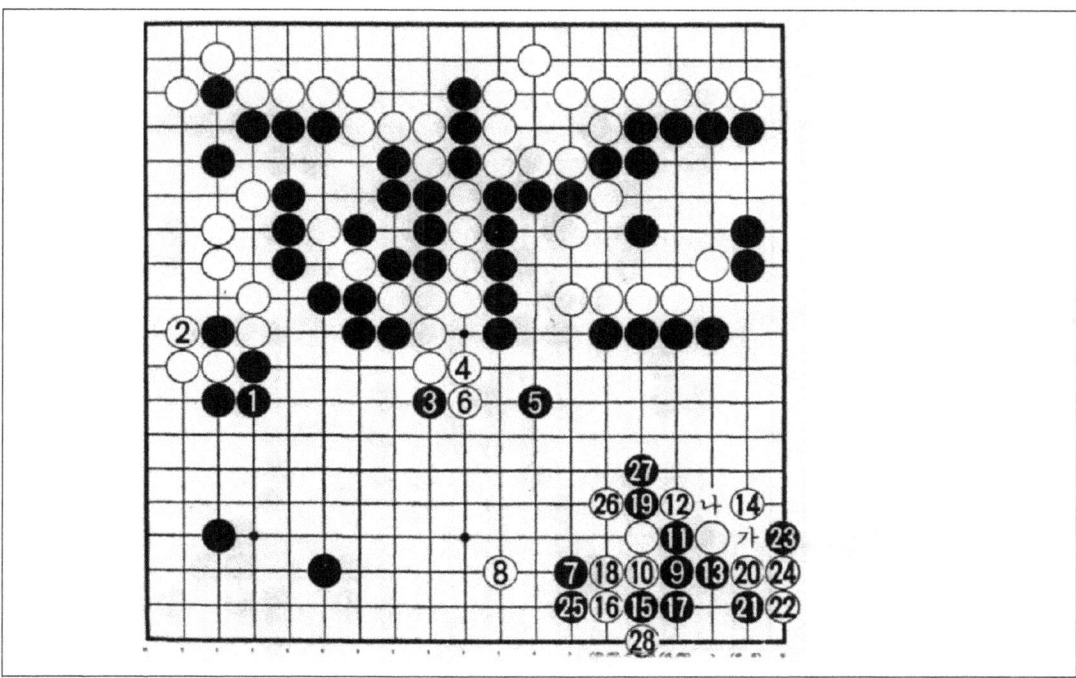

Diagram 1: (101-128)

is very promising for player A. So, A knows that t is promising for himself. In that situation, it is highly likely that player B knows that player A knows that t is promising for A.

As for the second desideratum, I think, this example shows that a strategy is not merely a tactics writ large. If it were so, insofar as this example includes two separate examples of sacrificial tactics, we would need to ask which tactic writ large it is. By incorporating this lesson, we may take a one small step toward the exact formulation of the second desiderata:

> A strategy must contain at least one salient tactics as its part, which crucially contributes to its success.

A bit confusing case would be found where there is only one tactics involved. Probably, that would be the cases, in which the strategy at stake might be counted to be a tactics writ large. In order to avoid confusion due to cases like this, it seems that we need to introduce some other factors such as the previous game situation that enable the particular tactics contribute to the strategy:

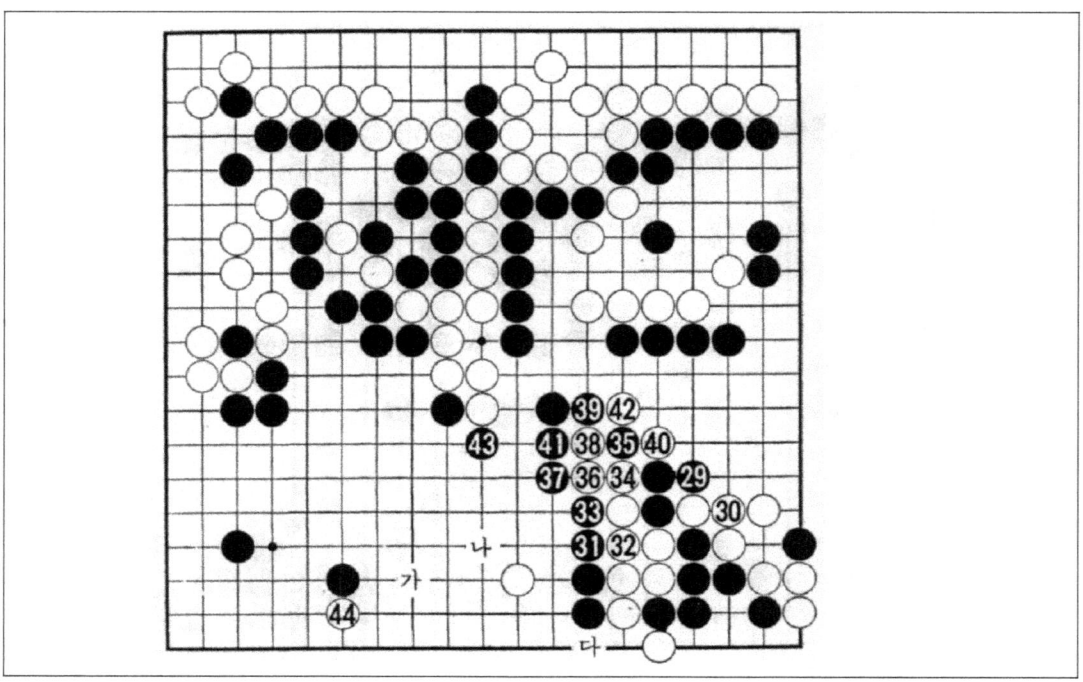

Diagram 2: (129-144)

A strategy must contain at least one salient tactics as its part, which together with other factors of the game situation crucially contributes to its success.

The point is that we are not distinguishing strategy and tactics in terms of the size but by certain inner dialectics within a strategy. There is no doubt that a further elaboration is needed. For example, we need to decide whether some problematic cases where there is only one tactics involved at the very early opening phase of the game should be excluded by the definition of strategy.

The only detail to fill in may be pointing out exactly when there is Inseki's strategy. And, as I intentionally insinuate here, this question as to "When is a strategy" could be multiply ambiguous. For, it may ask (1) exactly when Inseki conceived such a strategy, or (2) when the game record exemplify Inseki's strategy independent of his intention. I say "muplipy ambiguous", because both (1) and (2) can be subdivided. By using this double criteria, we can distinguish between four different possibilities. (See table 1) In principle, there is no reason why we cannot trace the inception back even further to the opening phase of the game. Even if

	earlier	later
intention		
independent of intenstion		

Table 1

that would be too far-fetched, Inseki may claim that at least at the time of giving up the three stones at the top center of the board he was scheming sa large-scale sacrificial strategy. At the time of the move W106, Inseki failed to block W's group. Probably for that reason, Jōwa allowed Inseki to play a sacrificial tactics at the right bottom corder, as shown in diagram 1. But, at the time of the move B143, it turn out that Jōwa underestimated the effect of combining the two examples of sacrificial tactics. W144 was the beginning of Jōwa's counter-attack to falsify Inseki's strategy. And, thanks to his ingenious move, i.e., W170, Jōwa won the game. But that has nothing to do withe value of Inseki's brilliant and overwhelmingly large-sclae sacrificial strategy. I would say that Inseki's strategy existed around B105 (in Diagram 1) to B143 (in Diagram 2).

4 Concluding Remarks

Naturalistic turn in the logical foundations of strategic reasoning opens entirely new possibilities for us. This paper is an attempt to demonstrate that in-depth study of strategic reasoning in Baduk can be extremely fruitful. One evidently nice outcome is that we secured a novel question: When is a strategy in games? By asking "When?" rather than "What?", we can give a fair hearing to the players in the formation, evaluation, and revision of the strategies. It is rather obvious that interdisciplinary, multi-disciplinary, or trans-disciplinary researches are required for this grand venture.

One of the most urgent question to raise must be this: Does AlphaGo have any strategy? No one contests the marvelous victory of AlphaGo over human Baduk players. AlphaGo was the winner in AlphaGo/Lee Sedol match: four to one. AlphaGo won sixty games with the top Baduk players in the world in a row. Finally, AlphaGo defeated Ke Jie, the current world champion of Baduk: three to zero. What should we learn from the glorious parade of AlphaGo? Even after DeepBlue's victory over human chess players, we thought that it would take a long time for computers to rival human Baduk players. How are we to explain such a rapid and unbelievable success of AlphaGo? What was the secret of success of Google DeepMind in developing AlphaGo? Of course, we need to examine carefully which component of AlphaGo was the crucial factor: Monte-carlo method, reinforcement learning, deep learning, machine learning, or what? But one thing evident is that even the DeepMind does not know exactly how AlphaGo achieved all the victories. If so, we have to ask whether AlphaGo has any strategy in playing games of Baduk. In answering this question, what I discuss here can be a point of departure.

I conclude with a disclaimer. The naturalistic turn in the logical foundations of strategic reasoning does not exclude all the other paradigms in game theory, epistemic logic, game logic, or strategic logic. I firmly believe that in innovation continuity is at least equally important as the discontinuity. In fact, for example, in any future study of strategic reasoning in Baduk, there are so many intriguing issues that desperately need the guidance and the assistance from the older paradigms. As an example, let me introduce another question: Could there be a strategy in a mirror game? What a stupid question could it be to classical game theorists! By applying the desiderata for a strategy suggested here, however, this question can be a blessing. For, by probing this question, we can synthesize what game theorists know about backward induction and what Baduk players know about the mirror games. There is a huge literature on backward induction. (For example, see [10, 13, 17, 25, 32, 43, 84, 94, 98, 104–107].) Though there seems to be no extensive research published on mirror games in Baduk, Baduk players and scholars have speculated about mirror game strategies all the time. If mirror game can provide us with a wonderful common ground for game theorists and Baduk players, why not tackle the closely related problem of common knowledge at the same time? Starting with Park [81], where I discussed Peircean abductive reasoning in Baduk, I have continued to examine some logical and philosophical problems raised by Baduk, such as thought experiment, belief revision and counterfactual reasoning in Baduk [81–83]. As is well-known, we can witness the Peircean abduction everywhere, including logic, philosophy of science, computer science, artificial intelligence, cognitive science, brain science, semiotics. Strategically speaking, "The role and function of abduction in the strategic reasoning in Baduk" might be an apt title for my next paper, which would incorporate everything I learned about abductive reasoning from my mentors and friends. (See especially, [60–62, 85, 94, 95, 117, 119–123].)

Acknowledgements

This paper was inspired by John Woods' key-note lecture at the International Workshop hosted by KAIST and KSBS (Korean Society of Baduk Studies). I am also grateful to Jeonghoon Kim, Lorenzo Magnani, Ahti-Veikko Pietarinen, and John Woods for their moral support.

References

[1] C. Alchorron, P. Gardenfors, and D. Makinson. On the logic of theory change: Partial meet contraction functions and their associated revision functions. " *Journal of Symbolic Logic*, 50:510–530, 1985.

[2] M. B. Andersen, T. Bolander, and M. H= Jensen. Conditional epistemic planning. In Herzig L. F. and Mengin A., editors, *del Cerro*, pages 94–106. vol. 7519, . Springer, Heidelberg, 2012.

[3] R. Aumann. Agreeing to disagree. *Annals of Statistics*, 4:1236–1239, 1976.

[4] R. Aumann. Backward induction and common knowledge of rationality. *Games and Economic Behavior*, 8:6–19, 1995.

[5] R. Aumann. On the centipede game. *Games and Economic Behavior*, 23:97–105, 1998.

[6] R. Aumann. Epistemic logic: 5 questions. In V. F. Hendricks and O. Roy, editors, *Robert Aumann*, pages 21–33. Automatic Press, 2010.

[7] R. Aumann and A. Brandenburger. Epistemic conditions for nash equilibrium. *Econometrica*, 63:1161–1180, 1995.

[8] M. O. L. Bacharach. The epistemic structure of a theory of a game. In M. Bacharach and M. al others Bacharach, editors, *Epistemic Logic and the Theory of Games and behavior*. Decisions, Kluwer, 1997.

[9] A. Baltag and S. Smets. The logic of conditional doxastic actions. In van Rooij and K. R. Apt, editors, *R*, pages 9–31. Amsterdam University Press, New perspectives on games and interaction, Vol. 4 of Texts in logic and games . Amsterdam, 2008.

[10] A. Baltag, S. Smets, and J. Zvesper. Keep 'hoping' for rationality: a solution to the backward induction paradox. *Synthese*, 169:301, 2009.

[11] P. Battigalli and G. Bonanno. Recent results on belief, knowledge and the epistemic foundations of game theory. *Research in Economics*, 53:149–225, 1999.

[12] B. Bernheim. Rationalizable strategic behavior. *Econometrica*, 52:1007–1028, 1984.

[13] C. Bicchieri. Self-refuting theories of strategic interaction: a paradox of common knowledge. *Erkenntnis*, 30:69–85, 1989.

[14] K. Binmore. Modeling rational players, part i. *Economics and Philosophy*, 3:179–214, 1987.

[15] K. Binmore. A note on backward induction. *Games and Economic Behavior*, 17(1):135–137, 1996.

[16] O. Board. Dynamic interactive epistemology. *Games and Economic Behavior*, 49:49–80, 2002.

[17] G. Bonanno. The logic of rational play in games of perfect information. *Economics and Philosophy*, 7:37–65, 1991.

[18] B. Bouzy and T. Cazenave. Computer go: An ai oriented survey. *Artificial Intelligence*, 132:39–103, 2001.

[19] A. Brandenburger. The power of paradox: Some recent developments in interactive epistemology. *International Journal of Game Theory*, 35(4):465–492, 2007.

[20] A. Brandenburger. Origins of epistemic game theory. In V. F. Hendricks and O. Roy, editors, *Epistemic Logic: 5 Questions*, pages 59–71. Automatic Press, 2010.

[21] M.: Shared cooperative activity Bratman. Philos. *Rev.*, 101:2, 1992.

[22] P. Calvo and A. : Gomila, editors. *Handbook of Cognitive Science*. Elsevier, Amsterdam, 2009.

[23] C. Chareton, J. Brunel, and D. Chemouil. Towards an updatable strategy logic. In F. Mogavero and others F. Mogavero al, editors, *Ist Workshop on Strategic Reasoning 2013*, pages 91–98. SR13, EPTCS 112, 2013.

[24] K. Chatterjee et al. Strategy logic. *Information and Computation*, 208:677–693, 2010.

[25] T. Clausing. Doxastic conditions for backward induction. *Theory and Decision*, 54:315, 2003.

[26] R. P. Cubbit and R. Sugden. Common knowledge, salience and covention: A reconstruction of david lewis game theory?? *Economics and Philosophy*, 19:175–210, 2003.

[27] A. E. Cudd. Game theory and the history of ideas and our rationality. *Economics and Philosophy*, 9:101–133, 1993.

[28] A Damasio. *The Feeling of What Happens: Body Emotion and the Making of Consciousness*. Heinemann, London, 1999.

[29] B. De Beruin. *Explaining Games: The Epistemic Programme in Game Theory*. Springer, Heidelberg, 2010.

[30] H. De Weerd, R. Verbrugge, and B. (2013) Verheij. How much does it help to know what she knows you know? *An agent-based simulation study*, 199:67–92.

[31] R. I. M. Dunbar and S.: Shultz. Evolution in the social brain. *Science*, 317:5843, 2007.

[32] R. Fagin, J. Halpern, Y. Moses, and M. Vardi. *Reasoning about knowledge*. MIT Press, Cambridge, MA, 1995.

[33] D. Fotland. *Knowledge Representation in the Many Faces of Go*. com/knowpap. txt, 1993.

[34] D. Gabbay and J. Woods. The reach of abduction: Insight and trial, a practical logic of cognitive systems. *Vol.*, 2, 2005.

[35] P. Galeazzi and E. Lorini. Epistemic logic meets epistemic game theory: A comparison between multi-agent kripke models and type spaces. *Synthese*, 193:2097–2127, 2016.

[36] P. Gardenfors. *Knowledge in Flux: Modeling the Dynamics of Epistemic States*. Cambridge, Mass., MIT Press, 1988.

[37] P. Gardenfors and H. Rott(1995). Belief revision", handbook of logic in artificial intelligence and logic programming. In *Vol*, pages 35–132. 4, Epistemic and Temporal Reasoning, edited by D. M. Gabbay et al., Oxford, Clarendon Press.

[38] S Ghosh. Strategies made explicit in dynamic game logic. In Pacuit and J., editor, *van Benthem*. Proceedings of the Workshop on Logic and Intelligent Interaction, ESSLLI 2008, pp. 74?81, 2008.

[39] S. Ghosh, B. Meijering, and R. Verbrugge. Logic meets cognition: Empirical reasoning in games. *In Proceedings of the*, 3:15–34, 2010.

[40] S. Ghosh, B. Meijering, and R. Verbrugge. Strategic reasoning: Building cognitive models from logical formulas. *J Log Lang Inf*, 23:1, 2014.

[41] P. Gochet and P. Gribomont. Epistemic logic. In Dov M. Gabbay and John Woods, editors, *editors, Logic and the Modalities in the Twentieth Century, volume 7 of their Handbook of the History of Logic , Amsterdam: North-Holland, 2006, pp*, pages 99–195. 99-195, 2006.

[42] M. Guo and J. Seligman. The logic of a priori and a posteriori rationality in strategic games. In *Grossi et al*, pages 218–227. LORI 2013, LNCS 8196, 2013.

[43] J. Y. Halpern. Substantive rationality and backward induction. *Games and Economic Behavior*, 37:425–435, 2001.

[44] S. O. Hansson. Ten philosophical problems in belief revision. *Journal of Logic and Computation*, 13:37–49, 2003.

[45] Sven O. Hansson. *A Textbook of Belief Dynamics: Theory Change and Database Updating.* Dordrecht, Kluwer, 1999.

[46] V. F. Hendricks. Mainstream and formal epistemology, new york: Cambridge university press. 2006, 2006.

[47] A. Herzig et al. Reasoning about actions meets strategic logics. pages 162–175. LORI LNCS 8196, 2013.

[48] J. Hintikka. *Knowledge and Belief.* Cornell University Press, Ithaca and London, 1962.

[49] J. Hintikka. What is abduction?: The fundamental problem of contemporary epistemology. *Transactions of Charles Sanders Peirce Society*, 34(3):503–33, 1998.

[50] R. v. d. Meyden Huang. X. *An Epistemic Strategy Logic.* 2014.

[51] S. Jeong. *Panorama on the Board.* Siwasahwe. (in Korean), Seoul, 1997.

[52] S. Jeong. *Understanding Modern Baduk.* Nanam, Seoul, 2004.

[53] S. Jeong and D. Trinks. The relationship between knowledge and strategic reasoning in terms of the aspire model of baduk. *Journal of Baduk Studies*, 13(2):29–43, 2016.

[54] Johnson Johnson(1997). To test a powerful computer, play an ancient game. *The New York Times (July)*, 29.

[55] F. Labra-Sprohne. The mind of a visionary: The morphology of cognitive anticipation as a cardinal symtom. In Anticipation: Learning from the Pasr and), editor, *Nadin (ed).* Learning from the Pasr, Springer, Anticipation, 2015.

[56] D. Lay. *Learning from the Stones; A Go Approach to Mastering China's Strategic Concept, Shi, Carlisle.* Army War College., 2005.

[57] Haebeum Lee. *Elements of Baduk.* Jeonwonmunhwasa. (in Korean), Seoul, 1997.

[58] D. Lewis. *Convention.* Harvard University Press, Cambridge, Mass., 1969.

[59] E. N. Luttwak. *Strategy: The Logic of War and Peace.* The Belknap Press of Harvard University Press, revised and Cambridge, Mass., enlarged edition, 2001.

[60] L. Magnani. Abductive cognition. In *The epistemological and eco-cognitive dimensions of hypothetical reasoning*. Springer, Berlin, 2009.

[61] L. Magnani. Naturalizing logic: Errors of reasoning vindicated: Logic reapproaches cognitive science. *Journal of Applied Logic*, 13:13–36, 2015.

[62] L. Magnani. *Playing with Anticipations as Abductions: Strategic Reasoning in an Eco-Cognitive Perspective??* this issue, 2017.

[63] L. Magnani, S. Arfini, and T. Bertolotti. Intelligence through ignorance. *An argument for ignorance-based chance discovery, International Journal of Advanced Intelligence Paradigms*, 2016. forthcoming.

[64] X. Mao. *The Thirty-six Stratagems Applied to Go*. Yutopian, 1996.

[65] B. Meijering, L. v. Maanen, H. v. Rijn, and R. Verbrugge. The facilitative effect of context on second order social reasoning. In *Proceedings of the 32nd annual meeting of the cognitive science society, . , PA, Cognitive Science Society*, pages 1423–1428. 2010.

[66] B. Meijering, H. v. Rijn, N. Taatgen, and R. Verbrugge. I do know what you think i think: Second order theory of mind in strategic games is not that difficult. *In Proceedings of the*, 33:2486–2491, 2012.

[67] B. Meijering, H. v. Rijn, N. A. Taatgen, and R. Verbrugge. *Reasoning about diamonds, gravity and mental states: The cognitive costs of theory of mind*. In Proceedings of the 35th annual conference of the cognitive science society, Berlin, 2013.

[68] B. Meijering, N. A. Taatgen, H. v. Rijn, and R. Verbrugge. In *Reasoning about mental states in sequential games: As simple as possible as complex as necessary*, Ottawa, 2013. In Proceedings of the 12th international conference on cognitive modeling.

[69] B. Meijering, H. van Rijn, N. Taatgen, and R. Verbrugge. What eye movements can tell about theory of mind in a strategic game. *PLoS ONE*, 7:9, 2012.

[70] F. Mogavero et al. A behavioral hierarchy of strategy logic. In N. Bulling and others N. Bulling al, editors, *CLIMA*, pages 148–165. XV, LNAI 8624, 2014.

[71] O. Morgenstern. Perfect foresight and economic equilibrium. In A. Scatter, editor, *Selected Economic Writings of Oskar Morgenstern*. Pages 169-183, New York University Press, New York, 1935. 1976.

[72] Y. Nagahara and R. Bozulich. *Strategic Concepts of Go*. The Ishi Press, Tokyo, 1972.

[73] Ch Nam. *Contemporary Go Terms: Definitions and Translations*. Oromedia, 2004.

[74] J. Nash. Equilibrium points in n-person games. In *Proceedings of National Academy of Science*, pages 48–49. U.S.A 36, 1950.

[75] J. Nash. Non-cooperative games. *Annals of Mathematics*, 54:286–295, 1951.

[76] Von Neumann. Zur theorie der gesellschaftsspiele. *Mathematische Annalen*, 100:295–320, 1928.

[77] Von Neumann J. and O Morgenstern. *Theory of Games and Economic Behavior*. Princeton University Press, Princeton, NJ, 1944.

[78] M. J. Osborne and A. Rubinstein. *A course in game theory*. MIT Press, Cambridge, MA, 1994.

[79] E. Pacuit. On the use (and abuse) of logic in game theory. *Journal of Philosophical Logic*, 44:741–753, 2015.

[80] R. Parikh and M.: Pauly. Game logic - an overview. *Studia Logica*, 75:165, 2003.

[81] W. Park. *Philosophy of Baduk*. Dongyun. (in Korean), Seoul, 2002.

[82] W. Park. Counterfactual reasoning in baduk: A preliminary survey. *Journal of Baduk Studies*, 1(1):125–143, 2004.

[83] W. Park. Belief revision in baduk: A preliminary discussion. *Journal of Baduk Studies*, 2(2):1–11, 2005.

[84] W. Park. Abduction and thought experiment in baduk. Unpublished paper, read at the International Conference on Applying Peirce, (Helsinki), 2007.

[85] W. Park. *Abduction in Context: The Conjectural Dynamics of Scientific Reasoning*. Sapere, Springer, Heidelberg/Berlin, 2016.

[86] W. Park. *Enthymematic Interaction in Baduk (WeichiGo)*. this issue, 2017.

[87] D. Pearce. Rationalizable strategic behavior and the problem of perfection. *Econometrica*, 52:1029–1050, 1984.

[88] C. S. Peirce. Perceptual judgments (1902). In J. Buchler, editor, *Philosophical Writings of Peirce, New*, pages 302–305. Dover, York, 1955.

[89] C. S. Peirce. The new elements of mathematics, vol. 4, c. In Berlin/New York: Mouton de Gruyter), Atlantic Highlands, and Nj: Humanities Press [Abbreviated as], editors, *Eisele (ed)*. /New York: Mouton de Gruyter; Atlantic Highlands, NJ: Humanities Press [Abbreviated as NEP, Berlin, 1976.

[90] C. S. Peirce. Writings of Charles S. In *Peirce: A chronological edition*. Indiana University Press [Abbreviated as W, C. J. W. Kloesel, Bloomington, Indiana, 1986.

[91] C. S. Peirce. The essential peirce: Selected philosophical writings. In *Vol*. Indiana University Press. [Abbreviated as EP, N. Houser and C. Kloesel, Bloomington and Indianapolis, 1998.

[92] C. S. Peirce. Collected papers, 8 vols., c. hartshorne and p. In *Weiss (vols)*. Harvard University Press [Abbreviated as CP], Cambridge, MA, (1931-1958).

[93] A. Perea. *Epistemic game theory: reasoning and choice*. Cambridge University Press, Cambridge, 2012.

[94] A. Perea. From classical to epistemic game theory. *International Game Theory Review*, 16(1):1–22, 2014.

[95] A.-V. Pietarinen. *Signs of Logic: Peircean Themes on the Philosophy of Language*. Springer, Games, and Communication, Dordrecht.

[96] G. Polya. *How to Solve It: A New Aspect of Mathematical Method*. Princeton University Press, Princeton, NJ, 1945.

[97] W. V. O. Quine. Two dogmas of empiricism. In *his From a Logical Point of View*, pages 20–46. Harvard University Press, Cambridge, 1953.

[98] W. Rabinowicz. Grappling with the centipede: defense of backward induction for bi-terminating games. *Philosophy and Economics*, 14:95–126, 1998.

[99] H. Rott. Two dogmas of belief revision. *Journal of Philosophy*, 97:503–522, 2000.

[100] H. Rott. *Change, Choice and Inference: A Study of Belief Revision and Nonmonotonic Reasoning*. A Study of Belief Revision and Nonmonotonic Reasoning, Oxford, Clarendon Press, Choice and Inference, 2001.

[101] M.: Norms Sergot. action and agency in multi-agent systems. In Governatori and G., editor, *Sartor*. DEON 2010. LNCS, vol. 6181, p. 2. Springer, Heidelberg, 2010.

[102] A. Schitter (ed.). *Selected Economic Writings of Oskar Morgenstern*. New York University Press, New York,

[103] P. Shotwell. *Go!!More than a Game,*. Tuttle Publishing, Tokyo, 2003.

[104] R. Stalnaker. On logics of knowledge and belief. *Philosophical Studies* 129: 169–199, 2006.

[105] R. Stalnaker. Knowledge, belief and counterfactual reasoning in games. *Economics and Philosophy*, 12:133–163, 1996.

[106] R. Stalnaker. Belief revision in games: Forward and backward induction. *Mathematical Social Sciences*, 36:31–56, 1998.

[107] R. C. Stalnaker. On the evaluation of solution concepts. *Theory and Decision*, 37:49–73, 1994.

[108] J. Van Benthem. Games in dynamic-epistemic logic. *Bulletin of Economic Research*, 53(5):219–248, 2001.

[109] J. Van Benthem. Epistemic logic and wpistemology: The state of their affairs. *Philosophical Studies*, 128:49–76, 2006.

[110] J. van Benthem. *Logical Dynamics of Information and Interaction*. Cambridge University Press, Cambridge, 2011.

[111] J. van Benthem. In praise of strategies. *J. v. Eijck et al. (Eds.), Games, Actions, and Social Software 2011, LNCS 7010*, pages 96–116, 2012.

[112] J. Van Benthem. *Logic in Games*. The MIT Press, Cambridge, 2014.

[113] J. Van Benthem. Logic of strategies: What and how? In [114], pp. 321–332, Berlin: Springer.

[114] J. van Benthem et al. (Eds.), *Models of Strategic Reasoning*, LNCS 8072, Berlin: Springer, pages 321–332, 2015.

[115] W. Van der Hoek, M. Wooldridge, and W. Jamroga. A logic for strategic reasoning. In *Proceedings of the Fourth International Joint Conference on Autonomous Agents and Multiagent Systems (AAMAS 2005), pp. 157?164. ACM*, 2005.

[116] R. Van der Meyden. The dynamic logic of permission. *J. Logic Comput.*, 6(3):465–479, 1996.

[117] Y. Venema. *Lectures on the modal mu-calculus*. 2007.

[118] R.: Logic and Verbrugge. *Logic and social cognition: The facts matter, and so do computational models*, volume Logic 38.

[119] J Woods. *Errors of reasoning: Naturalizing the logic of inference*. College Publications, London, 2013.

[120] J. Woods. Inconsistency: Its present impacts and future prospects, in c. In J. Woods and J. Spurr, editors, *Hewitt*. College Publications, Inconsistency Robustness, Milton Keynes, UK, 2015.

[121] J. Woods. *The logical foundations of strategic reasoning: Inconsistency management as a test case for logic*. This issue, 2017.

[122] J. Woods. *What Strategicians Might Learn From the Common Law: Implicit and Tacit Understandings of the Unwritten*. This issue, 2017.

[123] J.(2011) Woods. Recent developments in abductive logic. In *Studies in History and Philosophy of Science, 42(1),. [Essay Review of L. Magnani, Abductive cognition. The epistemologic and eco-cognitive dimensions of hypothetical reasoning. Heidelberg/,]*, pages 240–244. Springer, Berlin, 2009.

What Strategicians Might Learn From the Common Law: Implicit and Tacit Understandings of the Unwritten

John Woods *
University of British Columbia
john.woods@ubc.ca

"The common law ... has been built up, not by the writings of logicians and learned jurists, but by the summings-up of judges of experience to juries of plain men, not usually students of logic, not accustomed to subtle reasoning, but endowed, so far as my experience goes, as a general rule, with great common sense, and if an argument has to be put in terms which only a schoolman could understand, then I am always very doubtful whether it can possibly be expressing the common law."

Mr. Justice Percq, in

Smith v. Harris, 1939

Abstract

This paper is the companion piece of its predecessor in this issue, "The logical foundations of strategic reasoning: Inconsistency-management as a test case for logic." The present paper explores similarities between mainstream theories of strategy in military engagements and the jurisprudence of legal warfare at the criminal bar of common law justice. This exploration will be framed analogically. Given the implicities that are rife in the common law – the legal system that derives from English law – it only stands to reason that a legal scholar would want to know whether there is a logic and epistemology that reliably attests to the logico-epistemic *bona fides* of these features of the common law. My answer to that question is in the affirmative. Because the logico-epistemic fabric of strategic reasoning in military and diplomatic contexts is

---

*This paper, considerably revised, arises from a lecture at the Conference on Analogy in Philosophy and Law at the University of Konstanz in May, 2017. The title of the original was "Precedent and analogy in common law: a common misconception".

also awash in implicity and tacity, it is only natural to wonder whether the logic and epistemology that call the shots for law also call them for strategics. My answer is that it does. In particular, it is the only logico-epistemic theory that does justice to Luttwak's insistence that strategic reasoning is inherently inconsistent.

Parts A, B and C deal with law. In part D, a system of naturalized logic and causal-response epistemology is described, is advanced as a candidate for a theory that would work well for law. Part E reveals how the naturalized causal-response logic handles the inconsistency-management problem for law. Part F applies the analysis to how inconsistency pervades jury-room reasoning without, at the same time, destroying the logico-epistemic integrity of verdicts. Part G brings the paper to a close with an attempt to show that the new logic's accommodations for the implicit, the tacit and the inconsistent can be applied to the management of these same features of strategic reasoning. The central point in both applications is that to a striking extent, reasoning is transacted subconsciously and, in that regard, is itself implicit and tacit.

1 Judge-made Law

1.1 *Unwrittenness in the common law tradition*[1]

The common law, like all legal systems, is a principal means of regulating wars that are waged bloodlessly. In countries where vigilantes settle the score the legal system is at risk. All functioning systems of law have the same strategic objective. They aim to settle matters of legal conflict fairly, justly, accurately and decisively. The common law stands out from some of the others in its own appropriation of the measures of warfare in fulfilling its own regulatory objective: from war to peace without bloodshed. The centre of gravity of the common law's own warlikeness lies in its embrace of adversarial methods. One could say that warlike strategic procedure and purpose runs through the veins of the common law. A criminal trial at the common law bar of justice, is an especially illuminating exemplar of the strategies of warfare. Trials are bloodless wars and more or less politely fought ones, but it is a rare thing for a trial to end without creating harsh negative utilities for the loser. A good trial lawyer will have the instincts of a good general. A great trial lawyer will have the instincts of a Patton.

The common law, like other forms of civilized law, has as its chief strategic purpose the duty to administer the laws of the land in all their operational complexities

---

[1] Trial by jury arose in the pre-Norman county courts of the English shires, presided over by the shire's sheriff – note the etymological connection – and a diocesan bishop, exercising respectively both civil and ecclesiastical legal authority.

in the interests of true and honourable justice for all. Like the information systems underwriting SHEAF's strategic success in World War II, the common law's strategic successes require very large and complex information systems operating over many generations and centuries, in ways that give rise to vast swaths of detailed documentation. A good part of that large record includes verbatim transcripts of debates in Parliament. Also large, are the onsite transcriptions of everything said in trials and tribunals. Another considerable part is a written record of Parliament's legislative history, which chronicles the country's statutory law. A further and no less important part of the information upon which the common law crucially depend, is the information that underlies the making of laws *by judges*. Taken together the common law itself "is not to be found in the written records of the realm."

The English common law arose in the 12th century with Henry II's establishment of the secular tribuntls. (Strategics arose in 5th century B.C. in China) [2] The laws of Henry's secular tribunals would take force throughout the entire realm and so, in contrast to the regional systems of law that preceded it, would hold "commonly" across the land. The king's justices were required to attend to and respect each other's decisions, thus creating a unified law for all England. In that same century, common law would evolve a collectivity of judicial findings based upon tradition, custom and prior judgement. Centuries later in *R. v. Rusby*, 1801, Mr. Justice Kenyon wrote:

> "The common law, though not to be found in the written records of the realm, yet has been long well known. It is coeval with civilized society itself, and was formed from time to time by the wisdom of man."

One of the common law's most singular and least recognized features reposes in the epistemology it embodies. It is an epistemology of the implicit and tacit characteristics of judge-made and other forms of unwritten law (*lex non scripta*), which in turn is one of the common law's most singular legal features. Judges make laws by way of juridical findings that give rise to precedents. Precedents aren't given explicit formulation; judge-made law is unwritten law. Precedents provide that if the facts under consideration in a later case bear a sufficiency of relevant similarity to the facts in the precedent-making case, the later case must be decided in the same way as the precedent-originating one. "Sufficiency of relevant similarity" is undefined. Decisions would not be arrived at syllogistically or in any other deductively tight

---

[2] See Sun Tzu, *The Art of War*, http://ctest.org/art-of-war, Chinese English bilingual edition, Chinese Text Project [51]. Logic arose in the 4th century BC. Aristotle, *Organon*, in *The Complete Works of Aristotle: The revised English Translation*, two volumes, edited by Jonathan Barnes, Princeton: Princeton University Press, 1984. [6]

way, but by a form of reasoning known as "casuistry", in which findings in a given case would arise from consideration of earlier cases. Casuistic reasoning is case-based reasoning and carries none of the pejorative connotation of the word's present-day use.[3] Arising in the Middle Ages and extending throughout the Renaissance, casuity was a method of conflict resolution in moral contexts. It promotes case-based reasoning in which various cases would be compared to a paradigm case. Casuistically derived decisions are reached not by application of some universal axiom or theorem but rather, as Mr. Justice Cardozo would say some six hundred years later:

> "Common law does not work from pre-established truths of universal and inflexible validity to conclusions derived from them deductively .... Its method is inductive, and it derives its generalizations from particulars."[4]

This emphasis on cases prompted sceptical and bemused French legal theorists to quip, "la *superstition du cas*."[5] Lorenzo Magnani writes helpfully about this:

> "Indeed it is useful to remember that moral judgements and judgements *in general,* are not only "directly" and rigidly derived from well-established rules and principles, but may also be derived through complicated verbal argumentation of "practical reasoning", in part based on what cognitive scientists today would call "case-based reasoning", and in part characterized by careful attention to circumstances, concrete aspects, and possible exceptions. I think some aspects of casuists' methodology can be vindicated: [6] for example, the emphasis on circumstances and exceptions; the importance of "probable" opinion of and of the multiplicity of ethical "reasons; the role of analogical reasoning; and the need

---

[3] Despite its title, Jonsen and Toulmin's, *The Abuse of Casuistry*, is an *apologia* for its past effectiveness and a call for its revival in the present day. See Albert R. Jonsen and Stephen Toulmin, *The Abuse of Casuisty: A History of Moral Reasoning,* Berkeley and Los Angeles: University of California Press, 1988; reprint in 1990 [32].

[4] Benjamin N. Cardozo, "The nature of the judicial process", the *Storrs Lectures,1922 [13]*. See also Edmund Burke writing in *An Appeal from the New to the Old Whigs* : Whitefish MT: Kessinger Publishing, 2010; first appeared in 1791 [12]: "Nothing universal can rationally be affirmed on any moral or political subject."

[5] See here Lorenzo Magnani, *Abductive Cognition: The Epistemological and Eco-Cognitive Dimensions of Hypothetical Reasoning,* Berlin: Springer, 2009 [36], especially the section on "Manipulative Abduction", pp. 41-54; and Michael Polanyi, *The Tacit Dimension,* London: Routledge & Kegan Paul, 1966 [40].

[6] Vindicated in light of Pascal's criticisms in *Provincial Letters* of 1657 against the Jesuit's "prolix, jargon-laden writings about ethical cases [which] allowed the order to circumvent certain inconvenient abstract moral rules and principles" (in the words of Magnani, just below.)

to improve, to modify (*and to create new*) ethical knowledge when faced with puzzling cases."[7] (Emphases added)

Of course Justice Cardozo's differentiation of deductively valid and inductively secure reasoning comes nowhere close to capturing the peculiarities of good casuistic reasoning, but his notion of non-universal generalizations as "derived" from particulars seems closer to the mark. Think here of a kind of animal you've never seen, not even in photographs, ocelots say. Even so, when one of them is pointed out to you you'll know in a flash that

(a) Ocelots are four-legged.[8]

Generalizations of the ocelot kind are sometimes called "generic", as opposed to universally quantified.[9] There is reason to think that generic generalizations are free of quantifier clauses of all types – "nearly all", "most" and so on – notwithstanding efforts to find a place for them in contexts such as these. Generic generalizations are semantically interesting. Unlike universally quantified generalizations, they are not falsified by true negative instances such as Ozzie the three-legged ocelot.[10] This helps capture at least part of what Justice Cardozo may have meant in characterizing casuistic reasoning as inductive, not only nonmonotonic but also embedding true legal generalizations that are impervious to true negative instantiation. They are, so to say, paradigmatic. The inference to draw here is that the distinctiveness of unwritten laws is not fully captured by the casuistic nature of the reasoning that gives rise to them, and that their inexpressibility, though related, is a thing apart.[11]

---

[7] Lorenzo Magnani, *Morality in a Technical World: Knowledge as Duty*, New York: Cambridge University Press, 2007; p. 207 [35].

[8] Even if your first ocelot is three-legged, you won't in fact generalize on that anomaly. You will generalize to four-leggedness.

[9] Gregory N. Carlson and Francis Jeffry Pelletier, *The Generic Book*, Chicago: University of Chicago Press, 1995 [14]; Gila Sher, *The Bounds of Logic*, Cambridge, MA: MIT Press, 1991 [45]; Johan van Benthem and Alice ter Meulen, editors, *Generalized Quantifiers in Natural Language*, Dordrecht: Fortis, 1985 [50]; and Mark Wilson, "Generality and nomological form", *Philosophy of Science*, 46 (1979) 161-164 [53]. See also Kent Bach, "Default reasoning: Jumping to conclusions and knowing when to think twice", *Pacific Philosophical Quarterly*, 65 (1984), 37-58 [8].

[10] Consider, for example, while it is true that ocelots aren't three-legged, it is not true that no ocelot is three-legged. It might be that, owing to a congenital defect, Ozzie the ocelot has only three. Consideration of the falsification conditions for generic generalizations can be found in my *Errors of Reasoning: Naturalizing the Logic of Inference*, volume 45 of Studies in Logic, London: College Publications, 2013, reprinted with corrections in 2014; see especially chapters 7 and 8.

[11] Other forms of nonmonotonic judgement are typicality claims ("Birds [typically] fly", "Quakers are [usually] pacifists"), negation-as-failure inferences ("The departure board shows no flight from Vancouver to London after 6:00 p.m; [so there's no such flight]."), circumscription claims

It would be a mistake to overlook the extent to which scientific laws exhibit these same features, given the frequency with which they embed a *ceteris paribus* clause, which has the effect of converting a generalization in the form

(b) $\forall x \ (Fx \supset Gx)$

into something for which we'll need the notion of *type-aberration*. In the ocelot case, three-legged Ozzie is an aberration, having features that objects of his type aren't designed to have, whether by congenital defect or advantitiously by physical trauma. Ozzie, then, is an a-type-aberrant ocelot. This gives us the means to rewrite (a) in a way that captures the force of the *ceteris paribus* clause understood to apply there.

(c) $\forall x \ (Fx \land \sim\text{F-type aberrant } x \supset Gx)$.

Note, by the way, the resemblance between "Findings are binding" and "Ocelots are four-legged". Each permits true negative instances in the case of an aberration, cases in which things aren't working as they are meant to.

Judges who make laws, and traditions that engender the unstated underlying principles of the constitution, do so in ways that render its products intelligible to any hard-working neurotypical person. Bearing in mind that judges were making laws in England centuries before any parliament was enacting statutes, we shouldn't overlook the common law's historical but now much muted hostility to statute law. Writing in *The Law in the Meaning* [5], Sir Careton Allen observed in 1927 that the common law is living and human. Statutes have neither humanity nor humour".[12] Here to like effect is Mr. Justice Cockburn in the English case of *Wason v. Walton* 1868:

> "Whatever disadvantages attach to a system of unwritten law, and of these we are fully sensible, it has at least this advantage, that its elasticity enables those who administer [the law] to adapt it to the varying conditions of society, and to the requirements and habits of the age in which we live, so as to avoid the inconsistencies and injustice which arise when the law is no longer in harmony with the wants and interests of the generation to which it is immediately applied."

("All [normal] birds fly"), inheritance claims ("Mammals [typically] don't fly"), and default claims ("Birds fly [in default of indications to the contrary]"). Discussion of their pairwise independence or otherwise can be found in *Errors of Reasoning* in the places mentioned a footnote ago.

[12] Oxford: Clarendon Press, 1927; paperback in 1964; p. 302.

Writing in 1858 to opposite effect Mr. Justice Grier's found in the U.S. Supreme Court's *McFaul v. Ramsey:*

> "This system, matured by the wisdom of the ages, founded on principles of truth and sound reason, has been abolished in many of our States, who have rashly substituted in its place the suggestions of sociologists .... The results of these experiments ... has [sic] been to destroy the certainty and simplicity of all pleadings, and introduce on the record an endless wrangle in writing, perplexing the court, delaying and impeding the administration of justice."

1.2 A finding is binding

When arriving at its finding, sometimes a court subjects an expression, clause or section of a statute or regulation to an interpretation it has never had before. If the interpretation is integral to the finding, then the finding is binding. In the general case, the finding decides the case currently in play, and if the court's reason for judgement (*ratio decidendi*) is internally coherent, the finding plus the *ratio* creates a *precedent* which is binding on and on all courts below and subject to an expectation of juridical consideration and the possibility of influencing the proceedings of sister courts of the same appellate level elsewhere in the country. A precedent is a general rule of law, applicable with a wide reach in future sufficiently-like cases.

Once a precedent is created, it has both a license and a duty to travel. In so doing, it leaves decision-chains of points at which they've called the shots for subsequent cases. We might liken these points to the nodes of a patterning of decisions in Mintzbergian strategics. The longer the precedent's history, the more complex the patterning of its decision nodes. There are no algorithms for precedential decisions. Not only are the facts at each node of its decision-chain different, but precedents exhibit at least two kinds of operational looseness. Sometimes a judge-made law is pretty stupid (if I might resort to the vernicular). Mere stupidity does not, however, disturb the finding's force. Even so, the case-by-case application of a stupid precedent usually has a varyingly complex pattern history. When a judge or a court decides a case bound by a stupid precedent, there are two ways in which it might be approached. One has to do with the court's reckoning of the sufficiency of relevant similarity of present facts to facts in a precedent's pattern record, including of course the facts in the original case. If the court respects the precedent, it may interpret the similarities of prior cases rather generously. If it dislikes the precedent, it may interpret the similarities very narrowly. Relatedly is a court's reading of a precedent's *meaning* in law. If it likes the precedent, it may interpret its meaning rather broadly.

If it dislikes it, it could give it an interpretation which inhibits its license to travel. Although these flexibilities can serve similar objectives of precedent-management, they are nontrivially different. Both can be considered semantic flexibilities, but each concerns a different subject-matter. The first interprets the meaning of a sufficiency of relevant similarity between new facts and old. The second interprets the meaning of the precedent itself. These flexibilities impart to judge-made laws a degree of applicational indeterminacy which helps us see why precedents resist comprehensive expressiveness in any given documentary try.

It is a commonplace of law school how difficult it is for students to find the *ratio* of a case. The principal difficulty is finding the place in a court's reasons for judgement where the new law can be found. Some reasons for judgement are prefaced with a summary of its findings. They are natural places to look for the *ratio*. Otherwise, the new law is likelier to be placed towards the conclusion of the court's reasons. The elusiveness of precedents is not something students simply grow out of once they've been called to the bar. It is a problem that lingers even for those on the bench. We now know why.

> *Why judge-made law is unwritable*: Given a future court's semantic latitude, a precedent's applicability-range is intrinsically indeterminate to some unpredictable degree.

The phrase "*ratio decidendi*" is ambiguous in Latin, and also in the common but incorrect translation of it in English as "reason(s) for judgement." The most legally accurate English translation of the Latin phrase is "the rule in a decision". It is a regrettable ambiguity, inviting an unnecessary confusion.[13] *Every appellate court is required to give, in the ordinary everyday sense, its reasons for judgement. Trial judges are required to do the same when they are the finders of fact (in other words, when the trial is unjuried).*[14] *It is necessary to emphasize that reasons for judgement are not inherently or always precedent-setting and, to that same extent, not* rationes decidendi *in the legal meaning of that term. That the confusion is*

---

[13] One of the main reasons for law students' search difficulties is partly because instructors keep on misleading them by telling them that a *ratio* is the judgement's rationale.

[14] In a recent case, *R. v. Sliwka,* 2017, the Ontario Court of Appeals severely reprimanded a trial judge for failing to give written reasons for her finding, asserting that her omission "has frustrated the proper administration of justice". In an earlier case, *R. v. Cunningham,* this same judge failed to provide reasons to give reasons on a different matter until two years after her decision. The appeal court set aside the verdict and ordered a new trial, noting that the grossly delayed reasons appeared not to reflect the trial judge's actual reasoning processes *at trial,* but "were instead an after-the-fact justification for the result." This tells us something interesting about *rationes* in the general case. They must formulate a court's thinking at the time of decision.

not routinely noted and warned against either by law schools or in common law jurisprudence is an avoidable carelessness. Any room for pleading that the difference claimed here makes no real difference is foreclosed utterly by numbers of cases in which the opposite is provably so. Among the most notorious of such cases is the finding of the Supreme Court of Canada in Morgentaler, 1988, *which by a majority of five to two struck down s. 251 of the Canadian Criminal Code, which enshrined the entirety of Canada's criminal provisions for providing or submitting to the abortion of an unborn child.*[15]

S. 215 enacted two provisions, one defeasibly general, asserting that the performance of or submission to abortion were indictable offences in criminal law, and the other setting out ways of securing protection from criminal liability in certain health-related circumstances. Beyond question, the majority of five decided the case; s. 215 was gone forever. However, the majority justices rendered three different reasons for this decision, one a solo rendering and the two different ones each dual-authored. None of these *apologiae* had anything close to majority support, and none could present itself as the Court's *own* reasons for judgement. What, then, was the Court's reasons for striking down s. 215? It could only be found in the aggregate of the three majority reasons. The problem is that the majority reasons were in a number of respects incompatible with one another and were internally incoherent overall. From which we had it that Canada's criminal provisions for abortion were down and out until re-thought by Parliament (it has yet to happen). But there was no *ratio decendidi* (rule in the decision) discernible in the majority's scribblings. No precedent was set in *Morgentaler*, and no rule of law either.[16] The common belief that there is a constitutional right to abortion on demand in Canada is, as a matter of law, simply not so.

It should be emphasized that the precedent created in an originating case plays no *precedential* role in deciding it, notwithstanding that the case that led to the gen-

---

[15] Canadian jurisprudence is at its least helpful when it comes to the metaphysical status of the unborn. S. 223 (2) of the *Criminal Code* is headed "Killing child", and says "A person commits homicide when he causes injury to a child before or during its birth as a result of which the child dies after becoming a human being". S. 238 (1) is headed "Killing unborn child in act of birth", and says: "Every one who causes the death, in an act of birth, of any child that has not become a human being, in such a manner that, if the child were a human being, he would be guilty of murder, is guilty of an indictable offence and liable to imprisonment for life." However, s. 223 (1) says that an unborn child has not yet become a human being, despite the biological fact that the unborn child in question here is the unborn child of a human mother. Perhaps Canadian law allows for the possibility that the child of a human mother is an unborn turkey.

[16] Given Parliament's inaction, Canadian law is alone in the company of North Korea and China in standing mute on abortion. Parliament is so reluctant to consider this matter that it has yet to repeal the Criminal Code in response to the loss of s. 251. In my 2015 edition of the Code, s. 251 appears word for word in s. 287 (1) and (2).

eral rule also reflexively coheres with its provisions. The general rule of law applies precedentially in all further cases whose facts bear a sufficiency of relevant similarity to the facts of the originating case. It bears repeating that the critical clause "sufficiency of relevant similarity" is neither defined nor explicated in common-law practice, nor in its jurisprudence either. Certainly, for all practical purposes, both the law's general rule and its sufficiency clause are both inexpressible and only implicitly understood.

This would be a good place to say that judge-made law is not the sole specimen of inexpressible law. Many of the key provisions of statute law behave in this same way, and do so with no noticeable impairment of operational efficacy. Perhaps the best known example of a statutory provision that defies exposition is the proof standard in criminal trials. It is clearly enough laid out in the following simple-seeming words: "Guilty as charged beyond a reasonable doubt." The critical phrase, and the one that raises all the questions, is "reasonable doubt". In *R. v. Lifchus*, 1999, the Canadian Supreme Court foolishly succumbed to the temptation to issue a mock charge by a judge to a jury, according to which absolute certainty of an accused's guilt is too much to ask, and probable guilt isn't enough, and the sincere belief that he's guilty is not sufficient either. The proof standard was averred to lie somewhere within these extremes. Three years later, *R. v Starr*, 2002 weighed in with the view that it would be

> "... of great assistance for a jury if the trial judge situates the reasonable doubt standard between [these] two (sic) standards."

Let's call this elusive standard "Stan". Does any judge know where to find Stan? Does anyone know?

Had these jurists attended to Strong's *MacCormick on Evidence* at p. 517 [48], they would have read,

> "Reasonable doubt is a term in common use as familiar to jurors as to lawyers. As one judge has said, it needs a skilful definer to make it plainer by multiplication of words."[17]

A like sentiment is in instructions from the U.S. Seventh Circuit Court of Appeals in *U.S. v. Glass*, 1988:

> "Reasonable doubt must speak for itself. Jurors know what is reasonable and are quite familiar with the meaning of doubt."

---

[17] J. W. Strong, 5th edition, St. Paul, MN: West Group, 1999.

In England, it is now settled that reasonable doubt can neither be defined, uniformly understood, nor consistently applied. In Wyoming and Oklahoma, a judge's instruction on the meaning of Stan is automatic grounds for reversal, [18]

On the face of it, these features of the common law are epistemologically troubling. If judge-made law is unwritten and unspoken, and yet is understood by judges and lawyers, what would account for this grasp of the inexpressible? If the criminal proof standard is definable and beyond legally secure explication, and yet is understood by jurors, what again would account for it? If the established schools of epistemology were consulted here, it is all but certain that the law's apparent dismissals would meet with sharp resistance or worse. The same is true of books about strategy. The question I've set for myself here is whether there might be an alternative approach to human understanding and its like that is neither dismissible out of hand nor inhospitable to the common law's own understanding of the epistemic characteristics of common law, or to the Clausewitz, Luttwak, and Minzberg (CLM) approaches to strategic reasoning, of which more in section G. It bears repeating that if the answer were in the affirmative, a related question would then press. Is the epistemology that caters for judge-made and unwritten law an *ad hoc* manoeuvre equipped with but one-shot impact, or does it generalize in ways that show a greater appeal? I should emphasize that my purpose here is wholly descriptive. There is no time for normative evaluation. I want to present a descriptively accurate statement of the common law's doctrine of inexpressibility. Assuming the doctrine to be true, I also want to lay out a descriptively accurate account of an epistemology which, if sound, would lay credible claim to be the epistemology for the generality of humanity, not just for the juridical negotiation of precedents. In so doing, I seek the same kind of relief for the CLM approach to strategics.

In response to these questions, I will review two important decisions of the Supreme Court of Canada, one of which set a legal framework for arriving at precedent-making future findings, and the other gave rise to new law of precedential force. Once these reviews have been concluded, we'll return to epistemological matters.

2 Precedents

2.1 Reference Re Secession of Quebec 1998

The case was heard between February 16th and the 19th, 1998. It was a landmark decision, providing formidable security for the Canadian Union. The court's decision

---

[18] *Cosco v. Wyoming*, 1974, and *Pennell v. Oklahoma*, 1982.

was unanimous, and its reasons for judgement was unsigned. The Court held that

> "Quebec cannot secede from Canada unilaterally; however, a clear vote on a clear question referendum should lead to negotiations between Quebec and the rest of Canada for secession. However, above all, secession would require a constitutional amendment."

The Court gave no guidance on what a clear question and a clear vote would be. It left these matters undefined. It should be emphasized that in 1998 there was no actual case for the Court to decide.. The question submitted to the Court by the Governor in Council was a strictly hypothetical one, a what-if question. The question asked for some constitutionally expert opinion (*obiter dicta*) about what it would take, if anything at all, for Quebec to secede from Canada. The Court's advice was that it would require good faith negotiations between the two parties whose resolution point, if there were to be one, would require an amendment to the Canadian Constitution to bring into effect. Experts agree that the way in which the Constitution Act of 1982 is structured makes it preternaturally difficult to amend it even for well-liked improvements, and massively more so for matters as existentially fraught as secession of a province from the very country whose constitution it is. Given the remote likelihood that a future case would ever be brought for actual decision, some experts are of the view that the prospects of judge-made law on secession are vanishingly small. Why, then, would we take the time to consider this finding in a discussion of real cases actually settled by such laws? The reason why is that the common law's implicity and tacity manifest themselves in contexts other than precedent-making ones. Some of these contexts figured very prominently in the Court's advice regarding secession. I'll turn to this now.

The kernel of this matter is one of law and, according to the Court, one that requires the exposure of the most fundamental principles of the country's constitution, the "vital unstated assumptions undergirding Canada's Constitution".[19] We can see the Court's recognition of the influence of unwritten historical principles in interpreting Canada's written constitution:

> "What are those underlying principles? Our Constitution is principally a written one, the product of 131 years of evolution. Behind of the written word is an historical lineage stretching back through the ages, which aids in the consideration of the underlying constitutional principles. These

---

[19] See here Jerome E. Braybrooke's introduction to the Court's advisory at p. 38 of his edited work, *Canadian Cases in the Philosophy of Law*, 4th edition, Peterborough, ON: Broadview Press, 2007 [10].

principles inform and sustain the constitutional text: they are the vital unstated assumptions upon which the text is based. The following discussion addresses the four foundational principles that are most germane for resolution of this Reference: federalism, democracy, constitutionalism and the rule of law, and respect for minority rights. These defining principles function in symbiosis. No single principle can be defined in isolation from the others, nor does any one principle trump or exclude the operation of any other."

Perhaps those who are already familiar with the CLM approach will agree when I say that I see in these present considerations a ready likeness to the CLM approach to the historical line tracking from Alexander's strategic thinking in 334 BC to Eisenhower's strategic thinking between 1943 and 1945 AD.

These four unwritten elements of the constitution are enforceable in Canadian law, but the courts have yet to use them to override written parts of the constitution. So far, these provisions have been used to fill in gaps between text and interpretation.

From which we have it that

(a) The underlying principles of law are undefinable. For if none is definable in isolation from the others, and each of them is similarly constrained, none of them is definable.

(b) They are also unstated but influential, not usually in isolation from the others, also unstated and similarly constrained. Their influence therefore is often holistic.

(c) In the text of this *ratio*, we see the following citations of provisions or findings germaine to the Reference's findings

(1) s. 53 Supreme Court Act, 1985, (2) SCCR. v. Oakes, 1986, (3) *Manitoba Language Rights Reference*, 1985, (4) *Provincial Judges Reference*, 1997, (5) Charter of Rights and Freedoms, 1982.[20]

---

[20] The Supreme Court Act gives the government the authority to submit Reference questions to it. *R. v. Oakes* held that s. 8 of the Narcotics Control Act violated the right to the presumption of innocence under s. 11 (d) of the Charter and cannot be saved under its s. 1. *Manitoba Language Rights Reference* found that the Constitution Act of 1867 and the Manitoba Act of 1870 required the laws of Manitoba, Quebec and Parliament to be in both French and English, and that laws not so written were of no force and effect. *Provincial Judges Reference* held that there is a constitutional norm that protects the judicial independence of all judges. The Court also found that the constitutional protection of minority rights long precedes and is independent of the Charter's own attestations.

(d) One or other of the four underlying principles of law is invoked in several places in this *ratio* as influencing and helping direct the Court in giving shape to its advisory; and in certain instances, the invocation is specific. Given the Court's own attribution of the holistic and symbiotic character of the four underlying principles, singular invocation of them does not preclude their overall influence in all cases.

(e) The advice arising from the *Reference* is this:

"A democratic decision of Quebecers in favour of secession would put those other

relationships at risk.[21] The Constitution vouchsafes order and stability, and accordingly secession of a province 'under the Constitution' could not be achieved unilaterally, that is, without principled negotiations with other participants in Confederation within the existing constitution."

It bears on these reflections that clause 39 of the Magna Carta of 1215 provides that "no free man [shall be seized or dispossessed by the Crown] except by the lawful judgement of equals or by the laws of the land". Clause 40 undertakes that "To no one will we sell, to no one deny or delay, right or justice." Those provisions have been interpreted as imposing on rulers at least the following constraints, collectively known as "the rule of law": (a) Laws are binding on all, including their makers, (b) a law's application must be consistent with its meaning, (c) laws must be openly enacted and widely proclaimed, (d) laws should be formulated in clear language; and, in later interpretations, (f) laws must be democratically enacted, and (g) democratically enacted laws may not transgress certain rights and freedoms. It is worth noting that none of Canada's four underlying principles is discernible anywhere in the text of the Magna Carta. Nor is it clear that any of them originated from the findings of any given court on some given question. The respect for minorities principle is also enshrined in s. 35 of the Constitution Act, 1982, but did not arise there, as we have seen.[22]

---

[21] "In the 131 years since Confederation, the people of the provinces and territories have created close ties of interdependence (economically, socially, politically and culturally) based on shared values that include federalism, democracy, constitutionalism and the rule of law, and respect for minorities."

[22] Canadian law also recognizes the existence of "constitutional conventions". In the 1981 *Reference Re a Resolution to Amend the Constitution*, the SCC specified three factors necessary for the existence of a constitutional convention. They are (a) a practice or agreement developed by political interaction, (b) the recognition that the parties thereto have bound themselves to them, and (c) a purpose to be served by the practice or agreement. Although these are not enforceable as law, they too carry a court-recognized expectation of consideration and of such influence as may be given them in the exercise of a judge's reflective responsibility.

Had the Constitution Act of 1982 not made it practically impossible to amend the constitution, the security of the Canadian Union would have rested on a pair of considerations whose meanings had not been explained. Had the Court's opinion in 1998 been a finding rather than an opinion, it would have created a precedent, whose future interpretations by successor Courts could vary considerably from the implicit and tacit meaning it had for this Court in 1998. Yet no non-partisan legal scholar has said that those key expressions lacked intelligible sense then or lack it now.

Nowhere in the Court's reason for judgement is any of these four unwritten underlying principles of Canada's constitution expressly formulated. Even so, there is no need to think of these principles as hidden. Hey are wholly amenable to well-supported attribution. It is not precluded that a court might state in plain English its interpretation of its provisions for a given case with its own particular facts. In the Court's 1998 finding, federalism is given 15 lines of interpretation, democracy the better part of two pages, constitutionalism and the rule of law another close-to two pages, and respect for minority rights, one page. The *ratio*'s own conclusion runs to slightly over two pages.

2.2 Interpreting unwritten precedents

Let's briefly revisit the idea that some of our laws are unwritten and inarticulable risks the error of supposing that they cannot be understood, that they function in the dense fogs of inoperability. In fact, the reverse is true. Although the principle of federalism is not, in and of itself, open to comprehensive articulation, that it is interpretable in given ranges of cases is a clear indication of its intelligibility. Indeed, if we were to examine the common law's steady reliance upon the unformulable and unsayable provisions of our law, we could be left in no reasonable doubt about their intelligibility. Mind you, their intelligibility inherits in its own right the implicity and tacity of those same provisions. Though we understand them, we are unable to state the objects of our understanding with full generality. After all, how many different and incompatible ways are there for a state to federate its regions? What this tells us is that the correct approach to the jurisprudence of the unwritten underlying elements of Canada's constitution is much like the CLM approach to strategy. Given the full inexpressibility of each, the purposes served by these copious writings in each instance is provided by the causal stimulations of assured implicit and tacit understandings of them.

2.3 Saskatchewan Federation of Labour v. Saskatchewan

Until 2015, public service workers who performed essential services, such as police-protection and medical-emergency services were permitted to unionize but denied the right to strike. In 2015, the Supreme Court of Canada overturned that prohibition on constitutional grounds. The court's decision is binding on all public service arrangements in the country. It is now a constitutionally protected right in Canada for police officers, fire-fighters, ambulance drivers, medical emergency personnel to withdraw their labour in a strike. The Public Service Essential Services Act of 2008 (PSESA) assigns to employers the latitude to specify what services are essential to them. The court held that the PSESA unlawfully intruded upon the right to strike. It was declared unconstitutional because it violated the Charter's s. 2(d). S. 2(d) affords to everyone the constitutional guarantee of a fundamental freedom, namely "freedom of association".

In addition to s. 2(d) of the Charter, the dissenting judgement queries the applicability of s. 2(b) and s. 1. Section 1 provides that

> ="The Canadian Charter of Rights and Freedoms guarantees the rights and freedoms set out in its subject only to such reasonable limits prescribed by law as can be demonstrably justified in a free and democratic society."

Section 2(b) provides that "everyone has the fundamental freedom"

> "of thought, belief, opinion and expression, including freedom of the press and other media of communication."[23]

S. 2 (c) of PSESA defines essential services as follows. I quote from the Act:

"(i) with respect to services provided by a public employee other than the Government of Saskatchewan, services that are necessary to enable a public employer to prevent

    (A) danger to life, health or safety;

    (B) the destruction or serious deterioration of machinery, equipment or premises;

---

[23] S. 2 (c) asserts as a fundamental right "freedom of peaceful assembly." This subsection played no role in this case.

(C) serious environmental damage; or

(D) disruption of any of the courts of Saskatchewan; and

(ii) with respect to services provided by the Government of Saskatchewan, services that

(A) meet the criteria set out in subclause (i); and

(B) are prescribed.[24]

The reasons of the majority justices run to well over a hundred paragraphs, with the *minority* reasons taking but a scant six or so, terminating in its conclusion at para. 176. In the interests of time, I'll let the conclusion bring these reflections to a close. I quote it verbatim:

"176. Neither the PSESA nor the TUAAA[25] infringes s. 2(d) of the Charter. We would dismiss the appeal with costs. We would answer the constitutional questions as follows:

1. Does the Public Service Essential Services Act, S. S. 208 c. P-42. 2, in whole or in part infringe s. 2(b) of the Canadian Charter of Rights and Freedoms?

No.

2. If so, is the infringement a reasonable limit prescribed by law as can be demonstrably justified in a free and democratic society under s. 1 of the Canadian Charter of Rights and Freedoms?

It is unnecessary to answer this question.

3. Does the Public Service Essential Services Act, S.S 2008, c. P-42.2, in whole or in part, infringe s. 2(d) of the Canadian Charter of Rights and Freedoms?

No.

4. If so, is the infringement a reasonable limit prescribed by law as can be demonstrably justified in a free and democratic society under s. 1 of the Canadian Charter of Rights and Freedoms?

It is unnecessary to answer this question.

---

[24] In s. 2 (c): "prescribed" means "prescribed in the regulations".
[25] Trade Union Amendment Act.

5. Do ss. 3, 6, 7 and 11 of The Trade Union Amendment Act, 2008, S. S. c. 36, in whole or part, infringe s. 2(d) of the Canadian Charter of Rights and Freedoms?

No.

6. If so, is the infringement a reasonable limit prescribed by law as can be demonstrably justified in a free and democratic society under s. 1 of the Canadian Charter of Rights and Freedoms?

It is unnecessary to answer this question."

The court's *majority* finding was met with a wave of public disbelief. How can it be, people asked, that the right to associate implies the right of emergency-room surgeons to gather in a picket line? Surely, the protesters continued, the court had given to section 2 a meaning it simple doesn't have.

3 Meaning

3.1 Semantic coercion

This takes us now to what we may perceive as a clear difference between the common law's approach to the interpretation of meanings in legal discussion and the CLM's approach to the interpretation of meaning in the discourses of war-planning and war-making. The difference is stark and amounts to this. Judges often make meanings up by judicial *fiat*. Everyone with a working acquaintance with the abstracta of formal modelling will know the extent to which modellers supplant a term's existing meaning with one that it's never had before, this with the presumed stipulative authority of the formalizer. On the CLM approach, that would be ample reasoning to stifle the impulse towards the semantic stipulations of abstract formal theories.

The essential service employees of the public service of Saskatchewan are specified by type of service, supplemented by a services-like-them clause.[26] *For our*

---

[26] As set out in s. 2 (i) of the PSES Act:
(i) The Government of Saskatchewan
(ii) Crown Corporations
(iii) Regional Health Authorities
(iv) Saskatchewan Cancer Agency
(v) University of Regina
(vi) University of Saskatchewan
(vii) Saskatchewan Polytechnic
(viii) Municipalities

purposes here it suffices to concentrate on employees of five of these types, emergency paramedics and emergency ambulance drivers, and police, fire-fighters, and emergency-room surgeons.

If we consulted s. 2 (d), we'd find nothing in the natural meaning of "association" to justify the implication of the right of these police, paramedics and ambulance drivers to down tools. Of course, supporters will quite rightly say that, in its usage here, the meaning of "association" is contextually sensitive to a large body of labour law, whereby "associate" does indeed imply the right to unionize, which in turn implies the right to strike. While this might be so, it is another thing entirely as to what *Parliament* meant by "association" in 1982. But let that pass. Suppose, contrary to what many people believe, that s. 2 (d) did in fact imply the right of emergency crews and RCMP officers to down tools in a strike, what is there in s. 1 that wouldn't imply that the abandonment of the dangerously ill and of the public's necessary safety would demonstrably justify a limit on the right to strike in a free and democratic country? So in at least one of the two cases, the Court was simply making things up.

If in the case of s. 2 (d) the Court was giving it by force of law a meaning it did not have in fact, that would be a case of the semantic coercion of plain meaning. If in the case of s. 1 the Court was denying it a meaning it actually had, that too would be the semantically coerced suppression of actual meaning. Perhaps the best observations in English literature about semantic coercion are contained in the lines given to Alice and Humpty Dumpty by Lewis Carroll. "When I use a word", Humpty Dumpty said in a rather scornful tone, "it means just what I choose it to mean – neither more nor less." "The question is" said Alice, "whether you can make words mean so many different things." "The question is", said Humpty Dumpty, "which is to be master – that's all." Seen in Lewis Carroll's way, semantic coercion is Humpty Dumpty semantics, and that courts that practice it are Humpty Dumpty courts.

"Semantic coercion" is a term of my own coinage,[27] and has achieved no currency as of yet in common law jurisprudence. But the idea it expresses has an ample provenance there, usually in its reflections on judicial formalism and judicial activism.[28] In giving to plain words and phrases meanings they don't have and in

---

(ix) Police Boards

(x) "Any other person, agency, or body or class of persons, agencies or bodies that (A) provide essential services to the public; and (B) is prescribed."

[27] Woods, *Is Legal Reasoning Irrational?* [57]

[28] "Judicial formalism" is the Canadian term for what American jurisprudence calls "originalism". Originalism is not a uniquely American phenomenon. It was the received method of interpreting legislation in Canada for a hundred years or more after Confederation.

ignoring meanings they clearly do have, semantic coercion cleaves meaning into two kinds, *natural* and *legal*, the former arising from human usage and the later imposed by the force of law. Activists hold that judges have a duty in law to create legal meanings which conduce to the social betterment of the country, and formalists argue that, by the separation of powers, the duties of social betterment fall upon Parliament and legislatures alone. The debate that this unendlessly sparks is of undoubted and enduring importance for jurisprudence. However, my purpose in raising the matter here is its impact on the law's underlying epistemology. Before moving to that, I'll make a quick mention of two strictly legal points. One is that coerced legal meanings aren't generally discernible to the citizenry at large. A related one is that not knowing a law's coerced meaning impedes the likelihood of compliance. In each case, these impediments would appear to violate the rule of law.

Saskatchewan 2015 certainly created a precedent in ways that raise two levels of difficulty, one regarding what the precedent actually is, and the other regarding the extent of its applicability. It matters here how the bearers of the duties of essential service are specified. Saskatchewan boasts some of the country's most important military bases. The Canadian Armed Forces have a duty in law to provide emergency relief in conditions of natural distress, in addition to resisting and annulling armed threat from abroad. Does item (x) include the Canadian Armed Forces? The Saskatchewan Rough Riders are a celebrated community-owned franchise of the Canadian Football League, playing in a magnificent new stadium built with the substantial backing of the taxpayers of Saskatchewan. Are the Rough Rider players captured by clause (x)? (The very suggestion that they don't provide an essential service would be met with province-wide outrage in that fabled place.)

What, now, of other places? Alberta's universities are legally entitled to form Faculty Associations, having all the rights of public service unions except the right to strike. Does the precedent enacted from the law arising by *Saskatchewan* 2015 now extend to them? Does *Saskatchewan* 2015 apply to the garbage collectors of Drummondville, Quebec, or the forest-fire fighters of Chapleau, Ontario?

Whatever the answers to these questions might turn out to be, they won't be discernible in the *Saskatchewan* 2015's *ratio decidendi*. As a distinguished senior member of the bench once said to me in another but similar context, "We'll just have to see what happens, won't we?"

It's now easy to see why judge-made laws are unwritable. Before the 2015 decision, the precedent was that section 2 of the Constitution carried no meaning that implied an emergency-room surgeon to strike. Who at that time would have been able to predict that in 2015, it did carry that meaning after all? The very fact of future high-court semantic coercion, makes future findings unpredictable to a degree that would render any present full formulation of a precedent an inaccurate one.

Enough for now of semantic coercion. It is time to turn our minds more directly and generally to epistemology.

4 Epistemology

4.1 Causalizing knowledge and inference

We have now arrived at the point at which we can put our minds to the counterintuitiveness unattractiveness of my position on the permanent implicities and tacities of strategy and the common law alike. In philosophy, counterintuitiveness is often a matter of the distance of a thesis from the established paradigms of philosophical speculation. From the perspective of the long-enduring hard core of establishment epistemology, my characterizations of CLM strategics and the unvoiced epistemology of the common law are non-starters.[29] The whole focus of that paradigm is to bring the implicit and tacit into explicit written articulation. I want now to test the *bona fides* of the paradigm, not before making a small terminological point. I am a logician of the naturalized sort, and an epistemologist of the causal-response sort. The two go hand-in-hand. As a logician, I focus on entailment. As an epistemologist, I focus on belief. Entailments obtain or not in logical space, and requires for its proper understanding no help from psychology. On the other hand, consequence-drawing or inference occurs in psychological space, as a species of belief-revision. Accordingly, a full-bore logic of the consequence relation will be epistemology-free and psychology-free with respect to what obtains in logical space and yet will welcome both for what occurs in the psychological space of belief-revision.

The epistemology which I think best accommodates the laws' own views of implicity and tacity is the causal response model (CR) of a version of reliabilism, which arises in Alvin Goldman's classic paper of 1967.[30] The paper offers itself as an alternative to, and improvement upon, the still influential paradigm of knowledge as justified true belief (JTB). Goldman retains those conditions but radically reconstructs the justification condition. He does so by converting it from a forensic or case-making constraint on belief to causal condition on how beliefs are made to happen. The basic idea was that a belief is justified to the extent that it is caused by reliable belief-forming devices. Reliability, in turn, was transformed from

---

[29] Again, readers who have consulted my predecessor paper in this number of the journal will know what the CLM approach is, and those who haven't read the predecessor won't know what it is until section G of this one. It doesn't matter. What matters here are the ins-and-outs of causal-response epistemology.

[30] Alvin I. Goldman, "A causal theory of knowing", *The Journal of Philosophy*, 64 (1967), 357-372 [26].

the quasi-moral property of the truthfulness of people in a position to know to a more encompassing causal property of belief-forming devices that are in apple-pie condition and operating in the ways for which nature has built them. What had once been a forensic or case-making notion of justification is now, in Goldman's renewal, a causal condition. It is interesting that throughout a long and distinguished (and much admired) career, Goldman has shown scant interest in dropping the J-condition as a general condition on knowledge. In the large reliabilist literature that has subsequently arisen, it is striking how comparatively infrequently the 1967 paper pops up in the Goldman citations, not excluding citations by Goldman himself. I have a guess about this. In 1967 Goldman took the naturalistic turn in epistemology and, in so doing, beat Quine to the punch by nearly a year.[31] In those times, naturalism was in deep odour in some of the loftier reaches of analytic philosophy, in which philosophy's principled indifference to empirical matters was unquestioned holy writ. My guess is that in declaring war on that branch of analytic philosophy Goldman blinked. I have long regretted that what Goldman did in 1967 is not more widely hailed as a founding moment in the naturalization of epistemology, in spite of the fact that Goldman seems to have brought if off *malgré lui*. One of the discouragements of the naturalistic turn was the slightness of our understanding of how the causal mechanisms of human cognition actually work. We know more these forty years later, but much of what we want still eludes our grasp. If it were a problem for Goldman, it is less of one for us, but a problem even so.

CR is an extension of early Goldman, in which the J-condition is dropped as surplus to need, and the old definition of knowledge is replaced by a new characterization of knowing, which, in a first pass, is something like this for explicit knowing:

> *CR knowing-that*: A human individual S knows that p on information I when p is true, in processing I, S's belief-forming devices cause S to believe p, and those devices are in good working order and operating here as they should; I is good and properly filtered[32] information, and there is no interference caused by negative externalities.[33]

This characterization can be modified to accommodate implicit knowing.

---

[31] W. V. Quine, "Epistemology naturalized" [41], pre-printed in W. V. Quine *Ontological Relativity and Other Essays*, New York: Columbia University Press, 1969. First formal publication, in *Atken des XIV Internationales Kongresses für Philosophie*, 6 (1971), 87-103.

[32] See below in section 2 of this part, Harman's Clutter Avoidance Maxim.

[33] Needless to say, sometimes meeting the J-condition is essential for knowledge, for example, in high energy physics. But its force there remains causal, never mind its forensic character. In the absence of these case-making provisions, the scientist's belief-forming devices won't fire. But this is not a general condition on the formation of well-produced belief.

The motivating factor supporting the CR model is the plain fact that a human individual's knowledge is dominantly a matter of the states into which he has been put by his belief-forming mechanisms under environmental stimuli of various kinds. The qualification "causal" is intended to capture this being-put-into aspect of cognition, and "response" to reflect the factor of environmental impact. CR is a rival of the Command and Control (CC) model of knowing, of which JTB is a weighty exemplar. An important feature of the CC model is its attachment to the idea that knowledge is a state an agent puts himself into by the free exercise of his own intellectual powers, rather than by way of his cognitive devices. On the CC model, knowledge is dominantly down to the knower. On the CR model, it is dominantly down to the knower's devices. Each model recognizes in the other features which it should itself accommodate. The CC model recognizes the CR characteristics of perceptual knowledge. The CR model recognizes the CC characteristics of theoretical knowledge (*epistēmē = endoxon + aletheia + logos*). The CR model generalizes the *perceptual* paradigm to most of knowledge. The CC model similarly generalizes the *theoretical* paradigm. If the empirical facts true to lived cognitive experience cut ice here, the nod would go decisively to the CR model. As Timothy D. Wilson shrewdly observes, when it comes to knowing things we are rather massively "strangers to ourselves". A key feature of epistemology's naturalist turn is its readiness to do business with cognitive psychology and the other natural sciences of cognition. This is fine as far as it goes, but like other blooming courtships it demands mindful invigilation and well-judged chaperonage. The CC model has broad support among psychologists. Among the dissenters are Bargh, and Ferguson and Hassin, Ulemann and Bargh, who regard it as an uncritical assumption and a false one. They see it as embodying the "illusion" that cognitive processes are freely directed independently of causal mechanisms.[34]

Yet another feature of the CR approach, and the last one I'll mention in this preamble, is that it does for logic what Goldman and Quine did for epistemology. Bearing that logic was founded as a philosophical discipline for the investigation of real-time human reasoning, CR proposes to naturalize logic. In so doing, logic recovers an appropriately selective psychologism selectively rejected by Frege as immaterial to the foundations of arithmetic. Logic naturalized, like its epistemological sibling, would be properly but cautiously sensitive to empirical considerations.

---

[34] J.A. Bargh and M. L. Ferguson, "Beyond behaviourism: On the automaticity of the higher mental processes", *Psychological Bulletin* 126 (2000) 925-945 [9]; and R. Hassin, J. Ulemann and J. A. Bargh, editors, *The New Unconscious*, New York: Oxford University Press, 2005 [29]. See also Timothy D. Wilson, *Strangers to Ourselves: Discovering the Adaptive Unconscious*, Cambridge, MA: Harvard University Press, 2002 [54].

4.2 The human cognitive economy

A human ecology is an interactive set of arrangements between nature and its human habitants. A human economy is that part of an ecology which provides for the creation and circulation of wealth. A cognitive economy provides for the production and circulation of knowledge.[35] Like economies of the wealth-creating sort, cognitive economies place a premium on the careful husbanding of cognitive resources. They embed measures for the avoidance of what Harman calls "clutter". [36] Harman frames his requirement in these words: "Clutter Avoidance: One should not clutter one's mind with trivialities". The maxim is not, I think, a matter for mainly conscious implementation, but is better understood as a description of cognitive devices operating in the general case mainly automatically. Seen that way, these devices don't deviate from Harman's advice, but don't "obey" it either. My legal reasoning book postulates further filtration devices in addition to the irrelevance filter that keep clutter at bay. They are filters of inconsistency, misinformation, sayso unreliability, bad bias, premiss-conclusion insecurity and, of course, noise.[37]

While the human individual knows lots of things about lots of different things, he also makes lots of errors about lots of also different things. The frequency of error varies inversely with the successful operation of these filters. Especially vulnerable is the misinformation filter which, especially in conditions of war (on the field or in the court) is bombarded with erroneous information and often with disinformation, which is not only false but is designed to induce tactical miscalculation. Countering these vulnerabilities is a feedback mechanism which allows us to detect and correct error in a timely way. Another source of vulnerability is the premiss-conclusion security filter, which subdues the impulse to leap before we look. A related and equally essential capacity is the one we have for hypothesis-formation in the face of phenomena which puzzle us in ways that impede decisions to act. There are no algorithms for hypothesis-selection. Peirce has good things to say about this.

> "It is true that different elements of the [selected] hypothesis were in our minds before; but it is the idea of putting together what we never before dreamed of putting together which flashes the new suggestion before our contemplation."[38]

---

[35] See here R. Dukas, editor, *Cognitive Ecology: The Evolutionary Ecology of Information Processing and Decision Making,* Chicago: University of Chicago Press, 1998 [19], and R. Dukas and J. M. Radcliffe, editors, *Cognitive Ecology II,* Chicago: University of Chicago Press, 2009 [18].

[36] Gilbert Harman, *Change in View: Principles of Reasoning*, Cambridge, MA: MIT 1986; p. 12 [28]. See also his "Induction: A discussion of the relevance of knowledge to the theory of induction", in Marshall Swain, editor, *Induction, Acceptance and Rational Belief,* Dordrecht: Reidel, 1970 [27].

[37] See pages 203-209 of *Is Legal Reasoning Irrational?* [57].

[38] Charles S. Peirce, *Collected Papers,* Cambridge, MA: Harvard University Press, 1931-1958;

More on the epistemics of hypothesis-formation can be found in my "Reorienting the logic of abduction".[39]

Judges are human beings with all the abilities of the neurotypical person. Unlike counsel,

who receive considerable tutelage in pursuit of formal qualifications, there are no schools for judges and no formal academic qualifications either, beyond the ones required for their prior admittance to the bar. As with all the rest of us, in wide ranges of cases, they learn by doing. Even if his life depended on it, an experienced judge couldn't formulate the precedents he's created. Yet no one in the world's common law countries believes for a moment that precedential reason is inherently defective, still less irrational. What would it take to make it the case that judges understand and correctly apply precedents they can't formulate or explain? To answer this question, we shall have to turn our minds to how beings like us are cognitively structured.

The human animal is a being who extracts knowledge from the information he processes. In extracting this knowledge, he and his like get enough of the right things right enough of the time to survive, prosper and build the great cathedrals of mediaeval France. Let's turn now to one of the more important features of how we're as good as we are at extracting knowledge from information. To do that we'll have to say something about conscious awareness.

Consciousness has a very narrow bandwidth. The information processed in the sensorium, the place where information arrives from the five senses combined, is $\approx$ 11 million bits per second. Only $\approx$ 40 bits of those $\approx$ 11 million make their way into consciousness. If those 40 are processed linguistically, the rate falls to ≈ 16.[40]. We have it, then, that consciousness is highly entropic. Consciousness is a massive suppressor of information, and a thermodynamically costly state for a cognitive agent to be in. The reason for this is that knowledge is an information-thirsty state to be in, requiring more information than there is room for in consciousness. Therefore, most of what we know is known unconsciously. The cognitive system of the human being has a bicameral structure. It is a cooperative unification of two substructures. One is the cognitive up-above. The other is the cognitive down-below.[41]

5.181. Editors of volumes I-VI were Charles Hartshone and Paul Weiss; Arthur Burks edited volumes VII-VIII.

[39] In Lorenzo Magnani and, Thomasso Bertolotti, editors, *Handbook of Model-Based Reasoning*, pages 137-149, Berlin: Springer, 2017 [4].

[40] Manfred Zimmerman, *"The nervous system in the context of information theory"*, in R. F. Schmidt and G. Thews, editors, *Human Physiology*, 2nd edition, pages 166-175, Berlin: Springer-Verlag, 1989 [59].

[41] Woods,, *Is Legal Reasoning Irrational?*, chapter 4 [57], and Peter Bruza, Dominic Widdows, and John Woods, "A quantum logic of down below", in Dov M. Gabbay and Kurt Engesser, editors,

The descriptions of the preceding paragraph are ascribed to thermodynamic systems and are widely believed to be true of them. Human information systems aren't usually thought of as thermodynamic. Thermodynamic systems are closed, whereas human systems seem not to be. Even so, given the enormous complexity of the byplay of energy-to-energy transductions and energy-to information transitions,[42] and the daunting ambiguities of "information",[43] the characteristics ascribed above appear to be readily open to phenomenological adaptation to the lived truths of human cognitive economy. I say this in the spirit of Husserl and Brentano, whose influence on Wittgenstein early and late is abundant. The facts of lived cognitive experience provide ample phenomenological warrant to think of the management of human information in such terms independently of thermodynamic considerations. Think here of human memory, and of how little of its capaciousness is carried about by the conscious mind. Think too of the reams of background information, in whose absence even making the way on bus 33 from UBC to my house would be impossible. Notice, as well, how readily these hidden elements surface into productively conscious employment. Perhaps the most convincing and certainly the most universal example is all that a young child knows (and must know or die) before the acquisition of language. [44]

4.3 The cognitive up-above

Information-processing in the cognitive up-above has most or all of the following properties and does so in various degrees of intensity.

- *agent-centred:* centred in how human agents execute their intellectual execu-

---

Handbook of Quantum Logic, pages 625-660, Amsterdam: Elsevier, 2007 [?].

[42] Fred Attneave, *Applications of Information Theory to Psychology: A Summary of Basic Concepts*, New York: Holt, 1959 [7]; a golden oldie. More recent are Bargh and Ferguson, "Beyond behaviourism: On the automaticity of higher mental processes" [9], and Fred Dretske "Epistemology and Information", in Adriaans and van Benthem (2008), cited just below at pages 29-47 [17].

[43] See, for example, Peter Adriaans and Johan van Benthem, editors, *Philosophy of Information*, a volume in Dov Gabbay, Paul Thagard and John Woods, editors, *Handbook of the Philosophy of Science*, Amsterdam: North-Holland, 2008 [23]; especially Adriaans and van Benthem's introductory essay; and Luciano Floridi, *The Blackwell Guide to the Philosophy of Computing and Information*, Oxford: Blackwell, 2003 [21],

[44] See also Waismann for the view that in one of its uses the word 'understand' denotes no mental experience but rather a semantic *disposition*, in Friedrich Waismann, *The Principles of Linguistic Philosophy,* edited by Rom Harré, London: Macmillan and New York: St. Martin's Press, 1965; p. 347 f [52]. In a dictation to Waismann, Wittgenstein says that understanding the meaning of a word is the ability to apply it correctly, that an ability is not a disposition. The modified translation is in Alice Ambrose, *Wittgenstein's Lectures,* Cambridge 1932-35, from Notes of Alice Ambrose and Margaret MacDonald, Oxford: Blackwell, 1979; p. 92.

tive authority.

- *conscious:* lying in view of by the mind's eye.

- *controlled:* under the executive authority of the human processor.

- *attentive:* lying within the executive authority's durable focus.

- *voluntary:* subject to the free exercise of the agent's executive authority.

- *linguistically expressible:* subject to explicit formulation under the intellect's executive authority.

- *semantically loaded:* the information has semantic content. Spoken or not, it is propositionally structured.

- *linear:* the processing is temporally stepwise ordered.

- *surfacely contextualized:* the information has broken the surface into conscious awareness.

- *computationally weak:* the instruments of processing have scant computational fire-power; so the information cannot be too *complex.*

4.4 The cognitive down-below

The cognitive down-below is oppositely characterized. Information-processing in the cognitive-down below has most or all of the following properties in varying degrees of intensity:

- *mechanism-centred:* centred in how our cognitive devices operate.

- *unconscious:* the agent whose devices are doing the processing is consciously unaware of the information that's being processed.

- *automatic:* the processing is beyond the control of the person himself, and the devices operate "on their own".

- *inattentive:* the processing is not subject to the agent's executive inspection.

- *involuntary:* the processing is not something he freely decides to initiate or direct.

- *linguistically unformulated:* tacit and implicit.[45]

- *semantically inert:* the information being processed lacks semantic content and propositional structure.

- *parallel:* the processing is a multi-tasking one, performing several operations at once.

- *deep-down:* the information being processed is out of sight of the mind's eye, *beyond the reach of the heart's command, and unengageable by tongue or pen (or keystroke).*

- *computationally luxuriant:* the processing devices have vastly greater computational power than the human process does, and can therefore handle even very complex information.

Levels of upness and downness need not be uniform.[46] For example, in down-below processing the unconsciousness parameter might outweigh in parametic-value the value of the semantic inertness parameter. Equally, in the cognitive up-above, the consciousness parameter could carry a higher value than the computational weakness parameter. All parameters are subject to variations of intensity. Sometimes an agent has some conscious awareness of what's going on even when he's mainly unaware of it. Up and down are not rivals. They are equal partners in the cognitive economy, each indispensable to its good functioning in complex alliances of cooperation, involving a deal of cross-border traffic.[47] One of the unsettled questions about the varying weight of the parametric volume influences the interactions between the up-above and down-below. For example, is linguistically unstructured information subject to some or other degree of awareness? Some say not, for example, Brentano, Frege, Russell, Wittgenstein and Sellars.[48] Some say otherwise,

---

[45] See, for example, Zoltan Dienes and Josef Perner, " A theory of implicit and explicit knowledge", *Behavioural and Brain Sciences,* 22 (1999), 735-808 [16]; Gilbert Harman, *Change in View,* chapter 2, pp. 13-20 [28]; and Robert Brandom, *Making it Explicit,* Cambridge, MA: Harvard University Press, 1994 [11].

[46] Richard M. Shiffrin, "Attentional control", *Perception and Psycholphysics,* 21 (1977), 93-96 [46], and "Automatism and consciousness", in Jonathan D. Cohen and Jonathan W. Schooler, editors, *Scientific Approaches to Consciousness,* pages 49-64, Mahwah, NJ: Erlbaum, 1997 [47].

[47] An arrhythmic heart can retard the pace of its beat, and in so doing deny the brain the oxygen required to support consciousness.

[48]5 I wish to emphasize ... the denial that there is any awareness prior to, or independent of, the acquisition of language", Wilfrid Sellars, "Empericism and the philosophy of mind, in Herbert Feigl and Michael Scriven, editors, *Minnesota Studies in the Philosophy of Science,* volume I, pages 253-329, Minneapolis: University of Minnesota Press, 1956, section 31 [44].

say, Husserl and Romano.[49] We needn't try to settle this now. Let's mark it as a question of importance for future investigation by strategicians. The fact remains that most of the heavy lifting of human knowledge-production is done in the down-below. Why? Because, comparatively speaking, the cognitive resources of the down below are plentiful and cheap! The cognitive processes of the up-above are greatly encumbered by consciousness's supression of information.

It is necessary to emphasize the manic briskness of the ceaselessly nourishing traffic between the cognitive above and below. If we were drawn to the notion of supervience, we might venture to say in words modified from Quine's own physicalism:

> "Nothing happens in the cognitive up-above, not the flutter of a belief, not the flicker of an inference, without some redistribution of causal states in the cognitive down-below."[50]

Knowledge is an information-thirsty state to be in. It is the productive convergence of more information of various types than there is room for in any state that radically suppresses information. If the present characterizations of the cognitive up-above and down-below are accurate, we have it at once that

> *The dominance of the cognitive down-below:* Most of what we know at any time is known at one or other level of the cognitive down-below. Most of what we know we know subconsciously in some degree.

The idea that we know things, many of which actually matter, without knowing or being able to say why we do or how we came to know, is what Peter Struck calls "surplus knowledge".[51] By "surplus", he means beyond our ability to account for the fact that sometimes we "just know" things ("don't ask me how"). Just knowing is a cultural universal, which is problematic for present-day science. Struck locates the ancient phenomena in the divination rites of antiquity, and concentrates on the

---

[49] Husserl thinks that language is a redundant dimension of meaning, functioning as an unproductive supplement to pre-predicative experience. Romano holds that pre-predicative ability to read nature is a necessary condition of preconceptualized experience. Claude Romano, *At the Heart of Reason*, Michael B. Smith and Claude Romano, translators, Evanston: Northwestern Press, 2015 [20]. My own views can be found in *Truth in Fiction: Rethinking its Logic*, forthcoming from Springer in the *Synthese* Library series in 2018 [58]. See chapter 3, "What readers know".

[50] The original words are "nothing happens in the world, not the flutter of an eyelid, not the flicker of a thought, without some redistribution of microphysical states." W. V. Quine, *Theories and Things,* Cambridge, MA: Harvard University Press, 1981; p. 79 [43].

[51] Peter T. Struck, *Divination and Human Nature: A Cognitive Survey of Intuition in Classical Antiquity*, Princeton: Princeton University Press, 2016 [49].

early philosophical analyses of those practices as, for example, in Plato's *Timaeus*, the psychological works of Aristotle and the writings of Iambliclus. In his excellent notice of Struck's book, Brad Inwood writes:

> "It is reassuring, in an odd way, that on Aristotle's theory such [just know] insight is made possible by the limited intelligence of the diviner: too much rational analysis swamps the frail channels that open the diviner up to open up the diviner to the subtle causal influences from the world."[52]

He continues:

> "On Struck's unifying theory, ancient philosophers and modern cognitive scientists are doing similar work in coming up with theories to account for cognitive phenomena outside the reach of conventional epistemology."

Perhaps the less than conventional CR approach can help close the gap. Inwood astutely recognizes that the seeming similarities between ancient philosophers and present-day cognitive scientists could be imperilled if it turned out that the data to which their respective theories respond weren't actually the same data. This prompts Inwood to suggest that cultural anthropologists could be a welcome addition to the modern team. It may be of some interest that in writing *Errors of Reasoning* I recruited the services of a hypothetical team of a visiting extraterrestrial cultural anthropologists, who would assist in the collection and analysis of data on the host's ground that laid the foundation for the CR epistemology that the book developed for their accommodation.

In addition to knowing-that, CR encompasses knowing-what. Knowing-what is a powerful contributor to humanity's survival and prosperity and in many ways the paradigm case of implicity and tacity. As we have it now, knowing-what cannot be reckoned on the model of the CR characterization knowing that. Here is a first approximation:

> *CR knowing-what*: A human knower S knows what to do on information I when, in processing I, S's reactive devices causally induce S to behave in manner M, his reactive devices are in good working order and operating her as they're meant to and I is good and well-filtered information, all conducing to successful praxis.

---

[52] Brad Inwood, "Ancient gut instinct: On the theory of 'surplus knowledge' in the classical world", *Times Literary Supplement,* March 17, 2017; p. 16 [3].

Driving a car requires knowing what to do, as does getting home from UBC on the 33 bus and, all the more so, getting to Berlin before Stalin takes it all. Is anyone we know capable of making explicit what we know what we're doing when driving a car or making our way home from work? Why should it be any less so in the case of Berlin? What would we reckon to be the frequencies and range of just-knowings that and just-knowings what to do?

It is adjacently interesting to me that for Quine the core of logic is logical truth. Logical truth is predicated grammar, on the grammatical forms of canonical English. "Logic chases truth up the tree of grammar".[53] Looked at from a CR perspective, Quine's is an attractive position, although he certainly wouldn't have seen it that way. By assimilating logic to form and form to grammar, he opened a link (which he didn't click on) to the naturalization of logic, much in the way that he'd naturalized epistemology. Had he availed himself of that option, he might have tried to naturalize logic without having to commit Mill's psychologistic indiscretion. If so, logic could be naturalized without having to be an empirical science. I agree with the naturalism, but reject the equation of logic with Quinean form, and withal its dismissal from its ambit of the empirical sciences of cognition. But with this admonition: Like all grand alliances, this one between logic and empirical science must be more circumspect than heartfelt. A naturalized logic of human reasoning cannot flourish without a well-disciplined empirical sensitivity. But not anything we happen to like will do here. Some of the least attractive features of cognitive psychology have been borrowed from command and control epistemology, especially its embodiments in formal epistemology.[54]

We can now turn to the business more immediately to hand.

4.5 Implicity and tacity

Let's take belief and inference as examples. We could also have considered understanding, knowledge, doubt, presumption, and so on, but there isn't time.

> *S draws the implicit inference* that the dog is hungry to the degree that the information in the scope of the believes-that operator ("The dog is hungry") is semantically inert and propositionally unstructured, and the

---

[53] W. V. Quine, *Philosophy of Logic,* second edition with corrections, Cambridge, MA: Harvard University Press, 1986; p. 35 [42]. First published in 1970 by Prentice-Hall.

[54] See, for example, Vincent F. Hendricks, *Mainstream and Formal Epistemology,* New York: Cambridge University Press, 2006 [30]. See also Paul Gochet and Pascal Gribomont, "Epistemic logic", in Dov M. Gabbay and John Woods, editors, *Logic and the Modalities in the Twentieth Century,* volume 7 of their *Handbook of the History of Logic* pages 99-195, Amsterdam: North-Holland, 2006 [25].

cognitive devices that have brought this implicit belief about are in good working order and functioning here as they should.

It follows at once that implicit beliefs are *tacit*. The reason why is that human fact-stating language is itself semantically loaded and propositionally structured.[55] From which it also follows that if we applied semantically loaded measures of linguistic formulability to a belief or a principle that is semantically unstructured, we'd be guaranteed to get it wrong. This is an anchoring principle of the common law. Similarly and to like effect:

> *S implicitly infers* that the dog is hungry from information I when (1) his circumstances are such that, in processing I he is causally induced to believe implicitly that the dog is hungry, are in good working order and operating here as they should without interference, and (2) if asked what led S to believe (if he does) that the dog is hungry, any disposition to reply would be a disposition to cite information I (if he could).[56]

4.6 Precedents again

Very well. If the rules of law created by precedents are semantically inert and propositionally unstructured, they don't say anything. How, then, does a judge discern a precedent? How does he know whether he's doing what the precedent requires in the case he's now deciding? This is the hardest of the logico-epistemic questions posed by the common law. If we don't get a handle on this, we'll leave an important part of our present business undone.

When a judge writes his *ratio* in a precedent-making decision, the precedent is not stated. When a judge in a later case is bound by that precedent, he has an implicit and tacit understanding of it. This is partly because he has a perfectly explicit understanding of how he's required to decide this case. He is told that he is to decide the case before him in the same way, when there is a sufficiency of relevant similarity between the facts in the present case and the facts in the original precedent-creating one.

The important constant is how good judges are at spotting these sufficiencies, and how close to spot-on juries are in arriving at criminal convictions. Of course, in each case, they also do what they do implicitly and tacitly. There are no quantitative sufficiency-scales for computing the closeness of the new facts to the old ones. There

---

[55] Of course, this doesn't mean (and isn't true) that fully articulate statements can't carry implicit implications or proceed from implicit assumptions.

[56] These can easily be adapted to implicities and tacities of knowing-what.

is no calculus of sufficiency. Here judges are on their own, relying on their devices to get the matter right. The central fact here is that the cue provided by their procedural instructions doesn't take the precedent into conscious awareness.

The question whether implicit beliefs take truth-values (i.e. are either true or false) is a profoundly good one. My answer is that they do not. But they do take truth-values *implicitly*. What this means is that if the implicit belief were cued to surface into conscious awareness it would acquire the semantic content that enables it to have a semantically grounded truth-value. Notice that, on this same reckoning, rules of law have truth-values, or their legal counterparts, implicitly but sure-footedly.

The fact to keep in mind is that if, as the common law insists, these precedent-created rules of law are subject to an implicit understanding only, they are laws that lack semantic-content and therefore cannot accurately be put into words, no matter how learned. The unspeakability of judge-made law justice is one of the common law's greatest epistemological achievements.

4.7 Making it explicit

Some people are vigorously opposed to the implicity and tacity theses in the form in which I have stated them here. They distrust all such talk as smoke-blowing obfuscation of the plain facts of juridical practice, not the least of which is the making explicit the provisions of a precedent on a later set of particular facts. Obfuscation of such derring-do is thought to rival in its foolishness the foolishness of mystification. Nothing could be clearer, they say, than that judges do their thing by interpreting the meanings of statutes, codes and regulations, and of precedents too. I agree with most of this, but think that its last conjunct is a mistake. Certainly judges do interpret meanings as a matter of course, as did the justices in *Saskatchewan* 2015. In that case, they interpreted ss. 1, 2 (b) and 2 (c) of the Charter and, in so doing, created a precedent. The question to ask here concerns what a future court would do on a different set of facts bearing a sufficiency of relevant similarity to the facts of the precedent-making case. What for example, would it do if the Calgary Highlanders, a regiment of which I am an honorary member were to down tools in, say, Afghanistan. There is a constitutionally guaranteed right for essential services public employees to strike in all the applicable ways afforded by the applicable labour laws in circumstances having a sufficiency of relevant similarity to the circumstances of the original finding. Certainly things don't get more explicit than this, do they? What greater explicitness could we ask for? There is no doubt that the finding is perfectly intelligible, as is the *ratio* which supports it, merits of the case aside. This was never in doubt in my reflections here. The question that remains is whether formulation of the new rule law captures its full generality. The answer is that it doesn't. What is more, if this formulation did give the new law its most explicit expression, the further answer would be that it could not capture its full generality. The reason it doesn't is provided by the formulation's own *sufficiency clause*, which is a paradigm case of Waismannian open-texture. While there could be cases in which the *Saskatchewan* 2015 precedent is clearly applicable and others when it clearly isn't, there always remain cases in which there is no right answer. As remarked above, there is nothing paralyzing about this. When there is simply no applicational fact of the matter, a court will do one of two things. It will either

(1) deny the appeal

or

(2) find for the appellant under the protective cover of semantic coercion.

It is also *phenomenologically* apparent that on any given occasion, most of what we know we're not consciously aware of, not remotely close to it. It is equally apparent that on any given occasion most of what we seek to know, or is in our current interests to know, is when known at all not something we're consciously aware of. The same is true of knowing-what. What is known on these occasions is known, at some or other level, subconsciously in the cognitive down-below. An epistemology that disregards these facts of lived human experience, or dismisses them out of hand, begets scepticism on so grand a scale as to make naïfs of all of neurotypical humanity. Scepticism this aggressive warrants the appellation "big-box", outstripping by several orders the inhumanity of the Edmonton Mall and of the even much larger one in Minneapolis. If in the general case, none of us knows what he thinks he does and what evolutionary cognitivists ascribe to his subconscious, then that we survive, prosper and occasionally build great civilizations is an utter mystery. Its purported solution is enshrined in the hypothesis that mother nature has rendered its children so mindlessly ignorant precisely because had it not, humanity could not have made the evolutionary cut.

The main trouble with this is that there isn't the slightest independent empirical evidence from the most mature of the empirical sciences of cognition that lends the mass-ignorance hypothesis any credence. "Ah, yes," some philosophers will reply, "so much the worse for empirical evidence!" All we need in reply to this is the observation that only the scantest minority of the big-box crowd is inclined to slight the cognitive enlightenments of the climate science, cultural anthropology and paleontology, physical chemistry and empirical economics, to say nothing of evolutionary biology itself.

Needless to say, none of this settles the matter conclusively. But the point at hand should not be missed. CR epistemology offers plausible refuge from the big-box scorners of human cognitive fulfilment. Anyone achieving safe anchorage there can easily see that CR epistemology accounts for the empirical facts of humanity's lived experience without having to slight them. In its provisions for human belief and decision, it extends their reach effortlessly to the implicities and tacities of common law practice. This is further reason to favour the CR approach over its big-box nihilist rivals.

There is a considerable body of opinion to the effect that the implicities of sufficiencies of relevant similarity can be subdued by a good account of analogical reasoning. I harbour doubts about this. This is the subject a forthcoming paper, which contemplates analogy's place in strategics.

5 Inconsistency-management

It is helpful at the outset to mark a distinction logicians use for inconsistency. A system is said to be "negation-inconsistent" if and only if it has at least one properly derived sentence whose negation is also properly derived. A system is said to be "absolutely inconsistent" if and only if its every properly derived sentence has a properly derived negation. By a substantial majority, logicians are of the view that the two properties are equivalent. If true this would mean that a single instance of inconsistency condemns the system to total inconsistency. The statement giving one half of this equivalence is frequently known as *ex falso quodlibet,* which loosely translated means "From a contradiction [=logical falsehood], everything follows."[57]

Although *ex falso* enjoys majority support, there is a lively and growing minority which disputes it. There is considerable confusion in sorting out what this disagreement is actually about. True, it's about whether *ex falso* is true or false, but it is much less clear what the4 disputants think *ex falso* is true or false *of.* The majority of the majority think that *ex falso* is valid in the majority logics, including classical, intuitionist and modal logic. The majority of the minority think that *ex falso* invalid in the nonclassical logics they favour, including all varieties of paraconsistent logic, Hewitt's Inconsistency Robust Direct Logic, and dialethic logic.[58] There is a

---

[57] One of the meanings of "quodlibet" is "potpourri".

[58] See in addition to Schotch *et al.* 2009 and Hewitt and Woods 2015, Bryson Brown, "Preservationism: A short history", in Dov Gabbay and John Woods, editors, *The Many Valued and Nonmonotonic Turn in Logic,* pages 95-127, volume 8 of Gabbay and Woods, editors, *Handbook of the History of Logic,* Amsterdam: North-Holland, 2007 [24], and Graham Priest, "Paraconsistency and dialetheism", in Dov and John Woods, editors, *The Many Valued and Nonmonotonic Turn in Logic,* pages 129-204 [1], volume 8 of Gabbay and Woods, editors, *Handbook of the History of Logic,*

feature shared by all these logics, majority or minority. All their properties of note are defined over uninterpreted formal languages L. There are various views about how these formal properties relate to similarly named ones in natural language. On one approach, the formal properties map one-to-one to their counterparts in natural languages N and do so in ways that guide the development of the semantics of them. For this to work, there must be provable formal representability relations from the semantics of L to a dedicated fragment of N which preserve the properties of the semantics of L desired for a semantics of N. In yet another approach, the good done by a L-semantics for the semantics of N is done by bringing the grammar of N into line with the formal grammar of L in a regimentation procedure that rewrites N into canonical notation. In yet a third and more irrealist approach, there are no facts about the logic of N beyond the facts that hold in some other formal logic of instrumental interest to the logician.

The logic on offer here is not a formal one in the manner of those for uninterpreted formal languages. It is a disciplined naturalized empirically sensitive logic, *NL,* purpose-built for human-language reasoning, judgement and decision in real time. Formal models may or may not repay attention in various ways, but none of them calls the shots for *NL*. This makes how *ex falso* fares in those multiplicities of formalized structures irrelevant to its interests.

How, then, does *ex falso* fare in, say, English? It is provably *true* there. The proof is set out in section E of "The logical foundations of strategic reasoning", and there is no need to tarry with it here. Then reason why is that, even if perfectly true in English, it doesn't really matter. The reason for that has everything to do with the properties preserved under the closure of the logical implication in English. If *ex falso* is true, then any system harbouring a contradiction has the validly derived negation of any sentence derivable there. Would the omniderivability of such a system deny truth to all of its derivable sentences? Couldn't some be true and their validly derived negations not?[59]

Consider, for example, Newton's *Principia Mathematica* or more recently Frege's *Grundgesetze I. Principia* is inconsistent by way of its of "The logical foundstions of strategic reasoning", and there is no need to tarry with it here. The reason why is that, even if perfectly true in English, it doesn't really matter. The reason for that has everything to do with the properties preserved under the closure of the logical implication of all English sentences logically implied by a contradiction formulate calculus of infinitesmals, and *Grundgesetze* by way of the inconsistency abetted by its Basic Law V. Consider the wealth of knowledge about the laws of

---

Amsterdam: North-Holland, 2007 [24].

[59] Recall, *ex falso* does not say that any sentence of English is unambiguously true and false together. It says only what is validly derivable from any contradictory sentence.

motion and gravity in Newton, and all that Frege's students had learned from his pre-*Grundgesetze* lectures. Is there any independent empirically-based reason to support the claim that omniderivability precludes truth and knowability in some appropriately selective way. The great advances in astrophysics and mathematics in the aftermath of those notional setbacks speak for themselves. Omniderivability is not truth-wrecking, and not knowledge-exterminating either. Every sentence of those systems has a validly derived negation, but only some at most are true; and those that are are closed under truth-preserving consequence.

Perhaps it is not quite clear how beings like us are able to track the truth in a system in which every one of its sentences has a valid derivation. The CR approach has an answer in wait, a somewhat promissory and conjectural one, but better than a blank stare.[60]. Our ability to track truth in thoroughly inconsistent systems derives from a filtration device that helps keep false derivations at bay. As of now, it is a device of which we have a mainly implicit and tacit understanding.

6 The Inconsistencies of Verdicts

I come now to carriers of inconsistency in the common law. In its various ways the common law has a surfeit of inconsistency which, somehow or another, does not preclude it from true and just findings, not perfectly so, but so within an acceptably narrow margin of error. Of particular interest is how juries manage to come to unanimous verdicts.

6.1 Unanimity

When after a grueling two month trial a jury retires to consider its verdict, it is a wonder that one will actually be produced at all. It is easy to see why. Jurors have been required to pay careful but non-interactive attention to masses of complexly conflicted information, real or purported, to keep it all in mind without discussing any of it with anyone, especially with fellow jurors. All judgement having been postponed the jurors' world has now tilted madly. The whole burden of judgement now falls upon them in one fell swoop. It is not in the least unusual for a juror to enter the jury room in a haze, without any clear idea of what he should make of the trial he's suffered through for sixty days of sittings. Like an actor who "goes up" just before his entry on stage, the juror knows the panic of having forgotten

---

[60] Filters are briefly discussed in part E, section 2 of the predecessor paper, and more widely in Dov Gabbay and John Woods, "Filtration structures and the cut-down problem for abduction", in Kent A. Peacock and Andrew D. Irvine, editors, *Mistakes of Reason: Essays in Honour of John Woods,* pages 398-417, Toronto: University of Toronto Press, 2005 [22].

everything he's witnessed in court. Other jurors, far from having "gone up", emerge from the courtroom with their own settled theories of the evidence. His alone makes the point that, contrary to the bench's command not to rush to judgement, real-life human beings are incapable of withholding a provisional and still-open theory of what they've seen and heard in court. All the same, when pressed to say what that theory is, some jurors are at a loss for words, and still others can find the words to state it, but fewer ones to support it.

Beyond its duty to arrive at a verdict, the purpose of a jury's deliberation is to start the engines of each juror's cognitive down-below working interactively with the down-below, mechanisms of the others, so that together the multiagent that's the jury is enabled to render the *jury*'s verdict.

An especially tough requirement – tough and I would say unrealistic – is that juries are given one of only two pre-set verdicts to arrive at – guilty or not-guilty. There are no intermediate options.[61] What is more, there is no verdict, for or against, unless there is unanimous support for it. No other decision-method in criminal law is subject to so heavy a requirement, not even the highest court in the land when it decides the constitutionality of an existentially explosive and furiously disputed question on a five-to-four split. It bears on this that any matter put to a jury will likewise be an existentially stirring one, and the occasion of high adversarial strife. Indictable crimes are not easy things to be indifferent to. Neither are the deprivations of liberty and treasure, and the collateral ones of destroyed reputations, collapsed marriages, and deep and prolonged wretchedness. It is well known that the more a disputed matter is existentially fraught, the lower the likelihood – indeed the practical possibility – of arriving at an unanimous bimodal decision about it. Another complicating factor is the sheer size of the decision-instrument. Even big courts have nine judges at most, as for example, in Canada and the United States. Juries have twelve, which is another third more.

When a judge instructs a jury, they will be told that they must pay close and open-minded attention to all they see and hear in court and to what they will say to themselves when they retire to consider their verdict. This we may call the "total evidence". Total evidence incorporates several quite different elements. First and foremost is the evidence sworn by witnesses in answering the questions put

---

[61] In Scots law (which is not common law) a third verdict of not proven is permitted. Sometimes a common law jury does indeed have a third option of a kind. In certain instances, it can acquit the accused of the crime he's being tried for and convict him of a lesser contained offence. It is also true that a common law jury will sometimes to try to persuade the trial judge that they are incapable of reaching agreement. No judge will release a deadlocked jury before repeated exhortations to reconvene and break it on their own. Common law judges greatly dislike hung juries and take emphatic steps to prevent them.

to them by counsel, including their answers to questions put to them under cross-examination. Answers to these questions are "answers given in evidence". Given the adversarial character of criminal trials, it is entirely routine that the evidence of Crown witnesses will in some given particular or other be contradicted by defence witnesses. A second important component of the total evidence is provided by counsels' closing arguments, in which they develop their respective theories of the (witnesses') evidence. It is an inbuilt feature of the closing-argument stage that theories of the case will in a most central way contradict each other. There are still further elements of the total evidence, but there is no need to mention them here. We've said enough to show that the (total) evidence on which the jurors must base their decisions is inconsistent. If the decision is underlain by a big information system, it will be permanently and pervasively inconsistent. If *ex falso* is true, all of it will be inconsistent. When a judge tells a jury to attend to all they see and hear, he emphasizes the necessity of objectivity and the suppression of preconception and bias. He tells them they must use their common sense and their own shared experience of human life. Any juror heeding those instructions would be ill-disposed to give to his own decision on an inconsistent backing.

It would therefore appear that even though the total evidence is inconsistent and that juries must consider it all, an individual juror will reach his decision on some *consistent subset* of that total, corresponding roughly to which parts of it he believes with highest material confidence. As various commentators have supposed, if this were so, two things would have to have happened. Each individual juror would have divided the total evidence into all its consistent subsets, from which she would then have selected the largest consistent subset in which she reposes her most confident material belief. Two further problems are thereby occasioned. One is the question of how a jury's most-assured *belief* that there is no reasonable doubt of the accused's guilt rises to the *proof* of guilt required by the criminal standard. I shan't take up this question here, and want instead to turn our attention to a second difficulty.[62] It is actually two related ones.

One pertains to how the partitioning of an inconsistent evidence-set into all its consistent subsets is actually brought about. The other is how an evidence-set is to be scanned for consistency. Whatever the manner in which these objectives might be achieved in theory or principle, they are manifestly beyond the conscious powers of any human being doing the best that's humanly possible. While part of the problem is the daunting size of the evidence generated by a trial, that mightn't be the heart of the matter. Suppose that the neurotypical human's belief-system harboured a

---

[62] A fuller discussion can be found in chapter 21 of *Is Legal Reasoning Irrational?*[57], entitled "An epistemology for law".

scant 138 logically independent atomic beliefs. A consistency check of that slender set would require "more time than the twenty billion years from the dawn of the universe to the present."[63] This is a problem of long standing in research communities that investigate the computational capacities of real-life cognitive agents. There is a further problem which, so far as I have been able to learn, hasn't yet surfaced in the legal literature. It will help in framing the problem if we allow ourselves the hopeful assumption that somehow or other in some *subconscious* manner, the human reasoner actually does manage to achieve the partition of the inconsistent total evidence into each of its subsets, and is also able to run a consistency check on each. An even more simplifying assumption is that the partition task is surplus to need, and that what more plausibly happens is that an agent's *belief-forming* mechanisms scans the inconsistent evidence-sets and extracts the largest subsets that are reasonably believable. The consistency of these subsets would presumably flow from the reasonability of their belief-worthiness. All this is conjecturally abductive and should, as soon as practicable, be put to such experimental test as is currently or foreseeably available in the cognitive sciences.

The second and largely unnoticed problem flows directly from the logical structure of unanimity in hotly contested and existentially fraught decision spaces. Finding out will require the services of a unanimity logic, or would if we had one on hand. Consider a hypothetical case. When the jurors begin their deliberations, they are often doubly conflicted. They disagree about the accused's guilt and innocence – I mean, of course, his legal guilt or innocence as determined by what the jury makes of the total evidence. Moreover, even those jurors who agree on guilt and innocence, will frequently disagree on the evidential basis on which their respective conclusions rest. In the first instance, if the case has been contentious and strongly and capably fought by both sides, evidential subset-inconsistency is virtually assured. When this happens, the union of the twelve evidence-sets in which the verdict is grounded is inconsistent. Negotiation enters the picture in a way that I'll caricature in an over-simplified dialogue between two jurors.

> *Juror one*: Don't you see that even on your reading of the evidence, Guilty [not-Guilty] is the right verdict.
>
> *Juror two*: Yes. Thanks for your help in getting me straight about that.
>
> *Juror one*: Thanks in turn for being so open-minded.

---

[63] Christopher Cherniak, "Computational complexity and the universal acceptance of logic", Journal of Philosophy, 91 (1984), 739-758 [15]; pp. 755-756. Of course, "the present" of the Cherniak reference was 1984. These thirty-two years later, things have got no better.

Juror two: I was just doing what I'm supposed to do.

Juror one: Now that we've agreed on the verdict, there's only one other thing to clear up. Although it supports what we agree is the correct verdict, your reading of the evidence is in certain respects at odds with my own. Obviously, a verdict based on incompatible supporting evidence cannot be allowed to stand. So what will it take to bring your reading into line with mine?

Juror two: Look, we both want agreement on the verdict, but if the cost of having it is that I adopt your reasons for supporting it, I cannot in all conscience give the verdict we've just agreed to. For that to happen, I'll have to base it on my own evidence-set.

Juror one: Okay! Okay! I yield. Let's agree to convict (or acquit) and agree to disagree about why.

Juror two: After all, hasn't each of us reached agreement in an intellectually conscientious, though different, way?

There is nothing fanciful or tendentious about our imaginary example. Something similar frequently happens even in the courts above when some highly contentious issue has been decided by majority vote. Unless the majority justices have fully equivalent *rationes decidendi* (reasons for judgement), each must write a separate one. When this happens, and since the majority decision is *eo ipso* the *court*'s decision, the court will have decided a case for reasons it cannot agree on. This leaves it entirely open that one or more of the majority justices will have thought the *ratio* of other justices to have been an inadequate basis on which to reach the decision they all agree on. Nor can it be ruled out that one *ratio* is inconsistent with another. In some instances – think here of *R. v. Morgentaler* (1988) – the majority *rationes* not only differ but when taken together fail to be internally coherent, and in some instances contradict one another. As we saw, when this happens, the majority *decision* stands, but no precedent is set by the court's disunified *ratio decidendi*, hence no new rule of law.[64]

6.2 The composition fallacy

Suppose now that, in our hypothetical example, the jury decides to convict. On present assumptions, each juror convicts on a consistent subset of the evidence, but

---

[64] One of the majority justices in *Morgentaler*, found that the Charter's s. 7, which constitutionalizes everyone's right to "security of the person" immunizes abortion from criminal liability in all cases. Wo when writing the Charter in 1982 could have predicted that a high court judge someday in the future would say that this is what it meant?

the *jury itself* convicts on contradictory evidence. I take it that it is, at least tacitly, assumed by legal scholars that if the vote of each *juror* is consistently evidenced, the verdict of the *jury* is also a consistently based one. Were it otherwise, jurisprudential *angoisse* would have gone viral, and all over the place there would be distinguished named chairs in law schools devoted to inconsistency-expungement. The present comparative silence of legal scholars strongly suggests the commission of a special case of the fallacy of composition, according to which it is reasoned that if all parts of a whole have a given property, it follows necessarily that the whole entity has it too. Logicians and decision theorists who study the rationality of collective decision-making often take evasive action to preserve their thinking from this fallacy. One (the right one) is not to draw the compositional inference in the first place. The other (the dubious one) is to relativize the compositional inference to ideally rational agents, both collective ones and their component individual partners, largely to ease the theory's engagement of powerful and simplifying mathematical methods and higher prospects for some impressive new theorems. I happen to join with those who regret these idealized sleights of hand in so many precincts of rational decision theory. This is not the place to litigate that regret either. The point to note here is not that legal theorists fall into the same questionable habit with regard to inconsistently based jury verdicts, but rather that in a jury deliberation itself, any initial *verdict*-disagreement must be dispelled. However, in so doing, there is no requirement that all *grounding*-disagreement also be dispelled, never mind that the opposite might be what jurors will have tacitly assumed. Whatever the workings of an unanimity logic might turn out to be, the fact remains that the legal duty to arrive at verdicts without dissent or abstention is wholly dischargeable by eliminating verdict-disagreement. Whether grounding-disagreement is also expunged is entirely and, at best, a contingent collateral benefit of the resolution of the former. So let's repeat the central point:

> *Inconsistency-tolerance*: Removal of grounding-disagreement is neither a necessary nor frequently realized condition on verdict-disagreement removal.

Jury deliberations are said to be *negotiations*, and indeed the only ones that are permitted once a trial is underway. Sometimes counsel will negotiate issues out of court, as when a prosecutor offers the defence a reduced charge in return for a guilty plea. But none of this, beyond its formal announcement, happens at trial. Jury deliberations, on the other hand, don't occur in the courtroom, but they are fundamental components of criminal trials, and they usually take place in the Court House, not in a plush boardroom at The Ritz. The unanimity requirement puts the negotiation-space into a tight confinement. Between guilt (G) and innocence (not-G) there is nothing whatever to negotiate. Jury deliberations are also exercises in *mind-changing*. Jurors can change their minds about G and not-G, and they can change their minds about why-G and why-otherwise. We might think that the only rational and intellectually conscientious way of changing minds about verdicts is by changing minds about what parts of the evidence are the most probative and in which of its consistent subsets it is to be found. However, this is not in fact the case, and it is precisely the point at which some *bona fide* negotiation can conscientiously take place. In the hypothetical trial under present consideration, there are twelve consistent subsets of the inconsistent total evidence, some of which might with low likelihood be extensionally equivalent to one another. More typically, when the trial has been highly contentious and well-handled by opposing counsel, there is a nontrivial positive likelihood that each of the twelve will be inconsistent with at least one of the others. The fact that the very existence of this feature of juried decisions leaves no discernible footprint on jurisprudential scholarship might well be explained by the law's routine but tacit commission of the composition fallacy.

It is time to give the present problem a name. Let's call it what it is: the verdict-inconsistency problem (VIP). There will be plenty of logicians and decision theorists who won't be able to believe that experienced trial lawyers and judges are simply impervious to the likelihood that juries return inconsistently based verdicts with a notable frequency. Judging from what we know of jury room deliberations, based in large part on what jurors report after dismissal, these same theorists have the same difficulty in believing that none of them had cottoned on to the likelihood – indeed sometimes the fact – that their verdict had been crafted upon unresolved grounding inconsistency.[65] Suppose that sometimes these reservations are grounded in plain fact, and that it is simply not true that the VIP has escaped notice. When this happens, why wouldn't its spotters draw attention to the problem? Where, it might be asked, are the whistle-blowers, and why is it that they leave their whistles unblown?

Perhaps part of the reason is the public's substantial confidence in the criminal justice system, especially in politically stable countries whose governments aren't noticeably corrupt at their core. The man or woman at large in countries such as Britain, Canada and in most jurisdictions of the United States is likely to believe that false convictions, although profoundly regrettable, fall within an acceptably narrow margin of error. Let's also give a name to this. Call it the *verdict-confidence*

---

[65] In most common law jurisdictions such as England and Canada, it is a criminal offence for jurors subsequently to reveal any aspect of what occurred during jury deliberations. In the United States, there is no such general prohibition. Most of what Canadian scholars know of the ins-and-outs of jury rooms comes from American after-the-fact self-disclosure.

phenomenon (VCP). The question now is to sort out the tangled complexities of the relations, such as they may be, between VIP and VCP. My objective in this section has been largely exploratory. I want to call attention to what strikes me as an unnoticed – or anyhow undeclared logico-epistemic problem with verdicts in criminal trials. My hope is to have taken a small first step in passing to the research communities of common law jurisprudence the complexities of the VIP-VCP dynamic for their further consideration.

6.2.1 Criminal proof

Before we can say that the common law has anything of note to teach theorists of strategy, it would pay us to turn our minds to how, in these clouds of inconsistency, juries are able to arrive at true and just verdicts. The wording of the standard of criminal proof suggests that a jury's duty is to assess the respective success of the Crown's attempt to prove the guilt of the accused as charged, in light of the evidence adduced at trial, and the defence's attempt to rebut that proof. So construed, the jury is the counterpart of the referees of an accompanying learned journal, whose duties are to assess the respective merits of a submitted paper and a note to rebut it. If we attended to the facts of lived jury-experience, we'd see how off the mark this analogy is.

The best place to go to see why is by attending to how the proof standard is actually dealt with by real-life juries on the ground. There is a core sense in which prosecutors try to prove the accused's guilt and defence counsel try to prove his innocence, whether by proving it outright, or by proving that the Crown's proof is not a proof after all, hence that he is not guilty *as charged*. This sets up the expectation that the jury's job is to assess the success or failure of those proofs, by determining which, if any, meets the law's undefined standard for conviction.

This is not what juries do. Nothing will slow down a deliberating jury more than fruitless quibbles about what that standard is, and what it would take to meet it. So I surmise that in fairly quick order they stop doing it, and a good thing too. Consider now how a sensible judge instructs a jury about how it is to proceed in deciding the case. I take the liberty of pretending that he has had some exposure to CR epistemology and that this has helped shape his charge, as follows:

> "Ladies and gentlemen of the jury: If after having given careful and open-minded and fair consideration to all you've so far seen and heard at trial and to what will be said and heard in the jury room, and mindful too of all my instructions, you find that you cannot in all intellectual conscience convict the accused, then you must acquit him. If you find that under

these same conditions you cannot in all intellectual conscience acquit the accused, then you must convict him. Period."

We see at once the point of it all. It has to do with conscientious belief, induced with the indispensable impact by the working devices of the juror's fully engaged cognitive wherewithal. The down-below and the up-above converge on the decision point. The down-below bears most of the responsibility for what the juror implicitly and tacitly makes of the total evidence. The up-above weights in, making possible the juror's examination of his intellectual conscience, which discloses to a degree sufficient for action the durable state that he is now in.

For a juror to be in that state, it is necessary for him to track the patterning of events reflected in the totality of a trial which enacts the Crown's strategic purpose to put the accused in prison and the defence's counter-strategy to keep him out and free. To the extent that he is able to do that, he does it implicitly and tacitly. To the extent that this implicit and tacit tracking permits conscious engagement of the juror's examination of his intellectual conscience, the forces of down-below and up-above conduce to truth and justice at the common law bar of criminal justice.

At this point, we have concluded our search for a logic and epistemology that would make some of the peculiarities of common law reasoning and cognition matters of course. We have seen that the logic that works best here is a selective empirically-sensitive naturalized logic, working hand in glove with a causal-response epistemology, which is the epistemology that also works best for our purposes here. We have seen readily these naturalized approaches bend to the task of shedding real light on the mechanics of implicit and tacit cognition, especially on why it should so often resist full expression and articulation. Along the way, we've found it possible to expose inconsistency-management to theoretical treatments hitherto unknown to logicians and epistemologists. It remains to be seen whether these same measures will bear fruit with the peculiarities of strategical reasoning noted in CLM approaches to strategics. We turn to this now.

7 Strategics[66]

When in 1945 the World War II Allies arrived in Berlin before the Red Army could take it all, it was the conclusion of massively complex strategic planning, organization and execution involving literally large armies of war-fighters to bring off. In all its aspects, Berlin '45 was also both literally and figuratively the work of whole

---

[66] Again, in my companion piece in this issue of the *IfCoLoG Journal,* strategics is given a somewhat more expansive treatment than here. The companion paper is "The logical foundations of strategic reasoning: Inconsistency management as a test case for logic."

armies of interactive and concordant partnerships, each in its own turn a complex of smaller partnerships. Anyone with some military experience will know the words spoken by every soldier in human history: "The Army doesn't know what it's doing, and neither does the Government!" Even when taken as the jocular gripe of a tired and hungry grunt, it is a telling and insightful joke all the same. The Supreme Headquarters of the Allied Expeditionary Force (SHAEF) was established in 1943. It commanded the largest number of forces ever assigned to an operation on the Western Front, including the First Airborne Army, the British 21st Army Group (First Canadian Army and Second British Army), the 12th Army Group, and the American 6th Army Group (French First and American Seventh). Its strategic purpose was to launch a phase of Operation Overlord against occupied France. According to some experts, SHAEF's strategy is to be found in the *patterning* of nodes of complex and interacting decision-chains originating in General Sir Frederick E. Morgan's earlier plan, moulded into its final version in mid-March 1943 and executed on June 6th the following year.

SHAEF was a large and complex *multiagent*, a composite of its separate parts. A multiagent is an interactive cooperative aggregate of subagents, often themselves multiagents in their own right, working together collectively according to some operational agenda or in fulfilment of some conventional arrangement. Given the size and complexity of SHAEF, we would be right to think of it as a superagent. Multiagents pose interesting questions for epistemology and the philosophy of mind. SHAEF's causal impact on the European war is beyond doubt. Whether SHAEF had a mind of its own is not so clear. How could it have? Did it have a brain? Perhaps a virtual one?

In the Preface to *Strategy: The Logic of War and Peace,* [33]Edward N. Luttwak writes,

> My purpose ... is to uncover the *universal logic* that conditions all forms of war as well as the adversarial dealings of nations even in peace." (Emphasis added)

To this he adds:

> "... the logic of strategy is *manifest in the outcome* of what is done or not done, and it is by examining those often unintended consequences that nature and workings of the logic can best be understood." (Emphasis added)

In eschewing abstract theories for strategies, Luttwak's scornful words might well be borrowed by a jurist for a like rebuke of abstract theory in the jurisprudence of the common law. In *Strategy*'s Part I Luttwak writes of what he takes to be the "paradoxical character of strategy, indeed "the blatant contradiction that lies within":

> "Consider the absurdity if equivalent advice *in any sphere of life but the strategic*: if you want *A* strive for *B*, its opposite, as in "if you want to lose weight, eat more" or "if you want to become rich, earn less" – *surely we would reject all such.*" (pp. 1-2; emphases added)

It is not entirely clear that Luttwak's notion of blatant contradiction is intended to be understood in the way a logician would. It doesn't really matter. Luttwak certainly acknowledges the sheer magnitude of the information-systems of the strategic thinking that underlay the Allied victory in World War II. Information-systems this big would take multiples of millions of lines of code for software engineers to computerize, assuming that such a question could arise in any practicable way in the six-year interval from 1939 to 1945. Information-systems like this are what Carl Hewitt calls "big", such as those underwriting the Five Eyes security network, the climate sciences, the Peoples Liberation Army, and British Columbia's health care system. These systems have features that give orthodox logicians the vapours. They are perpetually, pervasively and ineradicably inconsistent in the logician's sense, notwithstanding their enormous (but not perfect) and often indispensable practical value. These inconsistencies are what Hewitt calls "robust".[67] A standard line in logic is that an inconsistent theory is a disaster, disabled for fruitful cognitive work of any kind.[68] The arresting thing about the systems in view here is that they are cognitively valuable and don't go off the rails.

It turns out that a system needn't be Five-Eyes big to be robustly inconsistent.[69] It is widely agreed that human deep memory is inconsistent. So, as we have seen, are the information systems animating the common law. This creates a serious and still largely unmet need for a comprehensive theory of inconsistency-management not just for strategics but far more widely.[70]

---

[67] Carl Hewitt, "Inconsistency robustness in foundations: Mathematics self proves its own consistency and other matters", in Carl Hewitt and John Woods, editors, *Inconsistency Robustness*, volume 52 of Studies in Logic, pages 104-157, London: College Publications, 2015 [2]. See revised edition in preparation.

[68] Apart from the knowledge that the system is no good.

[69] John Woods, "Inconsistency: Its present impacts and future prospects", in Hewitt and Woods, pages 158-194 [56].

[70] I have also written about this in section E of "The logical foundations of strategic reasoning: Inconsistency management as a test case for logic". See also my "How robust can inconsistency get?", *IfCoLog Journal of Logics and Their Applications,* 1 (2014), 177-216 [55].

Luttwak considers Carl von Clausewitz as "the greatest student of strategy who ever lived." (p. 267) yet he also says with evident approval that Clausewitz "was simply uninterested in defining things in generic [= universal] abstract terms; he regarded as such attempts as *futile* and *pedantic*. "(Emphases added)[71] Henry Mintzberg writes to similar effect:

> "Strategy making needs to function beyond the boxes to encourage the informal learning that produces new perspectives and new combinations .... Once managers implement this, they can avoid other costly misadventures caused by applying formal techniques, without judgement and intuition, to problem solving."[72]

If this matters for Clausewitz, it also matters for Luttwak because Luttwak models himself on Clausewitz. It would also matter for anyone who follows Mintzberg that he identifies a strategy as a "pattern in a stream of decisions" rather than as a kind of overt, articulated planning.[73] Mintzberg appears to be onto something important in a way that helps explain Luttwak's respect for Clausewitz. When he says that strategy is discernible in decision patterns, what Mintzberg seems to be suggesting is that, although discernible in a pattern of decisions, strategy is implicit and unvoiced in the interactive dynamics of decision-sequences on the ground, of which might be accessible to expression and consideration in a historian's or strategic analyst's theoretical speculations well after the fact. I have a different view. I think that what the successful military historian manages to do with his articulated formulations of opinion is to causally induce in his readers some tacit and implicit grasp of the pattern, which is just what the documentary record did for him. For ease of reference, I'll label this approach to strategics the *CLM approach*.

This would help answer some intuitively plausible questions. If strategies aren't amenable to full expressibility and are not matters that lend themselves to spoken presentation in the lecture halls of, say, the Imperial War College, and the likes of Clausewitz, Luttwak and Mintzberg could publish books which convey the doctrines

---

[71] Edward N. Luttwak, *Strategy, The Logic of War and Peace,* revised and enlarged edition, Cambridge, MA: The Belknap Press of Harvard University Press, 2001 [33]; first edition, also with Harvard, 1987. Carl von Clausewitz, *On War,* Michael Howard and Peter Paret, editors, Princeton: Princeton University Press, 1984 [31]; originally published as *Von Krieg* in 1832. See also Luttwak, *The Grand Strategy of the Roman Empire: From the First Century CE to the Third,* Baltimore [34]: John's Hopkins University Press, 1976 .

[72] Henry Mintzberg, *The Rise and Fall of Strategic Planning: Reconceiving the Roles for Planning, Plans, Planners,* Toronto: Free Press, 1994; p. 128 [39].

[73] Henry Mintzberg, "Patterns in strategy formation", *Management Science,* 24 (1978), 934-948 [37]. See also "The Design School: Reconsidering the basic premises of strategy formation", *Strategic Management Journal,* 11 (1990), 171-196 [38].

in view here and yet purport to render accounts in plain German or English of what strategies are and how they work? If an author believes that strategies are discerned but aren't amenable to articulate expression, wouldn't the words of these books on strategy fail this purpose outright? Why write books in the first place? We might note here that these same questions could be put to anyone writing about the implicity and tacity of *lex non scripta* in the common law tradition. My answer, which is conjectural, is that there exists a causal link that causally stimulates the theorist's implicit and tacit understanding of the documentary record of a sequence of decision points patterning efforts of a multiagent to achieve a common goal, which causally directs the production of his scholarly work. Given a strategy's inherent implicity and tacit, no such words can bring it to explicit written intelligibility, but the better the work is the greater will be the implicit and tacit understanding of its reader now have of the strategy in question. It is perfectly possible to arrive at a large knowledge of SHAEF's M45 strategy for Berlin. But the knowledge needn't itself be fully articulable or formulable to be *bona fide* knowledge. The first thing to learn is that the implicit and tacit are not impediments to knowledge. Needless to say, prospects of serious consideration of my causal-implicit conjecture depends on whether there is a credible epistemology that offers it independent support. This, as we saw, was undertaken in previous sections, where I supported a similar thesis for our understanding of the common law's implicities and tacities.

No one should be in the least doubt about the masses of documentation embodied in the decisions in virtue of which the Allies' just-in-time arrival in Berlin, or about the masses of historical and analytical writings thereafter. Equally, no one should shirk the extensive wordiness of the reasons for judgement penned in all precedent-creating decisions by common law courts, still less the libraries of volumes of scholarly discourse about what those findings reveal and how they call the shots for subsequent cases. Their value lies in the causal good they do.

I should be clear about the position I take on the CLM thesis. I am not saying that the thesis is true to what Clausewitz, Luttwak and Mintzberg intended, even though I think it might be. Neither am I saying that the CLM thesis is true, never mind what C, L and M may have intended, even though I think it is. What I offer here is an exercise in abduction in response to the following question: "If the CLM thesis *were* true, what would be the epistemology that best supports it?" This was also my intention in framing my interpretation of the common law doctrine of unwritten law. Masses of documentation underlie the practices and jurisprudence of the common law. This doesn't change the fact that parts of what practitioners and legal scholars know about the staggering complexities of case-law they know only implicitly and tacitly, and that much of it will remain unformulable and inarticulable for as long as the common law persists. The naturalized logic and causal-response epistemology in play here help explain these peculiarities. The same should apply to the peculiarities of strategic reasoning. Masses of documentation exist of the planning and execution of the Allied strategy to liberate Europe. This doesn't change the fact that parts of what military historians now know about that utterly tangled jumble of momentous events they, too, know only implicitly and tacitly, and that much of it will exceed their capacity for full expression and articulation. What's true of the historians, was massively more so of the participants. Moreover, as we saw, most of what we'll ever know is known subconsciously, and therefore is highly unlikely to surface into the light of consciousness formulation and articulation. If every individual human being is a cognitive stranger to himself, how could it not be massively more so for the superagent that liberated Europe between 1944 and 1945, never mind the absolute inconsistency of its data-sets?

Acknowledgements

I warmly thank the following people for their productive suggestions and helpful criticisms both during and after the Konstanz conference: Matthias Armgardt, Paul Bartha, John Beatty, Gottfried Gabriel, Thomas Müller, Jacob Rosenthal, Margaret Schabas, Chris Stephens, and some others whose names I didn't catch (apologies). I am also indebted for later correspondence to Eva-Maria Engelen, Gottfried Gabriel, Carl Hewitt and Dirk Schlimm. Lorenzo Magnani generously offered his fruitful advice on an earlier draft of this paper, and Woosuk Park has done me the kindness of suggesting improvements to the penultimate version. Christopher Mole, Jonathan Ichikawa, Matt Bedke and Paul Bartha were also of interest in their plucky resistance of my efforts to rationalize inconsistent verdicts. I also thank the journal's referees.

References

[1] Paraconsistency and dialetheism. In *The Many Valued and Nonmonotonic Turn in Logic*, volume 8 of *Handbook of the History of Logic*, pages 129–204. Amsterdam: North-Holland, 2007.

[2] Inconsistency robustness in foundations: Mathematics self proves its own consistency and other matters. In *Inconsistency Robustness*, volume 52 of *Studies in Logic*, pages 104–157. London: College Publications, 2015.

[3] Ancient gut instinct: On the theory of 'surplus knowledge' in the classical world. page 16, 2017.

[4] Reorienting the logic of abduction. In *Handbook of Model-Based Reasoning*, pages 137–149. Springer, 2017.

[5] Carleton Kemp Allen. Law in the making. 1964.

[6] Complete Works Aristotle. The complete works of aristotle.

[7] Fred Attneave. Applications of information theory to psychology: A summary of basic concepts, methods, and results. 1959.

[8] Kent Bach. Default reasoning: Jumping to conclusions and knowing when to think twice. *Pacific Philosophical Quarterly*, 65(1):37–58, 1984.

[9] John A Bargh and Melissa J Ferguson. Beyond behaviorism: on the automaticity of higher mental processes. *Psychological bulletin*, 126(6):925, 2000.

[10] Jerome Edmund Bickenbach. *Canadian cases in the philosophy of law*. Broadview Press, 1993.

[11] Robert Brandom. Making it explicit, 1994.

[12] Edmund Burke. *An appeal from the new to the old Whigs*. Bobbs-Merrill, 1791.

[13] Benjamin Nathan Cardozo. *The Nature of the Judicial Process*. Yale University Press, 1921.

[14] Gregory N Carlson and Francis Jeffry Pelletier. *The generic book*. University of Chicago Press, 1995.

[15] Christopher Cherniak. Computational complexity and the universal acceptance of logic. *The Journal of Philosophy*, 81(12):739–758, 1984.

[16] Zoltan Dienes and Josef Perner. A theory of implicit and explicit knowledge. *Behavioral and brain sciences*, 22(5):735–808, 1999.

[17] Fred Dretske. Epistemology and information. *Philosophy of information*, pages 29–47, 2008.

[18] R. Dukas and J.M. Ratcliffe. *Cognitive Ecology II*. University of Chicago Press, 2009.

[19] Reuven Dukas. *Cognitive ecology: the evolutionary ecology of information processing and decision making*. University of Chicago Press, 1998.

[20] Daniel Dwyer. Claude romano: At the heart of reason. michael b. smith and claude romano (trans.). *Husserl Studies*, 33(1):81–89, 2017.

[21] Luciano Floridi. Blackwell guide to the philosophy of computing and information. 2003.

[22] Dov Gabbay and John Woods. Filtration structures and the cut down problem in abduction. In *Mistakes of Reason: Essays in Honour of John Woods*, pages 398–417. Toronto: University of Toronto Press, 2005.

[23] Dov M Gabbay, Paul Thagard, John Woods, Pieter Adriaans, and Johan FAK van Benthem. *Philosophy of information*. Elsevier, 2008.

[24] Dov M Gabbay and John Woods. Handbook of the history of logic: The many valued and nonmonotonic turn in logic, vol. 8, 2007.

[25] Paul Gochet and Pascal Gribomont. Epistemic logic. In *Handbook of the History of Logic*, volume 7, pages 99–195. Elsevier, 2006.

[26] Alvin I Goldman. A causal theory of knowing. *The journal of Philosophy*, 64(12):357–372, 1967.

[27] Gilbert Harman. Induction: A discussion of the relevance of knowledge to the theory of induction. In *Induction, Acceptance and Rational Belief*. Dordrecht: Reidel, 1970.

[28] Gilbert Harman. *Change in view: Principles of reasoning.* The MIT Press, 1986.

[29] RR Hassin, JS Uleman, and JA Bargh. The new unconsciousness, 2005.

[30] Vincent F Hendricks. *Mainstream and formal epistemology.* Cambridge University Press, 2006.

[31] Michael Howard, Peter Paret, and Rosalie West. *Carl Von Clausewitz: On War.* Princeton University Press, 1984.

[32] Albert R. Jonsen and Stephen Toulmin. *The Abuse of Casuisty: A History of Moral Reasoning.* Berkeley and Los Angeles: University of California Press, 1988.

[33] Edward Luttwak. *Strategy: the logic of war and peace.* Harvard University Press, 2001.

[34] Edward Luttwak. *The Grand Strategy of the Roman Empire: From the First Century CE to the Third.* JHU Press, 2016.

[35] Lorenzo Magnani. *Morality in a technological world: Knowledge as duty.* Cambridge University Press, 2007.

[36] Lorenzo Magnani. *Abductive cognition: The epistemological and eco-cognitive dimensions of hypothetical reasoning*, volume 3. Springer Science & Business Media, 2009.

[37] Henry Mintzberg. Patterns in strategy formation. *Management science*, 24(9):934–948, 1978.

[38] Henry Mintzberg. The design school: reconsidering the basic premises of strategic management. *Strategic management journal*, 11(3):171–195, 1990.

[39] Henry Mintzberg. The rise and fall of strategic planning. *New York*, 450:67, 1994.

[40] Michael Polanyi. The tacit dimension. 1966.

[41] Willard V Quine. Epistemology naturalized. *Akten des XIV. Internationalen Kongresses für Philosophie*, 6:87–103, 1971.

[42] Willard V Quine. *Philosophy of logic.* Harvard University Press, 1986.

[43] Willard V Quine and Willard Van Orman Quine. *Theories and things.* Harvard University Press, 1981.

[44] Wilfrid Sellars et al. Empiricism and the philosophy of mind. *Minnesota studies in the philosophy of science*, 1(19):253–329, 1956.

[45] Gila Sher. The bounds of logic: A generalized viewpoint. 1991.

[46] Richard M Shiffrin. Attentional control. *Attention, Perception, & Psychophysics*, 21(1):93–94, 1977.

[47] Richard M Shiffrin. Attention, automatism, and consciousness. *Scientific approaches to consciousness*, pages 49–64, 1997.

[48] J.W. Strong. *MacCormick on Evidence.* MN: West Group, 1999.

[49] Peter T Struck. *Divination and Human Nature: A Cognitive History of Intuition in Classical Antiquity.* Princeton University Press, 2016.

[50] Alice Ter Meulen and Johan van Benthem. *Generalized quantifiers in natural language*, volume 4. Walter de Gruyter, 1985.

[51] Sun Tzu. *The art of war.* e-artnow, 2012.

[52] Friedrich Waismann and Horace Romano HARRÉ. *The Principles of Linguistic Phi-*

losophy... Edited by R. Harré. Macmillan; St. Martin's Press: New York, 1965.

[53] Mark Wilson. Generality and nomological form. *Philosophy of Science*, 46(1):161–164, 1979.

[54] Timothy D Wilson. *Strangers to ourselves: Discovering the Adaptive Unconscious*. Harvard University Press, 2004.

[55] John Woods. How robust can inconsistency get? *IfCoLog Journal of Logics and their Applications*, 1(1):177–216, 2014.

[56] John Woods. Inconsistency: Its present impacts and future prospects. In *Inconsistency Robustness*, pages 158–194. London: College Publications, 2015.

[57] John Woods. *IS LEGAL REASONING IRRATIONAL? AN INTRODUCTION TO THE EPISTEMOLOGY OF LAW*. College Publications, 2015.

[58] John Woods. *Truth in Fiction: Rethinking Its Logic*. Springer, 2018.

[59] Manfred Zimmermann. The nervous system in the context of information theory. In *Human physiology*, pages 166–173. Springer, 1989.

www.ingramcontent.com/pod-product-compliance
Lightning Source LLC
Chambersburg PA
CBHW081126170426
43197CB00017B/2760